BASE BLEED

BASE BLEED

First International Symposium on Special Topics in Chemical Propulsion

Editor

Kenneth K. Kuo
The Pennsylvania State University, USA

Associate Editor

James N. Fleming
Royal Ordnance, UK

HEMISPHERE PUBLISHING CORPORATION
A member of the Taylor & Francis Group

New York Washington Philadelphia London

BASE BLEED: First International Symposium on Special Topics in Chemical Propulsion

1 2 3 4 5 6 7 8 9 0   E B E B   9 8 7 6 5 4 3 2 1

Cover design by Debra Eubanks Riffe.
A CIP catalog record for this book is available from the British Library.

Library of Congress Cataloging-in-Publication Data

International Symposium on Special Topics in Chemical Propulsion (1st : 1988 : Athens, Greece)
Base bleed / First International Symposium on Special Topics in Chemical Propulsion : edited by K. K. Kao, J. N. Fleming.
p. cm.
Includes bibliographical references and index.

1. Solid propellants—Congresses. 2. Rocketry—Congresses.
I. Kuo, Kenneth K. II. Fleming, J. N. (James N.) III. Title.
IV. Title: Base bleed.
TL783.3.I54 1988
662'.666—dc20 90-27152
ISBN 0-89116-937-7 CIP

Contents

SESSION III: EXPERIMENTAL AND BALLISTIC PERFORMANCE OF BASE BLEED PROJECTILES AND GAS GENERATORS

Preface

This symposium was organized to review the state-of-the-art in the development of base bleed projectiles and its related research. It was conceived to fill the gap in the coverage of this new field of major technical and commercial interest. Although other meetings have included a few papers on base bleed, it has not previously been possible to devote any depth to this subject. Since base bleed projectiles have recently been seriously considered for production in several countries, and will likely become standard ammunition in the armed forces of many nations in the next decade, the need for such meetings is apparent.

This symposium was held in Athens, Greece, on 23–25 November, 1988. In total, 85 engineers and scientists, representing 20 different countries, participated in the conference. The meeting began with an invited paper given by Dr. Gerald V. Bull of Space Research Corporation. A total of 22 papers were presented. A collection of papers was bound into a volume and distributed at the symposium. A short workshop was also conducted for further exchange of information among the participants.

Following the symposium, each paper presented at the meeting, together with questions raised by the audience and replies prepared by authors, were sent out to three reviewers. The comments from reviewers and editors were incorporated into the final manuscripts by the authors for all accepted papers, which are printed in this symposium proceeding volume. Altogether, 20 papers were accepted for final publication.

We would like to thank Hemisphere Publishing Corporation for printing this proceedings and for making it commercially available for all interested individuals and organizations. We would also like to take this opportunity to acknowledge the financial assistance given toward this meeting by the European Research Office of the United States Army (London, U.K.) and by Space Research Corporation (Brussels, Belgium). Furthermore, we would like to thank Mrs. Ginny Smith for her great help in maintaining excellent records of conference papers, communication between various parties, and in the development of the meeting program as well as the proceedings volume.

It is intended that a series of International Symposia on Special Topics in chemical Propulsion be held in the future. The Second International Symposium was held in Lampoldshausen, Germany, 4–6

March, 1991. The subject of this meeting was Combustion of Boron-Based Solid Propellants and Solid Fuels. This second symposium was held under the chairmanship of Professor Kenneth K. Kuo of The Pennsylvania State University, with sponsorship provided by interested parties.

Shocking news was recently announced regarding the assassination of Dr. G. V. Bull in the Spring of 1990. Dr. Bull, an outstanding engineer in the propulsion field, will be long remembered for his technical contributions and warm friendship with many active researchers and engineers worldwide. His invited paper in the volume could represent the last formal publication of Dr. Bull's career. We are certain that all his friends will continue to appreciate his talented contributions.

Kenneth K. Kuo
James N. Fleming

Steering Committee

K. K. Kuo, Penn State University, USA (Chairman)

B. Zeller, S.N.P.E., France

W. C. Strahle, Georgia Institute of Technology, USA

G. V. Bull, Space Research Corporation, Belgium

J. N. Fleming, FCI, UK (Symposium Coordinator)

A. Gany, Israel Institute of Technology, Israel

N. E. Gunners, Defence Research Establishment, Sweden

Steering Committee

Base Bleed Technology in Perspective

G. V. BULL
Group President, SRC

ABSTRACT

The general fluid mechanic mechanism for flow about axisymmetric bodies (shell shapes) is reviewed, with reference to the very early experimental work (circa 1900) which resulted in determination over the subsonic–supersonic range of the drag–Mach number relationship.

A numerical example of drag component values for one type of modern shell is presented, to illustrate the major significance of the base drag component. The long range 210mm Paris gun shell of 1918 is discussed, being of interest as perhaps, an outer bound to the envelope of shell trajectories (apogee : 43 km, range 127 km).

The history of base injection studies is sketched briefly through early post–war work to the present systems. Various postulated wake flow models are discussed. Firing results for the minimum drag forebody projectile (commonly known as the ERFB projectile) are presented showing range increase with base bleed in the order of 30%.

INTRODUCTION

In scientific matters, as true in any general relating of human affairs, the versions of historical events often differ radically dependent on the historian. Often there exists a valid basis for the differences; unfortunately, and more frequently, there does not. The twisting of truth to perpetuate unfounded propaganda justifying past actions, or to inflate national or personal roles played, is the common stuff of history. The mixed blessing of alleged journalism, constituting the public media and often used as source references, generally relates more propaganda than reality. A short reference period in any archives reveals to the reader yesterday's "newspapers" filled with yesterday's lies. Probably more wars and chaos have been created by media propaganda than any other single source.

The above may seem somewhat irrelevant to the technical subject of this meeting. However I believe it appropriate in view of the rapid developments that have taken place in the application of the technique of Base Bleed to artillery in the last decade. These accomplishments in the practical application field tend, for a variety of reasons, to overshadow the lengthy historical base on which the technology rests. Since the papers in this conference are dedicated to the status of the applied development, it was felt appropriate to include at the outset a brief review of the history of the relative aerodynamic and aero–mechanical background with, of course, a look at the presently unresolved, complex problems.

FLOW ABOUT BODIES OF REVOLUTION

The General Flow Field

By simple application of the conservation of momentum principle, Sir Isaac Newton deduced the general relation that a flat plate inclined at an angle to the airstream would have forces acting on it parallel and perpendicular to the flow velocity vector of the general form :

$$\text{Force} = \text{Constant} \, . \, 1/2 \, \rho \, V^2$$

where ρ is the air density and V is the freestream velocity. (The constant for the vertical component being proportional to the sine of the angle of inclination to the flow stream).

This deduction failed to take into account the potential nature of flow fields wherein the disturbance to the flow by the body is propagated throughout the surrounding flow fields. For the case of irrotational, inviscid flow, the potential field defines streamlines which must satisfy the Laplace equation :

$$\nabla^2 \phi = 0$$

with the flow being affected significantly at large distances from the body. Early attempts by mathematicians to determine the effect on bodies immersed in irrotational flow fields resulted in D'Alembert's paradox (1744) that no net forces could be developed. The introduction of the concept of circulation (Helmholtz) implying vorticity and rotational flow led to the solution of the inviscid, incompressible flow equations yielding the relations :

$$L = \rho \, K \, V \, , \, D = 0$$

where L is the lift force, D is the drag force and K is the circulation.

The resolution of this paradox was accomplished by Prandtl (1904) by the introduction of the boundary layer, which essentially divided the flow about a body into two regimes. A thin layer adjacent to the body surface had to satisfy the no–slip condition, so that the fluid velocity went from a zero value on the wall surface to full free stream value a short distance away from the surface. This viscous flow layer, the boundary layer, had to satisfy the general Navier–Stokes equations. Outside this layer the field would behave as irrotational, inviscid flow.

As the flow progresses along a body, the boundary layer increases in thickness; against an increasing pressure gradient, the boundary layer flow momentum can fall below the value required to sustain the flow. Reversed flow inside the boundary layer can occur, with eventual separation of the boundary layer from the surface, such as the case of a wing profile at high angle of attack.

While the general aerodynamic problem is not directly relevant, at least in passing we should observe that the aircraft aerodynamicist expended enormous effort in overcoming boundary layer separation. This work included injecting energy into the boundary layer, boundary layer suction, etc. In this case, he was primarily interested in high lift–to–drag ratios and preventing the onset of stalling.

In relation to bodies of revolution, the axisymmetric flow can be divided into five distinct, but connected flow regimes if consideration is restricted to bodies at essentially zero angle of attack. These are illustrated in Figure 1(a) in the case of supersonic flight and may be noted as:

1. The inviscid flow field outside of the boundary layer.

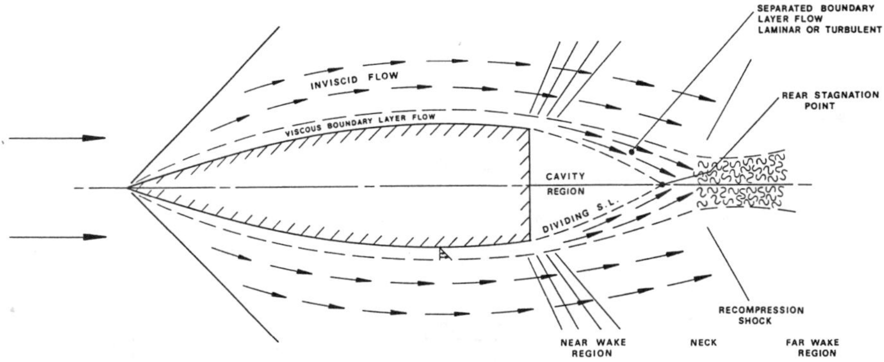

FIGURE 1(a). Simplified illustration of flow about an axisymmetric body at zero angle of attack.

2. The viscous boundary layer flow to the rear edge.
3. The "dead zone" or cavity region terminated by a rear stagnation point.
4. The separated boundary layer merging behind the rear stagnation point into a neck, with the turbulent wake flow after the neck carrying the fluid downstream and expanding, causing a recompression shock to form in the outer inviscid flow region.
5. The far wake region.

The flow illustrated in Figure 1(a) becomes considerably more complex when examined in detail, and of virtually unlimited complexity for the case of interest here, namely a yawing body with surface protrusions and a high spin rate.

The bow shock is curved through interaction with the expansion field about the ogival shape. The boundary layer grows as the flow proceeds over the body, its thickness and nature a function of the Reynolds number and Mach number.
The Reynolds number dominates the boundary layer characteristics, and is the ratio of the fluid momentum to viscous forces given by the dimensionless relation :

$$\frac{vl}{\upsilon} \text{ or } \frac{\rho vl}{\mu}$$

where v is the local velocity, l the characteristic flow length and υ is the kinematic viscosity, the viscosity, μ, divided by the density, ρ. Surface roughness, temperature, protuberances (driving bands, nubs etc) add to determining the boundary layer and general flow nature.

The Wake Flow

At very high Reynolds numbers, the boundary layer will be extremely thin; at low Reynolds numbers, it will be of appreciable thickness at the base of normal axisymmetric shapes of interest. A generally accepted, overall model of the wake flow is shown in Figure 1(b) for the case of no yaw. Restricting our interest to the supersonic Mach number regime, we should note that boundary layer flow will consist of numerous internal layers which as traversed will pass from supersonic adjacent to the free stream, through the transonic regime to subsonic flow in the region adjacent to the surface. As separation takes place the inner subsonic layer will turn, generating a Prandtl–Meyer type expansion throughout the supersonic portion of the flow field. Within the supersonic portion of the boundary layer, a lip shock may form due to boundary layer expansion and be propagated into the inviscid flow field.

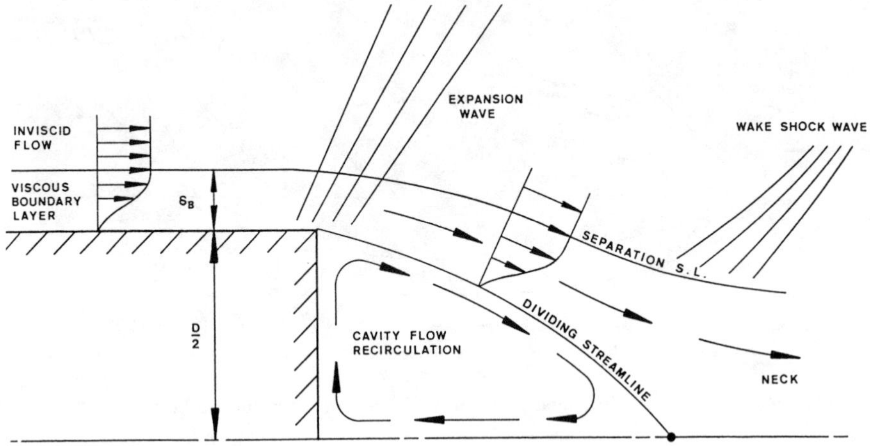

FIGURE 1(b). Details of postulated supersonic base flow.

The expanding boundary layer will be bounded by a separation streamline delineating the viscous region from the main inviscid flow on its outer surface and by what has been termed the Dividing Streamline (DSL) on its inner surface. The Dividing Streamline separates the inner cavity flow region from the viscous boundary layer, and terminates in the rear stagnation point which closes off the near wake cavity region adjacent to the base. Within the cavity region low pressure, low velocity circulatory flow is postulated.

Beyond the aft stagnation point a neck occurs in the boundary layer flow, with subsequent expansion into the far wake region. This causes the formation of the wake shock as shown.

The flow illustrated is in as simple a form as possible. In the more general case the boundary layer at the base prior to or even after separation, may be all turbulent, or a mixture of turbulent and laminar layers. After separation, the boundary layer may have viscous mixing regions, entraining gas both from the cavity zone and the inviscid flow zone.

Thus the detailed calculation of the base flow even for the case of zero yaw and spin, is beyond any generalized treatment. Clearly there will exist a significant scale effect, where δ_B/D, the ratio of base boundary layer thickness to body base diameter, is important.

DRAG OF AXISYMMETRIC BODIES

The total drag of a non–yawing axisymmetric body such as considered in Figure 1(a), is generally considered as consisting of three main components :

1. Profile (form or wave) drag
2. Skin friction drag
3. Base drag

The objective of the designer is to minimize each of these components to obtain maximum range for a given shell muzzle velocity.

The exact theoretical determination of all drag components for long range shell firings is not possible owing to the dynamic motion and flow regimes encountered. When the shell's dynamic modes are exactly known and the Navier–Stokes equations valid, current computer modeling can give good drag predictions for other then the transonic region. The real trajectory of the long range shell however, involves a spinning–yawing, axisymmetric body traversing large Reynolds and Mach number ranges. Semi–empirical codes have been developed which are force–fitted to experimental long range trajectory observations.

The aerodynamic problem embraces the most complex fields of fluid mechanics. Compressible inviscid flow solutions are rather well understood until we get into the region of "rarefied aerodynamics" or free molecular flow. For the boundary layer we can start with the classical text book of Schlicting (1), and note the incomplete but lengthy Bibliography of Reference (2), a reference to be discussed in more detail subsequently. The general fluid mechanic treatises of wakes may be considered as starting with the subsonic case studied by von Karman and leading to the now classical von Karman vortex street. But the general solution to the wake problem still presents a major hurdle in fluid mechanic analyses.

MINIMIZING AERODYNAMIC DRAG

Profile Drag

Starting as early as 1874, ballisticians began studying spark photographic techniques to measure the drag of small calibre bullets. By 1912 extensive data on the drag of various shape bullets existed as shown in Figure 2. In spite of this, the general major calibre artillery shell employed a rather blunt nose profile (approximately a 3:1 circular ogive) until October 1914 when Rausenberger and Eberhard of Krupp tried a 10:1 circular ogive nose cap on the 35 centimeter naval shell. This resulted in a range increase of approximately 50%, enabling the 38 cm Lange Max gun to reach Dover from Cap Gris

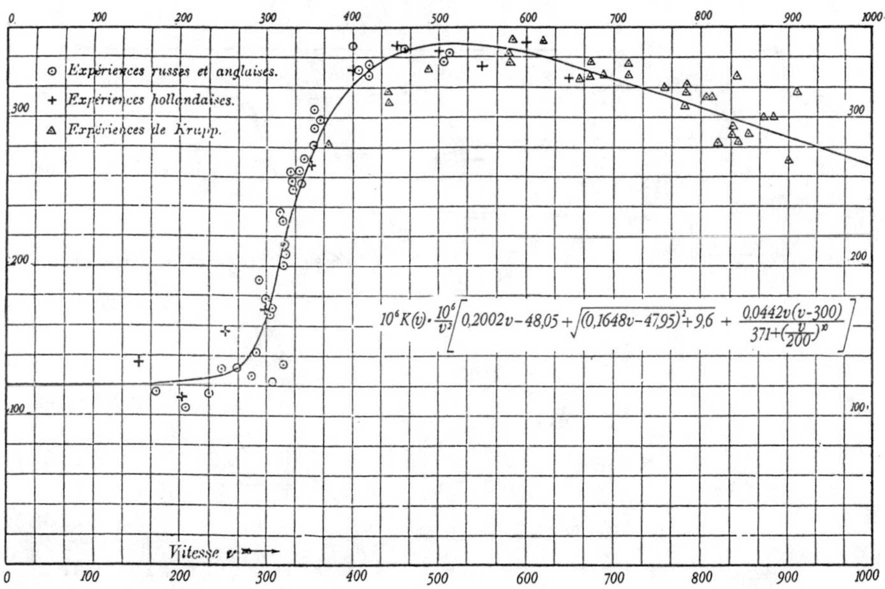

FIGURE 2. Experimentally determined drag–velocity variation, 1912, (From Reference (3)).

Nez (37 km).

Both theoretical studies and supersonic wind tunnel tests were conducted during World War 2 by the Germans (4) to determine minimum drag profiles. In general these are obtained by using an all ogival body. In practice this leads to problems. A front bourrelet must be provided either by canted nubs or by a discarding front sabot (bore rider). An interesting aside is that the World War 2 German artillery effort had started investigating nubs; an example may be seen in the museum at Picatinny Arsenal. The second problem is that of gyroscopic stability. The ogive cannot be completed at the aft end of the body without encountering stability problems that cannot be resolved in terms of permissible normal shell design parameters.

Skin Friction Drag

The skin friction drag is a function of the wetted area and any particular rough areas, such as in the vicinity of the driving band. In general, the designer has little control over this component of the total drag.

The Base Drag

This component represents an appreciable portion of the total drag of any shell configuration. The blunt rear end configuration shown in Figure 1 represents (normally) the worst case, or the case of maximum base drag. By boattailing the base drag can be reduced appreciably. However base drag remains an appreciable percentage of the total as shown by the drag coefficient break down for the nubbed, low drag profile boat–tailed round shown in Figure 3.

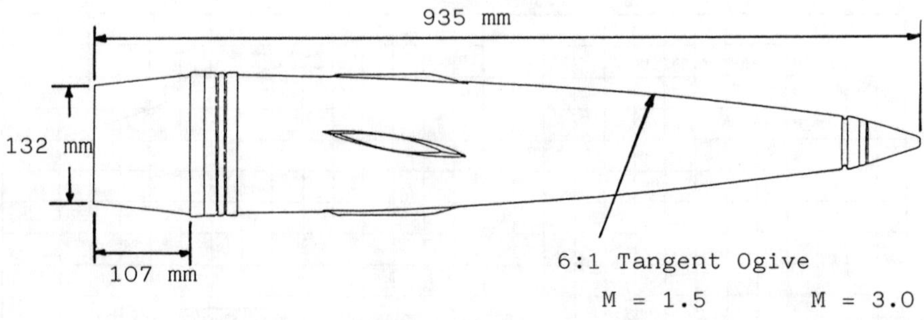

		M = 1.5	M = 3.0
C_D Skin friction	=	0.0468	0.0341
C_D Wave Drag	=	0.0844	0.0581
C_D Base Drag	=	0.1183	0.0647
C_D Total	=	0.2666	0.1683

Reference area : 155mm diameter circular cross-section

FIGURE 3. 155mm ERFB MK10 MOD2 drag coefficient components.

Reducing the Base Drag

Theoretical concepts for the reduction of base drag date back prior to World War 2. However serious interest may be considered to have started shortly after the war. The groups that sparked interest in shell base drag reduction were of course those working in ballistics. It was well known among external ballisticians that the velocity retardation of small calibre shells carrying tracers was less than the same shell without a tracer. It would be futile to enter into a debate as to whether German, French, Russian, English or American ballisticians were the first to observe this. As Figure 2 clearly illustrates, European small arms ballisticians were measuring shell retardation with differing shapes by the beginning of the present century.

After World War 2 the general interest in Western countries in major calibre ordnance declined in favour of other systems. The decision of the Westerveld Board of 1932 became established dogma. This cast field artillery into the role of a close support weapon, resulting in no serious requirement for long range firings. However the advent in the field of long range Soviet artillery such as the 130mm system (27.5 km) and the 152 system, created a NATO requirement of 30 km for the 155 system. With limitations placed on barrel length (39 calibres), this range could be obtained only by absolute minimization of shell drag or by rocket boosting after muzzle exit (RAP).

With the relative abundance of supersonic/hypersonic wind tunnels that appeared in North America in the early post–war years as a direct result of the wartime German developments, compressible aerodynamics experienced an enormous increase in scientific interest, both theoretically and experimentally. Inviscid, steady state compressible fluid flow mechanics was relatively well advanced and understood by 1950. Theoreticians then concentrated on the extremely complex areas of the compressible flow boundary layer and wake flows.

Theoretical concepts for the reduction of base drag date back well prior to World War 2. However serious technical effort on this problem may be considered to have started shortly after the war. Estimates of base drag from fitting observed firing range data were formulized as early as 1931 by Gabeaud (5) in the form :

$$C_{D_b} = 0.25 \text{ for supersonic flow}$$

while von Karman and Moore derived (1932, Reference (6)) a value:

$$C_{D_{b_{max}}} = 2/\gamma M^2$$

while Hoerner (7) showed that a good correlation with firing results in the supersonic range was given by

$$C_{D_b} = 0.7/M^2$$

These are effective total contributions of the base pressure distribution to the drag for blunt based bodies. C_{D_b} is the base drag coefficient based on base area, and M is the freestream Mach number.

Before considering the complex problem of examining base and wake flows in order to look at the general problems of base drag reduction, it is necessary to review the general aero– mechanical aspects of shell flight.

THE TRAJECTORY CHARACTERISTICS OF LONG RANGE SHELLS

General

A typical major calibre artillery shell leaves the muzzle of the gun with some angular (yaw) acceleration. This acceleration comes from mass and geometrical misalignments of the shell during bore travel and grows in magnitude with tube wear. The muzzle disturbance results in the excitation of two free–flight oscillatory modes, with the disturbance equally divided between them. The relatively low frequency processional mode is damped out normally after several cycles, with the maximum yaw usually occurring within 50 to 100 meters of the muzzle. The high frequency nutational mode is damped out much earlier than the processional mode, and is normally negligible at the point of first maximum yaw.

For well designed systems the maximum induced muzzle yaw generally is less than 3 degrees in the new barrel case. However as tubes enter the third and fourth quarter of their life, this initial yaw can be large. The effect of the initial yaw on range increases with the range of the shell, and for the very long range case, introduces a considerable correction factor. Clearly the reduction in range must then consider the shell at an angle of attack, with large Reynolds number and maximum Mach number values.

The shell proceeds through its flight trajectory with aerodynamic forces producing moments which result, through gyroscopic action, in a shell displacement angle in the plane at 90° to that of the applied aerodynamic moment. This displacement results in a gyroscopic moment causing the shell to maintain alignment with the flight velocity vector.

However as the shell traverses the large range of Reynolds and Mach number as it approaches apogee, permanent angles of yaw (in the side plane) can be present. In effect the density decreases to a value where gyroscopic moments dominate. On re–entry, the shell recovers through damped precessional mode oscillations as it traverses the increasing density atmosphere.

As interest increases in shells with increasing ranges, (ranges of up to 60 km are currently becoming of interest), the flight modes along the trajectory must be considered and it is obvious that effect of angle of attack on drag cannot be ignored. This of course, vastly increases the complexity of the aerodynamic analytical problem.

To illustrate the above comments two cases of long range shell trajectories will be presented.

The Paris Gun Shell Trajectory

This 210mm 106kg spin stabilized shell was developed in 1917, and fired against Paris from Crépy–en–Laon starting on March 23, 1918. During these firings it attained an altitude of 43 km and a maximum range of 127 km.

A recovered shell, with the nose cap replaced, is shown in Figure 4. A typical trajectory (taken from Reference (8)) is shown in Figure 5. It will be noted that the shell starts to develop significant permanent yaw (in the side plane) above 20 km, reaching an angle of repose of approximately 40° during traverse over apogee. The angular recovery mode is shown in the plot labelled (b) in Figure 5.

The 45 calibre ERFB MK10 MOD2 Trajectories

With a muzzle velocity of 897 meters per second and no base bleed, this shell attains a

FIGURE 4. Paris Gun 210mm shell.

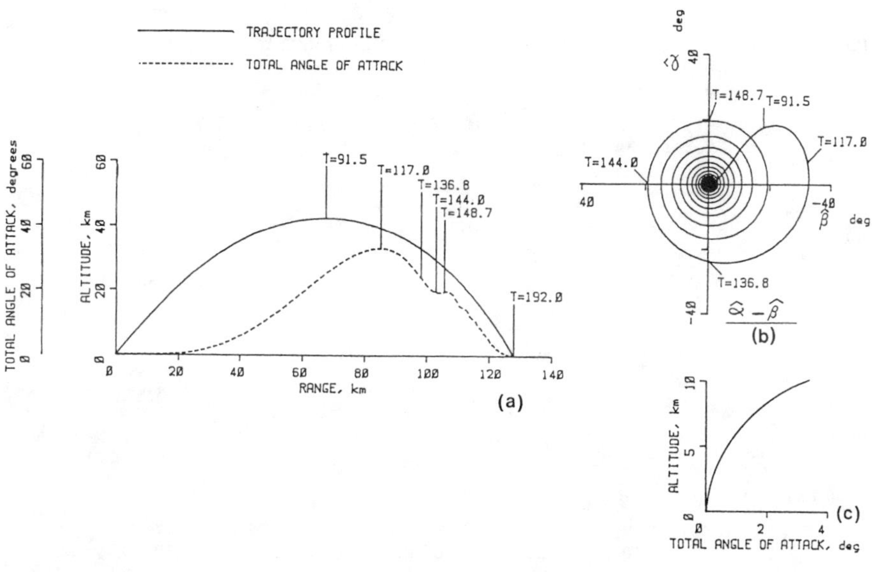

FIGURE 5. Paris Gun shell computed trajectory. Muzzle velocity = 1646 m/s, quadrant elevation = 50°.

range of 30 km. A typical trajectory plot for this case is shown in Figure 6(a), where it may be noted that the maximum yaw angle achieved is approximately 2 degrees. When the range is extended to 39km using base bleed the corresponding plot is shown in Figure 6(b).

The previous discussion illustrates the necessity of considering the entire trajectory when considering drag optimization for range. Among the complexities added to the simple consideration of axisymmetric body drag at zero degrees and a fixed Reynolds number – Mach number combination, must be added the drag component due to yaw and the optimization over the range of Reynolds number and Mach number while aerodynamic effects are still significant. The range of Reynolds number, Re, and Mach number, M, experienced by the two projectiles can be seen in Figures 6(a) and 6(b).

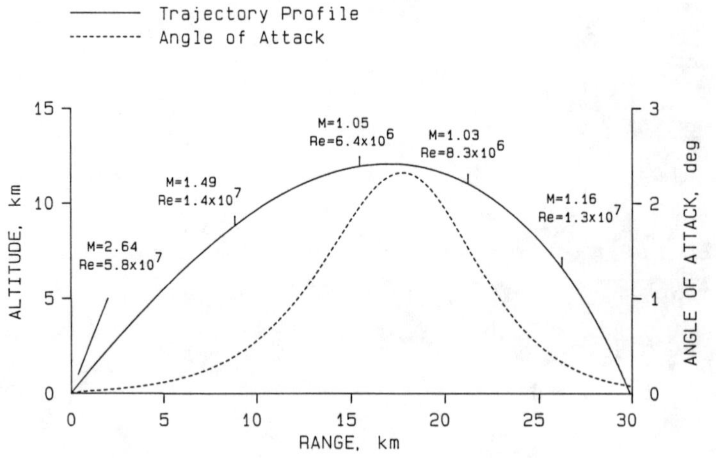

FIGURE 6(a). 155mm ERFB MK10 MOD2 maximum range trajectory.

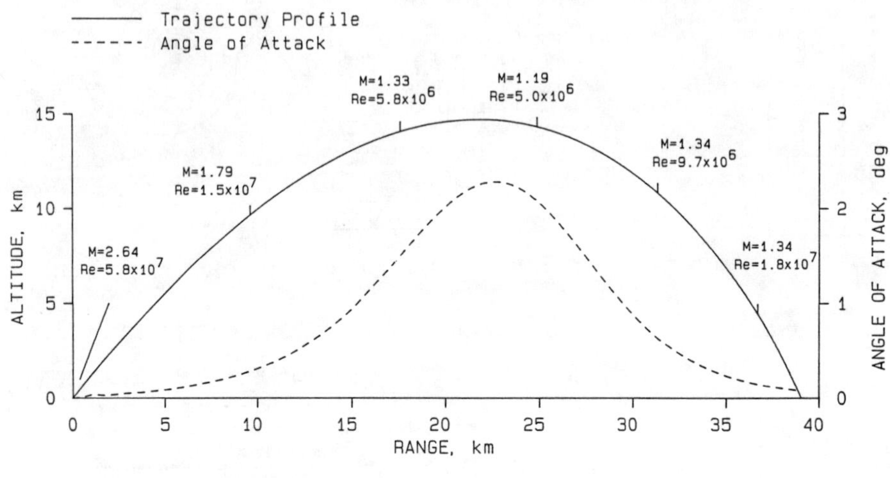

FIGURE 6(b). 155mm ERFB/BB maximum range trajectory.

REDUCTION OF BASE DRAG : BASE BLEED

Reference (9), dedicated to establishing the state of knowledge on wake flows and base drag reduction, published in 1976, must be studied as a base line work on this topic. It covers at least all the major work up to that time, including injection through the boundary layer ahead of the base, the effect on base drag of plates mounted downstream from the base on stings attached to the base, cold gas injection into the wake through the base, and base injection with combustion in the wake.

At the time this study was being compiled, no practical system had been devised (excluding tracers on small calibres) to apply this to major calibre artillery shells. The first successful system in fact, was then in the test and development stage in Sweden. It was brought to North America with the major firing series in Antigua conducted in 1978 (10). The grains used were carboxyl terminated polybutadiene/ammonium perchlorate manufactured to fit the SRC ERFB MK10 MOD2 projectile. The tests yielded the almost 30% increase in range predicted. Because of its assigned carcinogenic nature, the Swedes switched to hydroxyl terminated polybutadiene/ammonium perchlorate grains which for all practical purposes are identical in performance.

The method used in mounting these grains in the base of the ERFB MK10 MOD2 projectile is illustrated in Figure 7. This simply involved replacing the boattail by a conical section with a closed end cap and a central orifice of sufficiently large diameter to ensure subsonic flow into the cavity wake. The inside geometry was such as to accept the cylindrical shaped, sectioned grain, inhibited on all surfaces other than the central annulus and the section cut surfaces. A diaphragm seals the grain; as the pressure builds up to the shot–start value, the diaphragm is ruptured and the hot gun gas flows through the orifice and about the grain. Some indications exist that the grains are floating in the hot gas, and do not build up to the full spin rate of the projectile. The in–bore and muzzle exit conditions are of a complexity that while some effects can be observed,

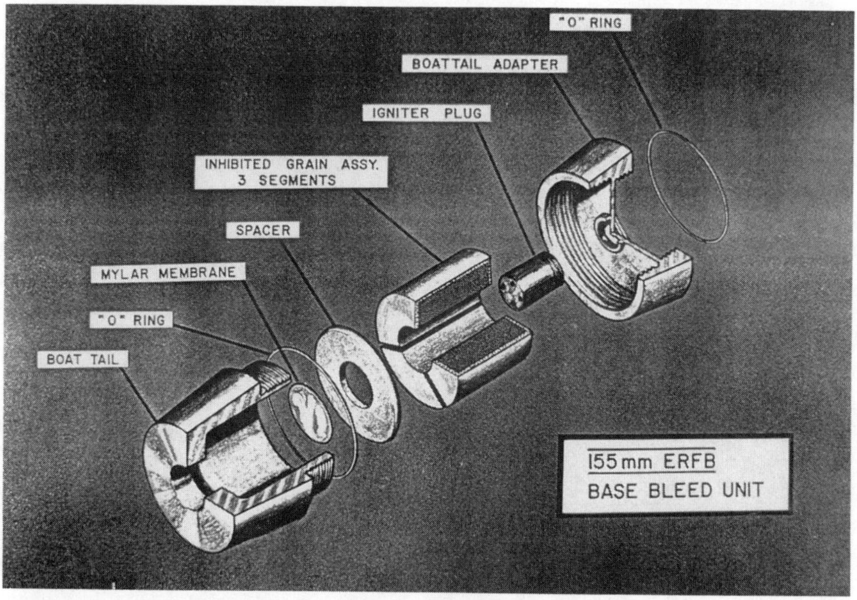

FIGURE 7. 155mm ERFB base bleed unit assembly.

detailed analysis is beyond the present state of the art. The grain and igniter begin burning probably at or near the seating position, and then experience the full in–bore gas pressure during shot travel down the bore. At exit, the sudden pressure drop causes quenching of the grain burning, but only sporadically. The igniter ensures continued grain burning after muzzle exit.

Returning to the flow models of the wake, the optimum condition would be when the momentum flux is such as to restore the base pressure to the full freestream value. The temperatures of the products of complete combustion lie in the range of 1000 to 2000°C, dependent on composition. Photographic evidence shows the presence of combustion in the wake during at least the early, and most critical part of the trajectory. Thus in practice, a satisfactory theoretical model must include combustion–heat transfer across the dividing streamline, through the separated boundary layer flow and into the main stream.

Various models can be postulated and solutions to the wake flow presented by making simplifying assumptions, particularly concerning combustion phenomena and heat transfer in the wake. Figure 8 sketches idealized conditions. In Figure 8(a) the wake geometry is sketched when the base bleed mass flow is less than optimum. In 8(b) the optimum case is sketched. The assumption in this case is that the momentum flux into the wake is large enough to cause the viscous throat to expand to full body diameter and the wake to be mass–filled to the point that the pressure is restored across the dividing streamline to the freestream value. The injected mass–wake mixing processes cause the boundary layer in this case to be simply a thin dividing shear layer between the mainstream flow and the lower velocity wake flow. Heat transfer across the shear layer to the main stream flow will add considerable complexity to the analytical model. Typical smear photographs illustrating a wide range of base bleed wake flow conditions are shown in Figures 9, 10 and 11.

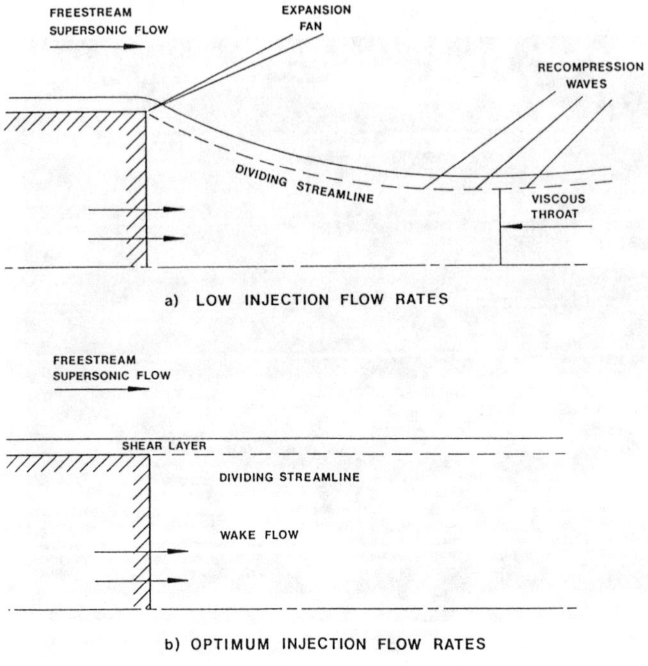

FIGURE 8. Schematic of base bleed flow with base bleed.

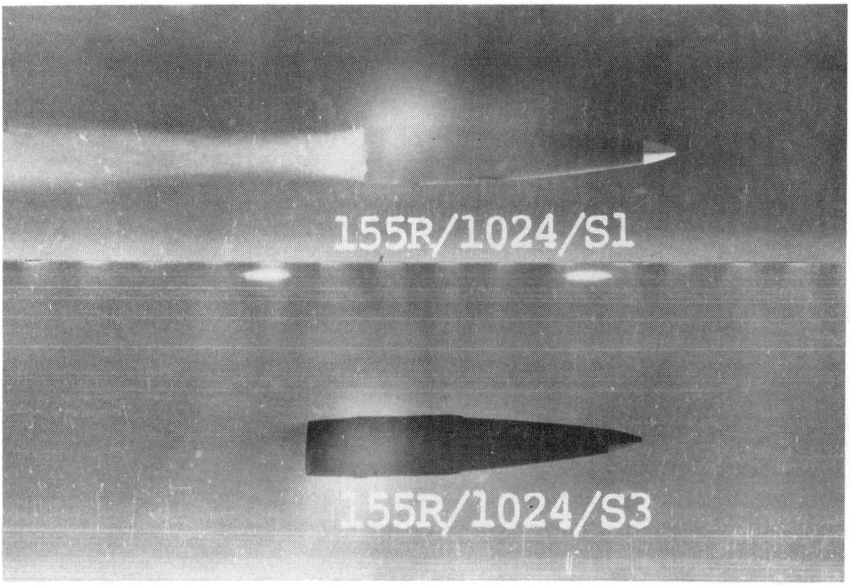

FIGURE 9. Smear photograph of base bleed projectile showing less than optimum base bleed mass flow.

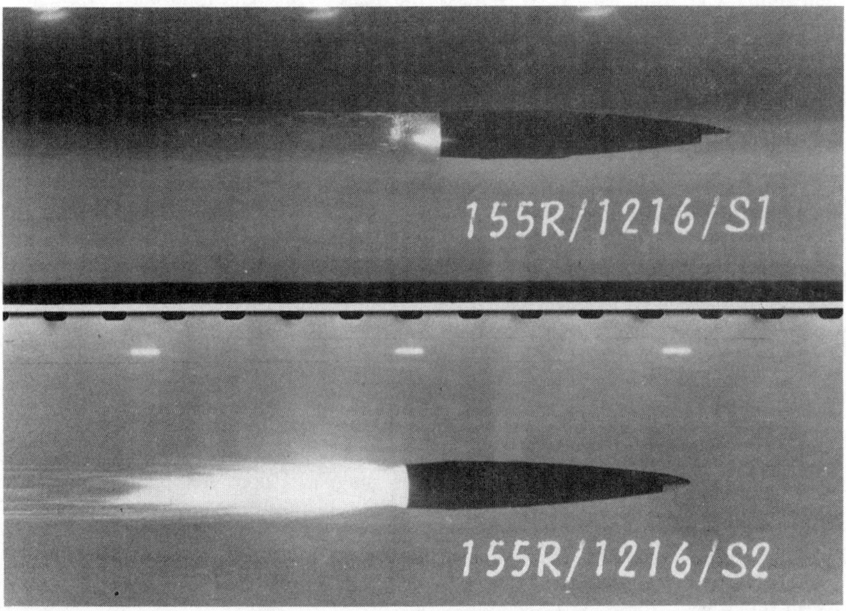

FIGURE 10. Smear photograph of base bleed projectile showing optimum base bleed mass flow.

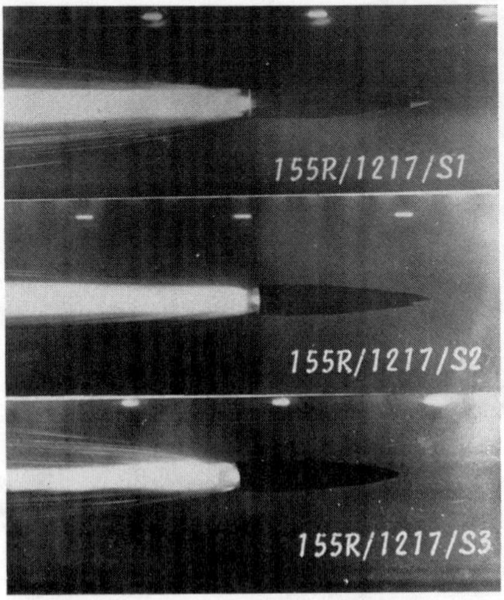

FIGURE 11. Smear photograph showing greater than optimum, supersonic type, base bleed mass flow.

TRAJECTORY OPTIMIZATION

In any long range trajectory, the projectile is traversing rapidly a large Reynolds number range during its ascent. Assuming that the optimum conditions could be achieved, this would imply a continuously varying mass–momentum flux from the base to the wake. Furthermore it would imply a detailed knowledge of base bleed – wake flow characteristics. To date the optimization is a trial and error proposition based on fixed grain properties and any variation in combustion rate caused by the change in the surrounding ambient air.

During the last decade, the base bleed concept has been applied to a wide range of projectiles in many different countries. One might say that the state of the applied field has advanced far ahead of the derivation of satisfactory fluid dynamic models. Contrary to the conclusions of Reference (11), range increases in the order of 30% have been achieved (12, 13), well exceeding the 10% estimate (matching RAP) the author concludes as a limit on base bleed systems.

Firing Results

Hoping not to appear parochial, but because of the simple reason that this is the only projectile for which the author has access to data, the performance of the base bleed version of the ERFB MK10 MOD2 projectile will be presented in concluding this presentation, and as a system currently deployed. At the outset, it should be noted that the base bleed system only reduces projectile drag, and does not apply thrust. Thus dispersion is not affected, other than malfunctions causing the base bleed drag reduction to vary. This results in increased range dispersion, but no variation in deflection dispersion. For fully developed systems, the reproducibility of drag reduction by base bleed is such that no increase is observed in range dispersion, the values for probable error in range using the data reduction techniques of Reference (14) lying between 0.3 and 0.45. Table 1 shows typical firing results of the ERFB MK10 MOD2 projectile (12), and of the base bleed version (13).

TABLE 1. Results of Typical Firing Series.

	Muzzle Velocity, m/s	Range,m	Probable error	
			Range	Deflection
ERFB	897.6	30577	.376%	.97 mils
ERFB/BB	899.9	39153	.334%	1.03 mils

CONCLUDING DISCUSSION

In no way can this paper claim to be more than a superficial scanning over the difficult subject of wake flows and trajectory analyses. However, hopefully, the papers to follow covering the current work in the many practical aspects of manufacture, supported by advances in theory, will at least allow it to qualify as an acceptable introduction to these more qualified, specific treatments.

REFERENCES

1. Schlicting, R., *Boundary Layer Theory*, 4th ed., McGraw–Hill, New York, 1960.

2. Murthy, S.N.B., and Osborn, J.R., Base Flow Phenomena with and without Injection, in *Aerodynamics of Base Combustion*, ed. S.N.B. Murthy, pp. 7–210, AIAA, New York, 1976.

3. Becker, K., and Cranz, C., Messungen über den Luftwiderstand für grosse Geschwindigkeiten, *Artilleristische Monatshefte*, Nr. 69, September 1912.

4. Ferri, A., Supersonic Tunnel Tests of Projectiles in Germany and Italy, NACA Wartime Report L5H08, October 1945.

5. Gabeaud, A., Sur la Resistance de l'Air aux Vitesses Balistiques, *Comptes-Rendus de l'Académie des Sciences*, No. 192, p. 1630, 1931.

6. von Karman, Th., and Moore, N.B., Resistance of Slender Bodies Moving with Supersonic Velocities, with Special Reference to Projectiles, *American Society of Mechanical Engineers Transactions*, vol. 54, p. 303, 1932.

7. Hoerner, S.F., *Fluid Dynamic Drag*, 2nd ed., published by the author, New Jersey, 1965.

8. Bull, G.V., and Murphy, C H., *Paris Kanonen - The Paris Guns (Wilhelmgeschutze) and Project HARP*, Mittler, Herford, 1988.

9. *Aerodynamics of Base Combustion*, ed. S.N.B. Murthy, AIAA, New York, 1976.

10. Anon, Data Analysis Report, 155mm ERFB MK10 MOD2 and ERFB Base Bleed Tests. Antigua Test Range, March 5–7, 1978, SRC–R–103, March 1978.

11. Hudgins, H.E., Range Increase of Projectiles by Heat and/or Mass Addition to Base or External Flow, in *Aerodynamics of Base Combustion*, ed. S.N.B. Murthy, pp. 495–521, AIAA, New York, 1976.

12. Anon, Qualification Report, SRC 155mm ERFB MK10 MOD2 Projectile, Volume 4 : Analysis of Test Data, Part 2 : Range and Dispersion over a Wide Range of Firing conditions, SRC–HE01–T–83330, August 1983.

13. Slack, A.D., Qualification Report, 155mm ERFB/Base Bleed Projectile, Volume 1: Analysis of Test Data – Range and Dispersion, SRC–TM–87615, April 1987.

14. NATO STANAG N° 4106, Procedure for the Determination of External Ballistic Performance of Shell.

SESSION I: CHEMISTRY AND IGNITION/COMBUSTION BEHAVIOR OF BASE BLEED PROPELLANTS AND IGNITER MATERIALS

Base Bleed Solid Propellant—Properties and Processibility for Industrial Solid Propellant

JACQUES GAUCHOUX, DOMINIQUE COUPEZ, and MAX LECOUSTRE
S.N.P.E., St Médard
33160 St Médard en Jalles
France

ABSTRACT

Composite solid propellants were studied in order to obtain optimized propellant with good properties and processibility required for base bleed applications. HTPB propellant with Ammonium Perchlorate oxidizer ratio from 70 up to 80 percent loading is selected after thermo-dynamical calculations . This paper describes the mechanical and ballistic properties of the optimized propellant and the importance of each parameter.

INTRODUCTION

The base bleed effect increases the shell range by reducing the projectile base drag. The system consists in injecting hot and reductive propellant gases at the shell base.
Before experimental investigations, the propellant specialist determines the burning gas theoretical composition and the temperature with the help of classical thermodynamical calculations.
During the following experimental propellant optimization, he looks more particularly :

- To obtain a propellant grain with appropriated mechanical properties in order to ensure structural integrity during gun firing tests.
- To give the required burning rate under sub-atmospheric pressure.
- To reduce production costs due to mixing time and tooling immobilization.

Then, the optimized propellant is incorporated in an industrial context to satisfy technical and economical requirements in order to get a base bleed generator with the best performances at the best cost.

THEORETICAL EVALUATION

Propellant formulation evaluations were performed initially with thermodynamical program[1] . Optimum oxidizer loadings were determined with used polymeric binder system. The optimization was based on the heat of combustion, density, and fuel ratio.
Propellants based on initial formulations were mixed, and Strand Burning rate, mechanical properties, and density determined.
From these initial formulations, additional modifications were introduced to evaluate the burning rate range capability and mechanical properties improvement. Working life evaluations were also performed.

THERMODYNAMICAL PROPERTIES

The type of composite propellant is composed of a polymeric binder (here a HTPB) loaded with an oxidizer, the ammonium perchlorate. Some additives like bonding agent, ballistic catalyst and curing agent and generally an amount of heat allow to obtain the final propellant material.

Composition of base bleed propellant (ex : Filler AP : 80 wt %)

Ingredient	Weight %
Binder, HTBP	18
Antioxydant	0,4
Bonding agent	0,1
Curing agent	1,4
Catalyst	0,1
Oxydizer	80

The propellant combustion delivers some gaseous, liquid and solid species that we are able to compute with a good level of accuracy [1].
The thermodynamical characteristics are :

- C : 13,63 At.G/kg
- H : 49,75
- O : 28,08
- N : 6,81
- Cl : 6,81

These significant products of combustion are gaseous and characterized by :
- Heat of combustion : T_b : 2370° (in chamber)
- Density : ρ : 1593 kg/m^3

The program delivers also some other parameters more commonly used for ballistic requirements :
- Isentropic coefficient : γ
- Flow coefficient : C_D.
- Characteristic velocity : C*

Where : $C_D = F(\gamma)\sqrt{\dfrac{R.T_b}{\rho}} = \dfrac{1}{C^*}$

with F (γ) : function of the isentropic coefficient γ
R : perfect gas constant

All these parameters mainly depend on the percentage of oxidizer and binder.

VISCOSITY OF COMPOSITE SOLID PROPELLANT SLURRY

The viscosity of composite solid propellant slurry is an important parameter to the processability and castability behaviours. It is affected not only by the binder, but also by the size, content, shape and surface properties of solid fillers in propellant.
Brookfield viscosimeter is used to measure the slurry viscosity. First it was evaluated for process temperature specified for end-of-mix.
The viscosity of the selected propellant is less than 15 Kp at the end of mix, for 40°C temperature. These low viscosity is remained over a eight hour period.

Figures 1 and 2 shows these results.

The viscosity of propellant slurry is significantly affected by the end-of-mix temperature, the ratio of particle size and the ratio of content between coarse and fine particles.

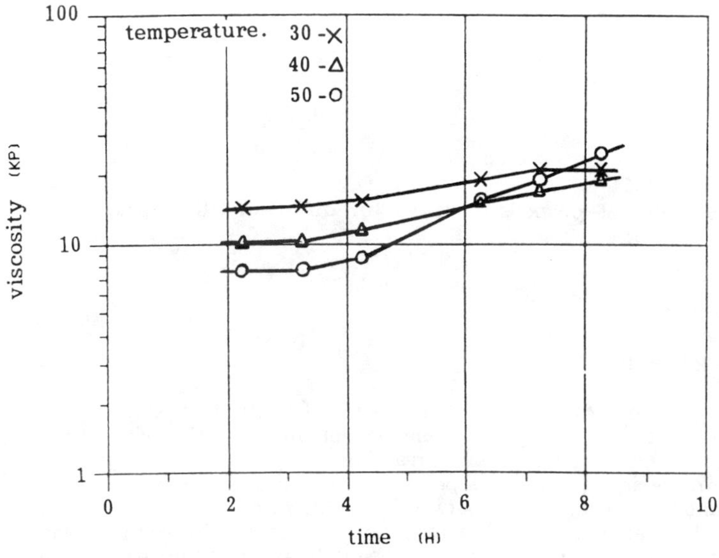

FIGURE 1 - Viscosity of the uncured propellant slurry versus time and different temperatures

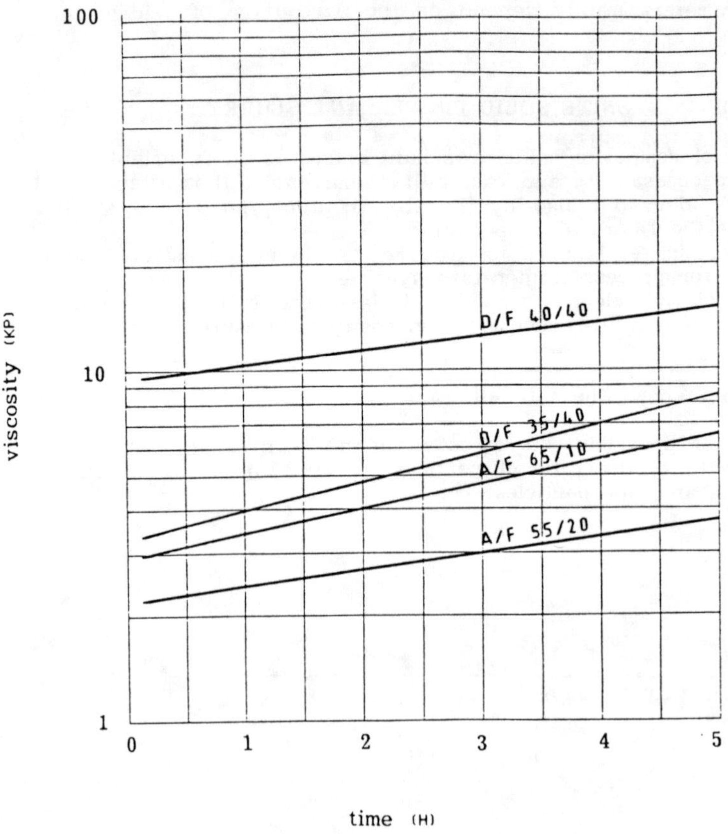

FIGURE 2 - **Viscosity of propellant slurry for different filler ratio and AP distributions**

MECHANICAL PROPERTIES

Due to the high acceleration of the projectile longitudinal strains are developed in the bore of the grain. The propellant must therefore have a high flexibility and ultimate tensile strain limit.
Mechanical characteristics mainly depend on the binder but also on the particle size and on the adhesion between particles and binder.
They vary with temperature and stress rate or strain in such a way that a time temperature equivalence has been determined for each type of binder. A shift factor a_T gives mechanical characteristics at every temperature T if they are known in standard conditions at 20°C. All mechanical properties (ultimate strenght. Sm, elastic modulus. E, deformation at maximum strength. em) are related to the number t/a_T which takes into account the time of application of stress t and the empirical shift factor a_T.

The tensile tests are carried out with JANAF specimen and the properties defined according to Figure 3.

Standard tensile test on standard test specimen at +20°C - Strain rate 50 mm/mn

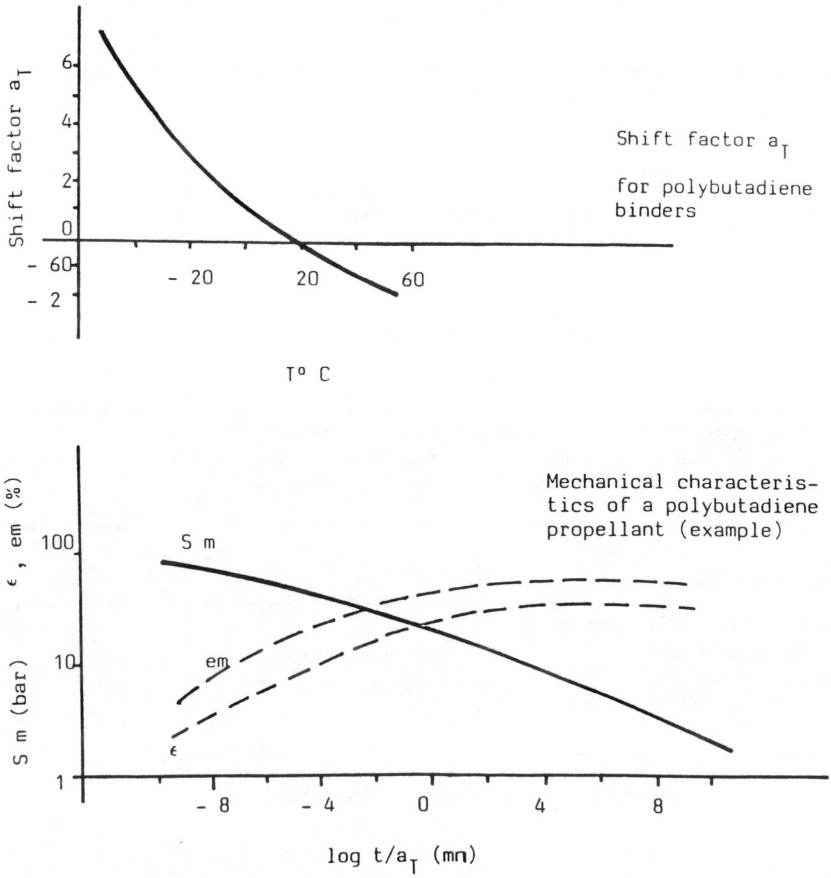

FIGURE 3 - **Example for mechanical behaviour of propellant**

MECHANICAL CHARACTERISTICS OPTIMIZATION

The combination of HTPB/AP is a current composite propellant composition, giving high performance, long service life and low cost. The nature of the fuel-binder allows high concentration of solid filler and good anti-aging characteristics has greatly extended the solid rocket technology. Therefore, the addition of a bonding agent is essential to ensure the improved interaction at the binder-filler interface and to have best mechanical property behaviour.

HTPB binder is hydroxylated polymeric material and, as such, is curable using isocyanates. To reduce the moisture sensitivity of the curing reaction, isophorone diisocyanate (IPDI) was selected as the primary curing agent to be evaluated. To achieve sufficient mechanical properties of the propellant the curing ratio (NCO/OH) ranging from 0,85 : 1 up to 0,95 : 1 were evaluated.

Laboratory testing resulted in the selection of the effects of :

Binder variation, including bonding agents.
Curing agent and AP oxidizer distribution.
Burning rate catalyst.

A series of formulations having the required burning rate were developed, and selected for evaluation in intermediate-scale propellant mixes. All these parameters were varied to assure selection of the formulation with the highest possible strain capability and optimum processing characteristics.
Propellant from this effort have been successfully scaled up to the 300 Gallon mixer, and the effect of mixer size and process on propellant processing, mechanical properties and balistic properties have been evaluated.

Some examples : optimization for some parameters. Effects of oxidizes particule size, curing ratio and bonding agent ratio. Tables 1, 2, 3.
We present also the mechanical properties versus temperatures and stress or strain rate on selected propellant. Table 4.

Characteristics	Curing ratio, NCO : OH			
	0,91	0,90	0,89	0,88
Maximum tensile strengh : Sm (MPa) at + 20°	1,4	1,4	1,4	1,4
Elastic modulus : E (MPa)	4	4	3,3	1,6
Strain at maximum tensile (%) : em	40	90	80	140

TABLE 1 - Curing ratio effects on mechanical characteristics

Characteristics	Oxidizer loading (%)					
	70	75			80	
Particle size distribution	D:30 F:40	D:35 F:40	A:65 F:10	A:55 F:20	b:53 F:27	D:40 F:40
Curing ratio (NCO : OH)	0,90	0,90	0,90	0,95	0,94	0,85
Maximum tensile strengh : Sm (MPa)	1,4	2	1,1	1	1,4	1,8
Elastic modulus : E (MPa)	4	8	8	8	11	7
Strain at maximum tensile (%) : em	90	50	50	24	31	42

TABLE 2 - Oxidizer loading and particule size distribution on mechanical characteristics

Characteristics	Bonding agent ratio (%)			
	0,5	0,3	0,1	0
Maximum tensile strengh : Sm (MPa) at + 20°C	2	1,6	1,4	1,2
Elastic modulus : E (MPa)	8	7	8	8
Strain at maximum tensile (%) : em	50	47	36	32

TABLE 3 - Bonding agent ratio on mechanical characteristics

Temp. (C)	R (mm/mn)	Sm (MP_a)	Eps (%)	E.tg (MP_a)	em (%)	er (%)
- 60	1	6,46	8,3	77,76	16,9	20,3
- 60	50	8,30	6,2	133,71	11,8	12,3
- 60	500	9,77	4,0	243,76	8,0	8,7
+ 20	1	1,40	30,3	4,64	38,1	38,9
+ 20	50	1,93	31,6	6,10	52,2	56,2
+ 20	500	2,36	28,9	8,16	56,5	66,8
+ 40	1	1,07	29,8	3,59	39,2	40,0
+ 40	50	1,51	32,7	4,62	52,1	55,2
+ 40	500	1,84	30,9	5,97	54,5	66,2

TABLE 4 - Mechanical characteristics versus temperatures and strain rate for a selected propellant

BALLISTIC PROPERTIES

Ballistic properties are defined by the burning rate of the propellant which depend mainly on the particles ratio and sizes and on the quantity of ballistic additives (accelerators).
Besides, the burning rate is also proportionnal to the initial temperature of the propellant.

The equation of the burning rate is :

$$V_b = ap^n$$

Where a and n are the characteristics of the propellant.
The burning rate of the propellant, in our case, is measured by the "Strand Burner" method, in nitrogen, and in a large range of atmospheric and subatmospheric pressures.
At the ambient temperature, the range of the burning rate must be

$$0,9 < V_c < 1,5 \text{ (mm/s)}$$

This range is optimized for different propellant grain dimensions.

The experimental propellant, in atmospheric pressure and T° = 20°C, is selected for V_c = 1,22 mm/s.
Figure 4 shows experimental results burning rate versus pressure.

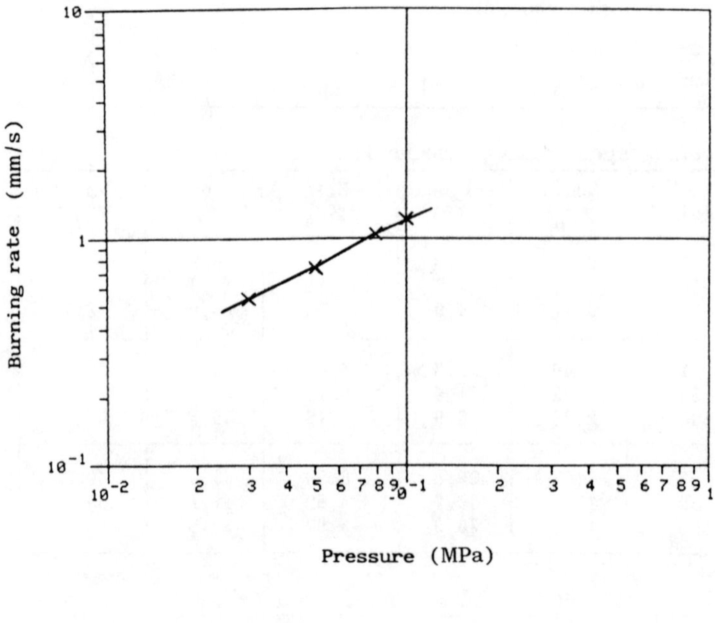

FIGURE 4 - Burning rate versus pressure

BURNING RATE OPTIMIZATION

In general, the binder has a relatively small influence on the burning rate.
The burning rate is governed more particularly by the ammonium perchlorate (AP) oxidizer.

The investigations on the oxidizer are :
- oxidizer ratio (Filler ratio)
- particle size distribution
- particle size diameter ($Ø_m$)

Some additives govern also the burning rate :
- Ballistic catalyst
- Carbon Black

If the ballistic catalyst is a burning rate adjuster, the carbon black is an opacifier, which decreases reflectivity and increase radiative absorption at propellant surface. It cannot be recognized to be an active catalyst particularly at subatmospheric pressures.

. Effect of the filler ratio

Filler ratio is chosen, essentially, on the base of the thermodynamical calculations. To obtain the required burning rate it is necessary to optimize the other parameters.
- Particle size distribution
- Balistic catalyst ratio

Table 5 shows the results for constant burning rate.

Characteristics	Filler ratio (%)					
	70	75			80	
Particle size distribution	D:30 F:40	D:35 F:40	A:65 F:10	A:55 F:20	b:53 F:27	D:40 F:40
Burning rate (mm/s)	1,25	1,26	1,22	1,27	1,25	1,22
Ballistic catalyst ratio (%)	0,75	0,2	4	2,5	0,5	0

TABLE 5 - Effects of filler ratio, particle size distribution and ballistic catalyst ratio at constant burning rate

. Effect of particle size distribution

Particle size distribution is optimized for give the required burning rate, in first, but also for a good processability. For a constant burning rate if it is directly dependant of the average particle size, it is also in relation with ballistic catalyst ratio. See results table 5.

. Effects of ballistic catalyst ratio versus particle size distribution

Particle size distribution used in this propellant is a bimodal blend using 10 and 90 µm or 10 and 200 µm.
The results are shown on figure 5.

For the required burning rate, all these parameter were correlated and used to determine along the constant burn line the best distribution which gives the minimum mix viscosity and longer pot life for better casting of the grain.

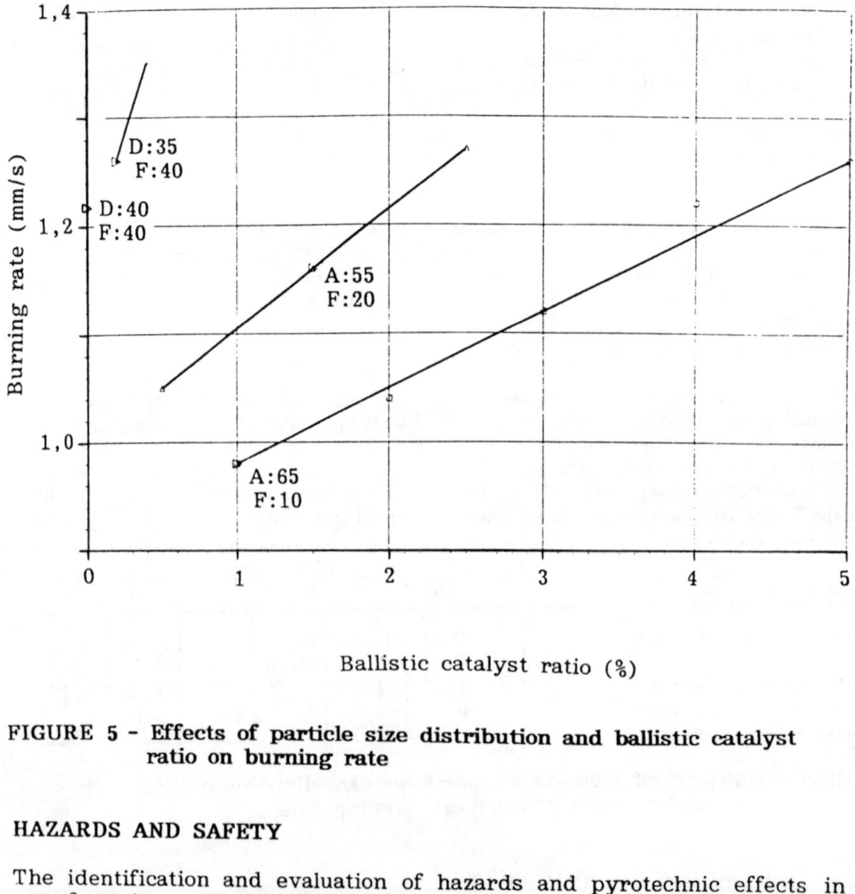

FIGURE 5 - Effects of particle size distribution and ballistic catalyst ratio on burning rate

HAZARDS AND SAFETY

The identification and evaluation of hazards and pyrotechnic effects in manufactoring solid propellants are necessary.
Consideration of the ignition consequences suggests modifications of process conditions or plant instruction to minimize risk to personnel, facilities, and product. Attempts to estimate the probable hazards of handling a new propellant were based on correlation of available test data with that of materials which had been processed successfully.

The results of different tests made on the optimized propellant are shown in table 6.

. Autoignition temperature	278°C
. Impact sensitivity (High energy Shock)	> 4 meters
. Impact sensitivity (Julius Peters Appartus)	11 Joules
. Friction sensitivity	203 Newton
. Regression Speed (burning rate) at Patm	< 1,6 mm/s
. Detonability	< 1 card
. Electrostatic sensitivity	No sensitive

TABLE 6 - Safety results

. Process and processibility

The composite solid propellant manufacturing system is shown in table 7. It is a typical block flow diagram. Noting that processability is an important factor in propellant formulation which processibility determination is a function of the rheology of the propellant.
The different steps in propellant processing are binder preparation, oxidizes preparation, propellant mixing, casting and curing. Processing differences between size mixes are a source of propellant variation. Frequently a new mixing cycle is optimized for each size mixer or type (Sigma or Vertical). After optimization of the required specifications propellant, for one mixing cycle adapted to one size mixer, the propellant processing parameters are selected.
In our application and with different size industrial mixes since 5. Gallons to 300 Gallons we have developed and demonstrated the feasibilite of an HTPB propellant for Base-Bleed applications, and with a low cost owing to the fact that we obtain many thousan grains by batch.

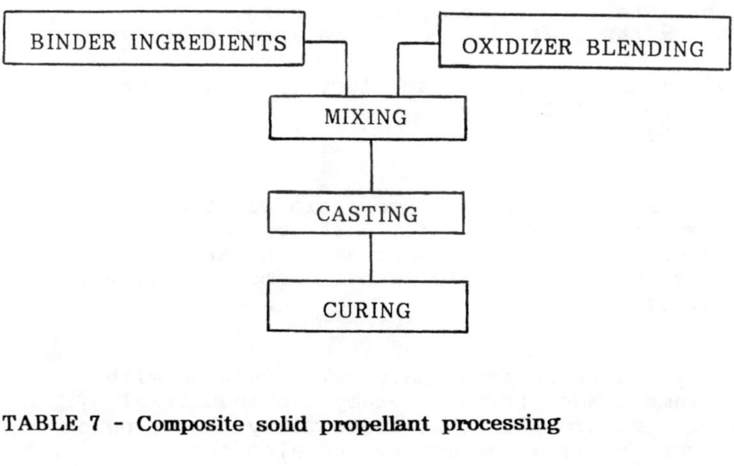

TABLE 7 - Composite solid propellant processing

CONCLUSION

Based on the initial formulations and the success of their foregoing evaluations a large-scale mix was prepared from horizontal mixer for 1000 kgs and Vertical "DAY" mixer for 2400 kgs.
From these initial formulations, additional modifications were introduced to evaluate process and processibility of the propellant. Rheological properties, burning rate range capability and mechanical properties improvement were evaluated.
High volume manufactoring have been conducted to provide reproductible process required for the production of large quantities of base-bleed composite grain at low cost.

REFERENCE

[1] S. Gordon, B.G Mc Bride
Computer program for calculation of complex chemical equilibrium composition - rocket performance - incident and reflecting shocks and CHAPMAN JOUGET detonation.
NASA Lewis research center (1971)

COMMENTS

B.Bolinder, Nobel Chemicals, Nobel Industries Sweden, KARLSKOGA . SWEDEN. In your Table 2, you designated the different particle size distributions by the letters A, D, F and b. What do these letters stand for ?

Author's Reply. Each particule size distribution is designated by a letter. In the present case :
AP : A : 200 µm (AP.b similar to AP : A about particle size)
AP : D : 90 µm and AP : F : 6 µm. That is an internal normalization fo SNPE Company.

J. Fleming, Fleming Consulting International, BUCKS SL 90 JJ UNITED KINGDOM. Please clarify what type of mixers you use to produce. B.B. propellant at SNPE. Are horizontal mixers used (as this is not normal practice in other companies).

Author's Reply. SNPE is principaly well equipped with horizontal sigma blade mixers. Thereby the industrial production of B.B. propellant use essentialy this type of mixers. However, if it is necessary, we also the possibility to use vertical mixer (300 Gal.)

S. Dart, Space Research Corp S.A. BRUXELLES. A shift factor, a_T, is mentioned in your paper. Can this factor be used to estimate the mechanical properties for short-time, responses corresponding to high strain rates ? Is there a correlation between sample temperature and strain. For example, can mechanical properties at low temperature be related to the properties at high strain rates ?

Author's Reply. The shift factor mentioned in our paper and computed according to the theory of W.L.F, can be used to estimate the mechanical properties for short time or high strain rates with a very good correlation. Also good relation appears between mechanical properties at low temperature and properties at high strain rates.

T. Wijker, Muiden Chemie. MUIDEN. NETHERLANDS. Is there significant change in your shift factor a_T between a solid filling of 70 % and 80 % AP particles ? Can you differentiate the mechanical properties under strain rate of 50 mm/min at -40°, +60°C with respect to + 21° for both 70 % and 80 % AP base bleed propellant samples ? What would be the most optimum mechanical properties for a B.B. propellant grain, when represented by the JANNAF - dogbone at + 21° C at a rate of 50 mm/mn ?

Author's Reply. An average shift factor a_T is eveluated for a large family of propellant. i.e Base bleed propellant between 70 an 80 percent of AP solid fillers. It's suffisant for a primary evaluation. With a good compromise between the magnitude of the effects of each formulation parameters (i.E; particule size and distribution and curing ratio) and mechanical properties, the "master-curve" can be simular in a large range of time-temperature correspondence for both propellant loaded to 70 or 80 percent AP. For better optimization of shift factor a_T it is necessary to reevaluate the "master-curve" and time-temperature correspondence for each typical propellant. For a Base Bleed propellant grain the best mechanical properties between two propellant with same modulus are those from the propellant having the bigger stress and the maximum elongation.

Study of Pyrotechnic Compositions as Base Bleed Propellant

GONGPEI PAN, CHANGJIANG ZHU, and ZHAOQUN WANG
Department of Chemical Engineering
East China Institute of Technology
Nanjing, PRC

ABSTRACT

This paper describes pyrotechnic compositions as base bleed propellant for increasing the range of projectile.

The wind tunnel test, the dynamic simulation test, the live projectile flight test are carried out with $Sr(NO_3)_2$-Mg-CaRes(calcium resin) and $Sr(NO_3)_2$-Mg-NH_4ClO_4 additives as a center, and better results are obtained. The paper shows that it is feasible and prospective to use pyrotechnic compositions as base bleed propellant.

INTRODUCTION

The current bleed propellants used in base bleed projectiles are composite propellants mainly of hydroxyl terminated polybutadiene(HTPB) and ammonium perchlorate(AP).[1] On the basis of pyrotechnics, we choose pyrotechnic compositions as base bleed propellants with the intention of improving ignition, increasing range, simplifying technological processes and reducing cost of the projectiles.

TEST CONDITIONS AND METHODS

1. Samples

The samples for the live projectile flight test are machined to the product structure. The sample used in laboratory is shown in Figure 1.

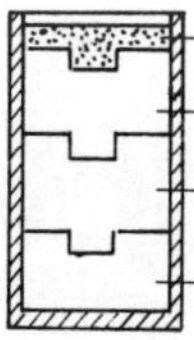

FIGURE 1. Sample charge

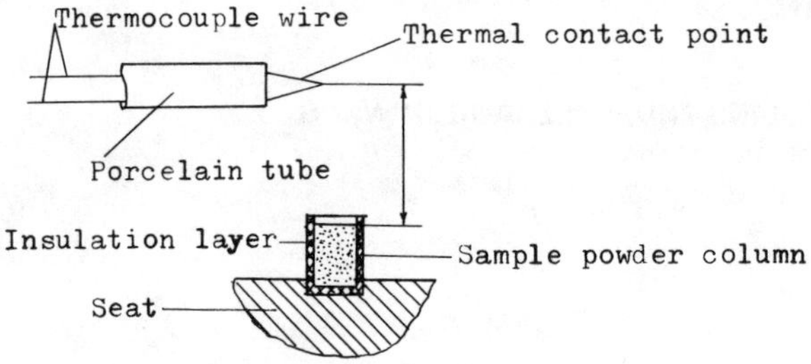

FIGURE 2. Illustration of temperature measurement

2. Temperature Measurement

The measuring device having W-Re thermocouple as its sensor is shown in Figure 2, by which the temperature on the specific point in flame area can be measured.

3. Gas Measurement

The gas-measuring device is shown in Figure 3.

4. Dynamic Drag Reduction Simulation Test

Live firing from 14.5mm machine gun and flight velocity measuring by velocity radar are made in the test. And the drag reduction is judged from the attenuation of the velocity.

5. Live Projectile Flight Test

A flight comparison test is carried out under the same conditions, of small-caliber and medium-caliber projectiles between both with base bleed composition and without it. The drag reduction and the range increase are determined.

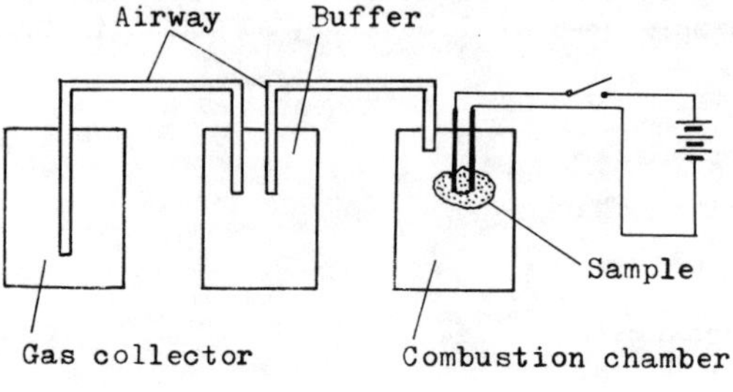

FIGURE 3. Device of measuring gas value

TABLE 1. Formulation Test of $Sr(NO_3)_2$-Mg-CaRes

No.	Formulation (%)				n	K	Combustion temp. T(°c)	Combustion velo. $\bar{u}$(mm/s)
	$Sr(NO_3)_2$	Mg	CaRes	DOP				
1	69	20	10	1	-15.0	0.63	1905	2.77
2	64	25	10	1	-20.2	0.54	1942	2.80
3	61.5	27.5	10	1	-22.8	0.50	2058	2.96
4	59	30	10	1	-25.4	0.47	2100	3.28
5	54	35	10	1	-30.5	0.40	1942	3.26
6	49	40	10	1	-35.7	0.34	1835	2.59

TESTS AND RESULTS

1. Formulation Test

Test result of $Sr(NO_3)_2$-Mg-CaRes formulation. Test results of combustion temperature and velocity of the formulation under static conditions are listed in Table 1. In Table 1, K is the oxidant availability coefficient of the composition. It is ratio of oxidant in the composition to the oxidant necessary for complete combustion of the total combustibles in the composition, while the oxidant dfference n is the difference between both.

Table 1 tells that the formulation with K=0.47 has higher combustion temperature. The measured gas of the formulation is 348.8ml/g. The average molecular weight is 33.15.

The wind tunnel test of the K=0.47 formulation shows poor ignition at low pressure and low drag reduction(see Table 2).

Adjustment test of $Sr(NO_3)_2$-Mg-CaRes formulation. This test is carried out with the addition of AP and the removal of DOP(dioctylephthalate) to keep the fluctuation of oxygen balance within a small range(K=0.40-0.48). The test results are shown in Table 3.

The test proves that combustion of the adjusted formulation is bettered, the measured temperature is increased, especially, that of No.2

TABLE 2. Result of the Wind Tunnel Test of $Sr(NO_3)_2$-Mg-CaRes formulations

No.	Stable pressure in the wind tunnel (Kg/cm^2)		Pressure after combustion (Kg/cm^2)		Remarks
	Excess pressure	Absolute pressure	Excess pressure	Absolute pressure	
1	-0.4433	+0.5567	-0.3645	+0.6355	
2	-0.4531	+0.5469	-	-	unignited
3	-0.4630	+0.5370	-	-	unignited
4	-0.4433	+0.5567	-	-	unignited
5	-0.4334	+0.5666	-0.3448	+0.6552	
6	-0.4531	+0.5469	-	-	unignited
7	-0.4433	+0.5567	-	-	unignited

TABLE 3. Adjustment Test of $Sr(NO_3)_2$-Mg-CaRes Formulation

No.	Formulation (%)				n	K	Measured temp. T(°c)	Combustion time t(s)
	$Sr(NO_3)_2$	Mg	AP	CaRes				
1	41.4	23.9	19.7	15	-32.85	0.40	1700	7.7
2	35.0	35.0	20.0	10	-29.31	0.41	2975*	3.4*
3	33.5	28.8	25.7	12	-29.11	0.42	1852	7.4
4	30.6	32.2	27.2	10	-26.67	0.44	1990	6.1
5	29.4	33.3	27.7	9	-25.04	0.45	2034	5.2
6	28.2	35.6	28.2	8	-24.21	0.46	2168	4.9
7	26.0	39.0	29.0	6	-21.24	0.48	2218	4.95

* Data obtained under the condition of rotating at 18,000 rpm

formulation under rotating condition is considerably raised. Rotation increases the mass burning rate, causes the atmosphere surrounding the flame to flow, with the second combustion efficiency being increased due to abundant oxygen.

The measured gas of the formulation is 448ml/g. The average molecular weight is 29.46.

Test of AP-Mg-CaRes formulation. To reduce the molecular weight still further, $Sr(NO_3)_2$ is entirely replaced by AP. The curves obtained in the test are shown in Figure 4(a),(b).

The measured gas of the formulation is 570-690ml/g. The average molecular weight is decreased below 25. Meanwhile its mechanical sensitivity is rather high, so it is inconvenient to process.

Test of $Sr(NO_3)_2$-Mg-AP-additive formulation. The test is carried out with the intention of increasing the combustion temperature, decreasing the average molecular weight, and especially forming the secondary combustion of the product.

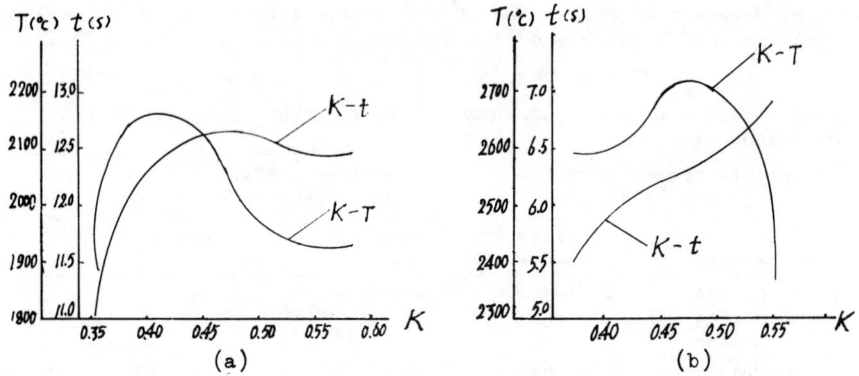

FIGURE 4. Characteristic curves of the relationship between combustion temperature, time and ingredient ratio of the AP-Mg-CaRes composition, (a) static combustion; (b) spin combustion

TABLE 4. Amount of the Base Drag Reduction in Wind Tunnel Test

No.	Mass flow rate (Kg/s)	Mach number M	Jet parameter I	Rate of base drag reduction $\Delta\bar{C}_{Db}$(%)
1	2.70×10^{-3}	1.72	1.58×10^{-3}	50
2	4.05×10^{-3}	1.72	2.35×10^{-3}	61
3	5.90×10^{-3}	1.72	3.43×10^{-3}	69.9
4	6.15×10^{-3}	1.72	7.68×10^{-3}	78
5	8.26×10^{-3}	1.72	10.00×10^{-3}	88.9

Proper selection of additive will make the formulation a bleed propellant burning at a low rate. The wind tunnel test results of naphthalene are listed in Table 4. In Table 4,

$$I=\frac{\dot{m}}{\rho_\infty u_\infty A}$$

where $\dot{m}$—average mass burning rate of the compositions, ρ_∞—free stream static pressure, u_∞—free stream velocity, A—area of the burning surface and I—injection parameter.

Figure 5 shows the result of the present formulation and the HTPB-AP composite propellant in the wind tunnel test with the Mach number being 1.3, 1.73 and 2.5. Obviously, when M=1.3, the drag reduction of pyrotechnic compositions is much better than that of composite propellants as I is increased fuether. The greater the Mach number, the better the drag reduction of $Sr(NO_3)_2$-Mg-AP-additive formulation. Figure 6 shows the variation of projectile base pressure with combustion time of the additive naphthalene formulation with M=1.3 and M=2.5.

2. Dynamic Simulation Test

The test is carried out in 14.5mm machine gun, and the velocity curve of the flying projectile is determined by velocity radar. In the test, $Sr(NO_3)_2$-Mg-AP-additive formulation is pressed into the simulation projectile base. The test result is shown in Figure 7. The curve B in

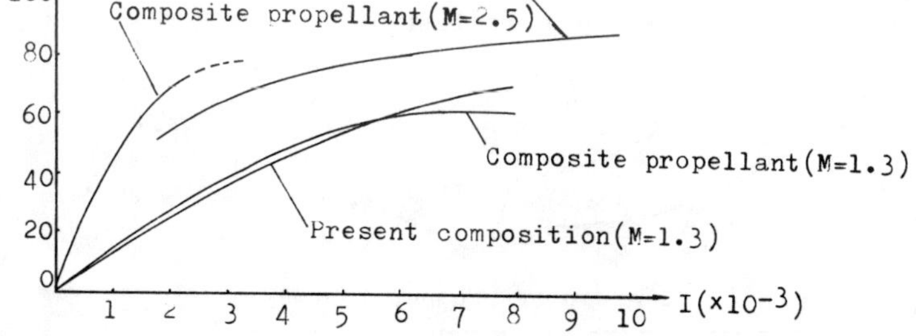

FIGURE 5. The test result in the wind tunnel

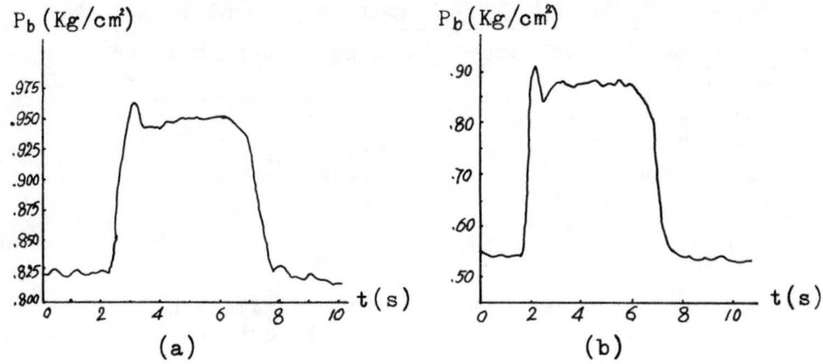

FIGURE 6. Curves of change of the base pressure with the combustion time when (a) M=1.3; (b) M=2.5

the figure refers to naphthalene additive.

3. Live Projectile Flight Test

Test result of small-caliber projectile. 37mm aerogun is chosen to perform the test. The test result shows high ability to remain velocity (see Table 5) and fine drag reduction (see Figure 8) of 37mm base bleed projectile loaded with present pyrotechnic composition.

Test result of medium-caliber projectile. 122mm howitzer is chosen to perform live projectile flight test. Its result is shown in Table 6.

DISCUSSIONS

1. According to the design principles of base bleed propellant[2]: (a) relatively low combustion velocity or mass ejection rate, (b) relatively high temperature of ejected gas, (c) relatively low molecular weight of ejected gas, (d) combustion in the wake, i.e., the secondary combustion, $Sr(NO_3)_2$-Mg-AP-additive used as the base bleed propellant can broadly conform to the above-mentioned principles. Its combustion rate is low, while the measured temperature is high because of the

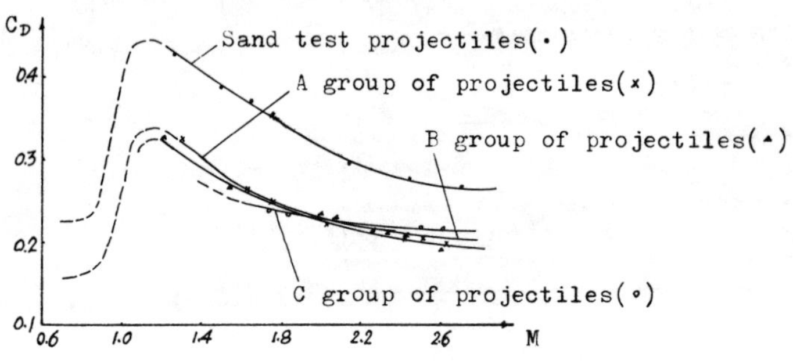

FIGURE 7. Projectile drag curves in dynamic simulation test

TABLE 5. The Remaining Velocity of Flying Live Projectiles

Type of proj.	37mm proj. without base bleed agent		37mm proj. with base bleed agent		35mm proj. (foreign)	
Proj. weight(g)	740		640		550	
V_i(m/s)[a]	1000		1162		1175	
Slant range(m)	t(s)[b]	V_r(m/s)[c]	t(s)	V_r(m/s)	t(s)	V_r(m/s)
2000	2.67	633	2.09	796	2.18	720
3000	4.75	446	3.48	658	3.78	530
4000	7.70	324	5.15	543	5.96	385
5000	11.75	244	7.18	444	9.15	292
5500			8.46	391		
6000			9.71	357		

a) V_i—Initial velocity; b) t—Flight time; c) V_r—Remaining velocity

presence of SrO, MgO in product. Mg with low boiling point can be easily evaporated (the melting point 651°C, the boiling point 1100°C), and the low molecular weight (M=24.3) of Mg after being evaporated can easily form the secondary combustion in the wake, which can not be accomplished by HTPB-AP. The advantage of $Sr(NO_3)_2$-Mg-AP-additive over HTPB-AP can be seen both in the wind tunnel test and the live projectile flight test.

2. It is possible to reduce the average molecular weight of combustion product through the use of the pyrotechnic formulation as the bleed propellant in the range of compositions. Table 1 and Table 2 refer to the same formulation. In Table 1, the average molecular weight of formulation 4 is 33.15, while in Table 2 the average molecular weight of the adjusted formulation 2 goes down to 29.46. If all the oxidizer is replaced by AP, the average molecular weight of the formulation can be reduced to less than 25.

3. It is feasible to use the pyrotechnic compositions in increasing the range of projectile. The pyrotechnic compositions can greatly in-

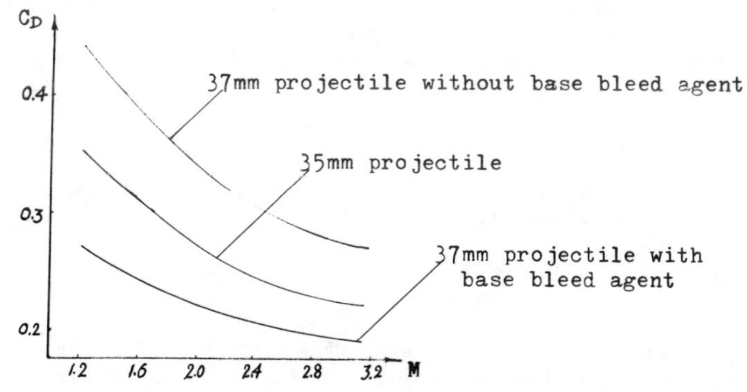

FIGURE 8. The C --M curves of small-caliber projectiles

TABLE 6. Range Increase of Live Projectile Dynamic Flight

Kind of base bleed agent	V_i (m/s)	Range(m)		Precision		Range increase (%)
		Proj. A[a)]	Proj. B[b)]	Deflection(m)	Range	
$Sr(NO_3)_2$-Mg-AP-additive	550	11008	13031	7.3	1/237	18.4
Ditto	520	12877	15027	3.1	1/233	16.7
Ditto	690	17500	21887	12.7	1/279	25.1
HTPB-AP	550	11008	12328	17.3	1/220	12.0

a) Projectile without the base bleed composition; b) Projectile with the base bleed composition

crease the range, simplify the technological processes and reduce coat under certain conditions. As the bleed propellant, they hold the bright promise of much more to come.

REFERENCES

1. Nils-Erik Gunners, Rune V. Hellgren and Torsten Liljergren, Gun Projectile Arranged with a Base Drag Reducing System, U.S. Patent 4,213,393, Jul. 1980.

2. Ward., J. R., Baltakis, F. P. and Pronchick, S. W., Wind Tunnel Study of Base Drag Reduction by Combustion of Pyrotechnics, BRL-R-1745, AD-B000431, April 1974.

COMMENTS

drs W. Pell, NL Ministry of Defence, Netherlands. Refer to Table 3. Did you calculate the adiabatic flame temperature? If so, what is it? What mechanism do you propose for a higher T, as we expect only higher burning rates?

A. J. Stomp, 2287 VT Rijswijk, Netherlands. In Table 3, test No. 2 is held under rotating conditions. It seems to have great influence on the combustion temperature and combustion time. As these rotating conditions are normal for a rotation-stabilized shell, why are not all the tests mentioned in these table held under rotating conditions? Comparism of test 2 with the other tests 1, 3, 4, 5, 6 and 7 seems impossible.

Author's Reply. We have calculated the combustion temperature of present base bleed compositions according to

$$t = \frac{Q - \Sigma(Q_s + Q_k)}{\Sigma C_p}$$

given by A. A. Shidlovskiy in "PRINCIPLE OF PYROTECHNICS"(refer to AD/A-001859 pp142-146). The obtained results are not actual because the C_p changes with the change of temperature and the reacting product is complicated which it is difficulty to determine.

The calculated results are not mentioned in our paper and the given temperature in the tables are all measured by using a W-Re thermocouple. It is only the temperature on the specific point in flame area and not the flame temperature. The further research on the temperature field will be carried on.

About the reason of the higher temperature under rotating conditions, our views are in the following.

(1) The increase of the combustion velocity under rotating conditions makes the exothermic quntity increase in a unit time, thus geting a raise in temperature.

(2) Under the rotating condition, the bundary of the flame and air is expanded and the secondary combustion is promoted. As a result, the reaction produces larger amount of heat, which makes the temperature raise.

Ignition and Combustion Behavior of MTV Igniter Materials for Base Bleed Applications

B. L. FETHEROLF, D. M. CHEN, T. S. SNYDER, T. A. LITZINGER, and K. K. KUO
Department of Mechanical Engineering
The Pennsylvania State University
University Park, Pennsylvania 16802, USA

ABSTRACT

The pyrolysis and ignition characteristics of an igniter material composed of magnesium, Polytetrafluorethylene, and Viton A have been studied using a high-power CO_2 laser as the radiative heating source. Methods of analysis and the instrumentation employed were high-speed direct and schlieren photography to visualize the flame structure and gas-phase dynamics, near-infrared photodiodes to obtain ignition delay data, and probe sampling and a gas chromatograph/mass spectrometer to analyze the gaseous products evolved. Important test variables were incident heat flux, chamber pressure, and oxygen concentration. In addition, the effect of boron addition on the ignition behavior of the igniter charge was studied.

Ignition occurred on the sample surface under all test conditions. Experimental results indicated that the ignition delay time decreased monotonically as the incident heat flux was increased. Also, ignition times decreased as the ambient pressure or the ambient oxygen percentage was reduced; it is believed that this decrease is caused by competitive oxygen and fluorine reactions. The direct and schlieren images showed that the burning behavior was also distinctly different as the chamber pressure or oxygen partial pressure was reduced. At low pressures (~ 0.1 atm), no evident luminous plume evolution and gas-phase dynamics were observed during combustion, and the igniter material decomposed in large fragments. In an inert atmosphere, pure PTFE pyrolyzed into many fluorocarbon compounds and Viton A pyrolyzed into a variety of fluorocarbons and hydrofluorocarbons. However, for combustion of the composite MTV material in air and in an inert helium atmosphere, only hydrocarbon compounds, and CO and CO_2 for combustion in air, were observed. The addition of boron or Viton A to a basic Mg/PTFE composition significantly lowered the ignition delay time and the addition of boron also greatly enhanced combustion.

1. INTRODUCTION

The development of a reliable and efficient base bleed system to increase the range of artillery projectiles is currently of great interest. This system is designed to inject gases into the wake behind the projectile, increasing the base pressure and reducing the profile drag, which accounts for 50-75% of the total drag. This reduction of the drag will increase the range capabilities of the projectile. To insure sustained ignition of the base bleed propellant during the transition from the severe gun barrel environment to the ambient air and also during the subsequent flight, a supplementary, redundant igniter system is required. Thus, the base bleed igniter propellant must be able to sustain burning over a wide range of pressures, temperatures, and gaseous environments. It must also be durable enough to withstand the tremendous forces (~ 10,000 G's) of setback acceleration and the centrifugal forces of high-speed rotation (16,000 rpm).[1]

Most of the work to date on base bleed has concentrated on the aerodynamics and selection of candidate propellants. Several researchers have done theoretical calculations of the effect of base bleed systems on the aerodynamics of a projectile.[2-6] Experimental simulations have also been conducted to validate the effectiveness of base injection.[7,8] Kloehn and Rassinfosse have written a paper on the manufacture of base bleed propellants.[9] However, very few papers have been published to date on studies of the performance of practical base bleed motors and their supplementary redundant igniter systems.

Magnesium/PTFE/Viton A (MTV) igniter materials have been considered in solid rocket motor igniters and are now being tried experimentally in base bleed igniter systems. Peretz[10] stated that the use of magnesium as the metal counterpart is beneficial because it ignites and burns readily, thus promoting efficient combustion, and requires small amounts of oxidizing agent to burn. Viton A $(-C_5H_{3.5}F_{6.5}-)_n$, a copolymer of vynilidene fluoride and perfluoropropylene, is often added to Mg/PTFE formulations to increase homogeneity and facilitate fabrication. Because of its chemical similarity to Polytetrafluoroethylene [PTFE → $(-C_2F_4-)_n$], significant amounts of Viton A can be added without altering the combustion characteristics.

Peretz[11] listed the following advantages of the MTV formulation for use as an igniter material: high energy content, high degree of safety in preparation, low temperature and pressure dependence of the burning rate, ease and low cost of igniter pellet or grain fabrication, favorable aging characteristics, and stable burning at low pressures. He also listed several advantages of metal-fluorocarbon compositions in comparison with metal-hydrocarbon materials: higher volumetric heat release, higher rate of reaction, much higher density, excellent thermal stability (due to the high C-F bond strength), higher activation energy, and much lower production of soot due to the lower content of carbon. For magnesium, the standard gravimetric and volumetric heats of fluorination [producing $MgF_{2(s)}$] are almost exactly twice as much as those for oxidation [producing $MgO_{(s)}$].

The thermal decomposition of PTFE has been studied by Griffiths et al.[12] and Kubota and Serizawa.[13] Both groups used simultaneous DTA/TG with heating rates of 20 K/min[12] and 10 K/min.[13] Griffiths et al. observed that the decomposition of PTFE was endothermic in argon and nitrogen but was exothermic in air. However, Kubota and Serizawa observed the decomposition of PTFE in argon to be exothermic. Griffiths et al. also observed that the decomposition process became increasingly exothermic in all gaseous environments as an increasing amount of magnesium was added to the PTFE, and the nature of the decomposition in argon changed from purely endothermic for PTFE to purely exothermic for a Mg/PTFE (3:2) mix.

Kubota and Serizawa have published two of the more comprehensive papers on experimental studies of the combustion of Mg/PTFE materials.[13,14] Their material also contained 3% Viton A. Chen et al.[15] have conducted studies of the thermophysical properties and combustion of metal-based solid fuels in a strand burner, including the material that is the subject of this paper. They found that the burning rate was consistently higher in a nitrogen atmosphere than in air, and they also measured the thermal diffusivity to be 1.6×10^{-2} cm^2/sec. However, to the authors' knowledge, no experimental papers have been published to date on the ignition behavior of MTV igniter materials.

Characterization of the ignition behavior of the MTV material under a wide range of conditions is critical for determining the feasibility, applicability, and subsequent performance of the igniter. Radiative ignition has been often used in the studies of ignition behavior of solid propellants. The high-power CO_2 laser is now the prevailing choice for the radiant energy source due to its advantages of precise control of lasing time and energy flux, and a high level of beam intensity that simulates energy fluxes found in igniters and propulsive devices.

The overall objective of this research was to investigate the ignition behavior of a Mg/PTFE/Viton A igniter material under various ambient conditions, to gain a better understanding of the processes governing ignition, and to correlate this data with conditions experienced by a base bleed igniter. Specific objectives were: (1) to investigate the flame structure and gas-phase dynamics during the ignition transient under various pressures and ambient oxygen concentrations, (2) to study the variation of ignition delay time as a function of pressure, ambient oxygen concentration, and incident heat flux, and (3) to analyze the gaseous species produced during pyrolysis and ignition of the MTV material and during pyrolysis of PTFE and Viton A individually.

2. EXPERIMENTAL APPROACH

A diagram of the overall experimental set-up is given in Figure 1. The radiative energy source was a high-power Coherent Super 48 CO_2 Laser, capable of producing 800 watts of power in the continuous wave mode and 3500 watts in the pulsed mode with precise control of the power output and lasing time. A beam profile, uniform within ± 10% across the propellant sample surface, was obtained using an aperture with a diameter of 7 mm to allow only the most uniform center section of the beam to irradiate the surface. For all tests, the actual heat flux was measured with a calorimeter. A cubic Plexiglass test chamber, 25 cm on a side, with a stainless steel top cover was used for ignition delay tests and schlieren visualization. Two high-quality glass windows were installed in opposite sides of the chamber for the schlieren flow visualization. For the species sampling tests, in order to have considerable species concentrations, a smaller Plexiglass test chamber, 10 cm long by 5 cm in diameter with aluminum end caps, was employed. A septum, installed in the side of the chamber, allowed for gas sampling with a syringe. The top covers of both chambers were fitted with potassium chloride windows, which are highly transmissive in the infrared range, for beam entrance. Pressure and composition of the initial gas in the chamber was controlled using a vacuum pump, a pressure/vacuum gauge, and an inert gas supply system.

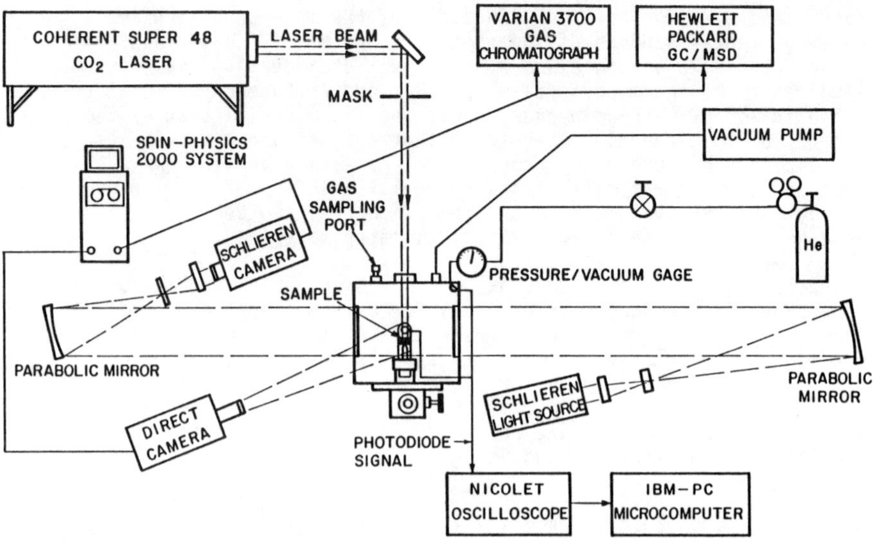

FIGURE 1. Experimental setup for CO_2 laser pyrolysis and ignition

Both the direct and schlieren images were recorded with a Spin Physics 2000 video recording system. As seen in Fig. 1, the system consists of a controller, video monitor, and two camera heads. Maximum recording speed is 12000 pps, but a 2000 pps recording rate was employed for this research. The controller can display both the schlieren and direct images simultaneously, allowing direct comparison of the flame structure and gas-phase dynamics. The schlieren system is a "Z-type" configuration, employing 15 cm parabolic mirrors to direct a parallel light beam through the test chamber from a 100 watt continuous tungsten-halogen lamp. The beam is directed and shaped by lenses and rectangular apertures at each end as it travels from the light source, through the chamber, and into the video camera at the other end.

The MTV igniter material studied in this research consisted of 50% magnesium particles of 325 mesh size and 99.8% purity. The balance of the material was composed of a granular resin of PTFE particles (dia. ~ 30 μm) and Viton A particles. The boron powder used in the boron/Mg/PTFE formulations was composed of 99% pure amorphous particles with a 0.5 μm average size.

Ignition delay times were determined using two near-infrared photodiodes (spectral response: 0.35-1.15 μm) to sense the first light emission from a flame or heat production from chemical reactions. One photodiode was positioned seven cm from the sample and aimed directly at its surface while the other one was positioned in an upper corner of the chamber and aimed 2.5 cm above the surface to observe the gas-phase infrared emission. To signal "zero-time" on both the photodiode traces and the direct video image, a strobe light was employed. The data was recorded on a Nicolet oscilloscope and transferred to an IBM PC for processing.

After a test, samples of gaseous products were drawn from the test chamber with a one milliliter syringe. These gaseous product species were then analyzed using a Hewlett/Packard gas chromatograph/mass spectrometer. Most of the samples were analyzed with a Carbopack B/5% Fluorocol column run isothermally at 50°C. A column with Porapak Q&R in series, used to obtain data on hydrocarbon species, was run isothermally at 50°C for the first 5 minutes and then ramped at 15°C per minute up to 200°C for the remainder of the run.

3. FLAME STRUCTURE AND GAS-PHASE DYNAMICS

Figure 2 is a sequence of images depicting the ignition transient for a sample ignited in air at 1 atmosphere. Both the direct and schlieren photographs of the ignition sequence are shown. The times below each pair of images represent the time elapsed from the onset of laser heating. In each pair of images, the sample surface is at the same height; however, the magnification of the direct image is slightly less than that of the schlieren image.

Ignition occurs in Fig. 2a after a delay time of 19 milliseconds. The faint light emission seen on the sample surface in the direct image corresponds to the abrupt rise in the photodiode signal (within 2 milliseconds) that was used as the criterion for ignition. The initial evolution of a gaseous sphere on the schlieren image occurs slightly off the center of the sample due to minor variations in the laser beam intensity profile. The gaseous sphere expands upward in Fig. 2b and a ball of flame is established within the dark outer gradients of the sphere. The flame and evolved gases continue to propagate through Figs. 2c-2f. A mushroom-shaped gas-phase structure can be seen inside the outer gradient in the schlieren picture in Fig. 2e. The cap of the mushroom is due to the formation of a vortex resulting from the

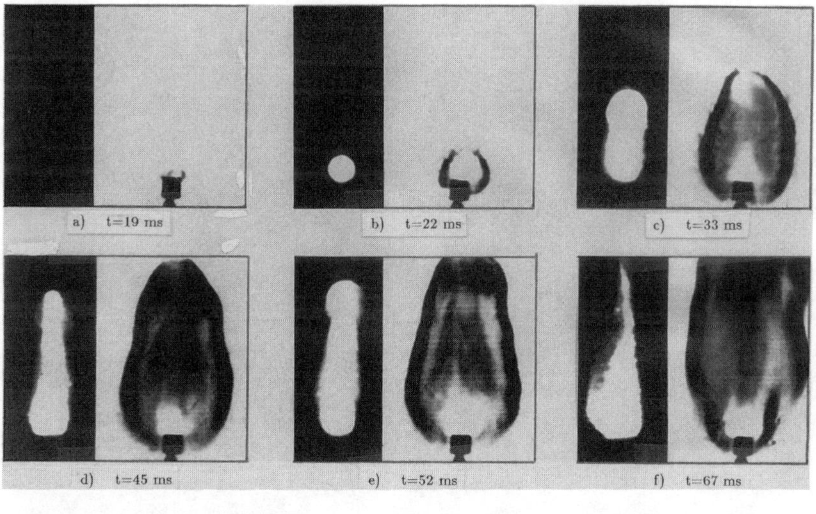

FIGURE 2. Ignition sequence of the MTV igniter material in air at 1 atmosphere ($q'' = 640$ W/cm^2)

propagation of the initial sphere of gases evolved, and the stem is formed by the continued evolution of a stream of gases normal to the surface. The actual steady-state flame shape has not yet been established in Fig. 2f, but it is quite similar in size to the inner flame region that can be seen in the schlieren image. Large particles can be seen in both the direct and schlieren images in this figure.

An ignition test conducted at a low pressure of ~ 0.1 atmosphere is displayed in Fig. 3. Clearly, this ignition process is much less vigorous than the rather explosive dynamics observed in the higher pressure ignition. The first emission of visible light does not occur until 108 milliseconds, but the photodiode signal abruptly increases at 32 milliseconds. Evidently, the rise in the photodiode signal was caused by a significant amount of heat (infrared emission) being evolved well before the visible light of a flame was observed. Thus, the ignition at low pressure is correlated with the significant evolution of heat from chemical reactions; the heat evolution does not directly correspond to the onset of a visible flame.

Two distinct phenomena observed in Fig. 3 are the abundance of very large particles and fragments evolved from the sample and the lack of any gas-phase structure in the schlieren pictures. Figures 3b and 3c depict the evolution and transport of two large fragments from the surface, and Fig. 3d shows the first sign of a substantial flame zone established on the surface. Then in Figs. 3e and 3f, huge amounts of very large particles are evolved as the sample "burns." In this test, the side surfaces of the sample were not inhibited, and particles evolved horizontally in Fig. 3f are a result of side burning. After the test, these particles fully coated the schlieren windows 15 cm from the sample and blocked the schlieren image by the time the sample had been completely burned.

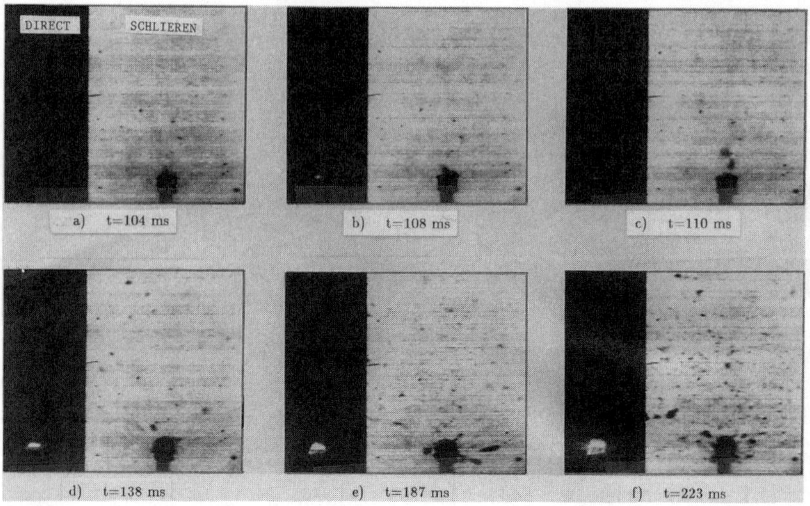

FIGURE 3. Ignition sequence of the MTV igniter material in air at ~ 0.1 atmosphere (q" = 430 W/cm^2)

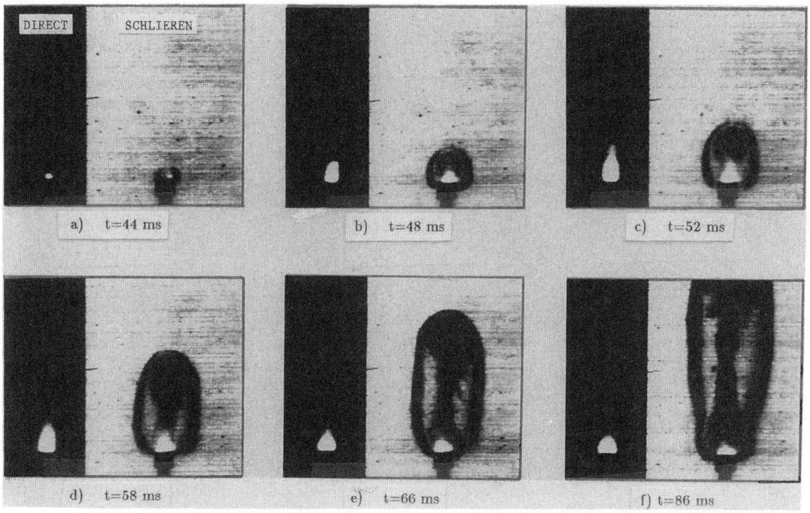

FIGURE 4. Ignition sequence of the MTV igniter material in an inert N_2 environment at 1 atmosphere (q" = 430 W/cm^2)

A set of images depicting ignition in an inert atmosphere of nitrogen (~ 0.5% O_2) is given in Fig. 4. The gas-phase dynamics follow roughly the same progression as those for tests in air. The gradients are noticeably darker, though, indicating stronger density and/or temperature gradients, and the flame structure is obviously quite different from the ignition tests in air at one atmosphere. The flame initially propagates to a height around 1.0 centimeter above the sample surface and then recedes to the steady-state shape seen in Fig. 4f. Comparing Figs. 2 and 4, it can definitely be concluded that oxygen plays a significant role in the reactions that produce a large luminous plume during ignition. The final steady-state flame is also larger for the case of air.

Finally, Fig. 5 displays two schlieren images with different levels of magnification that were obtained using different focal lengths of collimating lenses in front of the camera. The lenses used for Figs. 5a and 5b provide magnifications of 1.8 and 4.5, respectively, of the images displayed in Figs. 2-4. Both images in Fig. 5 represent ignition in air at 1 atmosphere. Measurements taken from all the images, irrespective of the level of magnification, corresponded very well, confirming the accuracy and reproducibility of the schlieren system. These measurements showed that the primary flame zone established on the sample surface is 1.2-1.5 mm in height.

4. IGNITION DELAY BEHAVIOR

The ignition delay behavior of the MTV igniter material was investigated as a function of pressure, heat flux,and ambient oxygen concentration. The ignition delay times were determined from the photodiode traces as the elapsed time from the initiation of laser heating to the distinct point where the signal rose very abruptly, evidenced by the typical

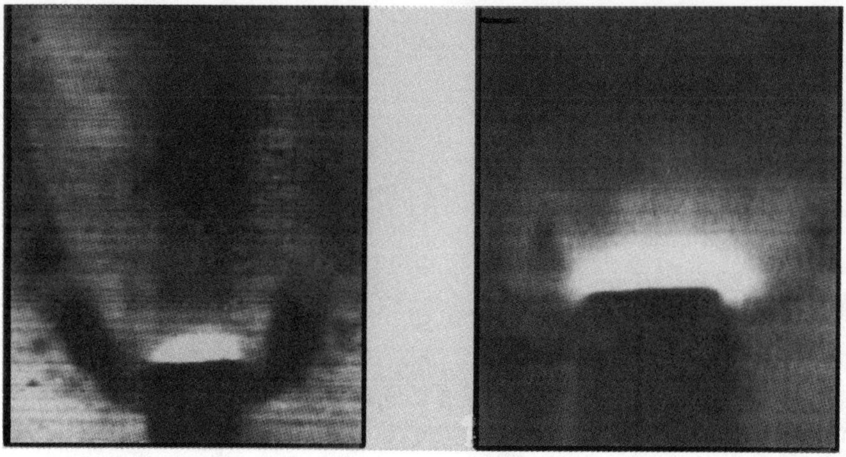

FIGURE 5. Primary flame zone attached to sample surface; two different magnifications of the size of the schlieren image in Figs. 2-4

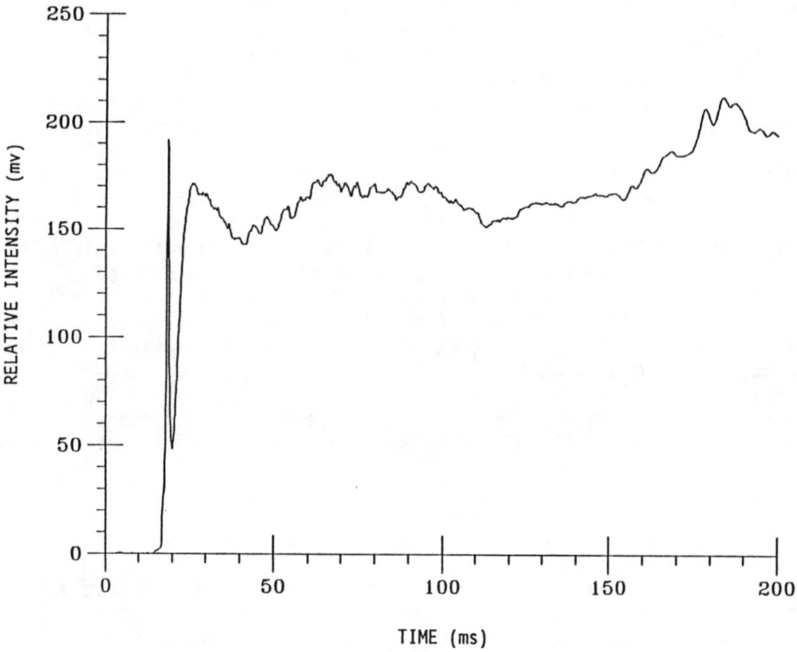

FIGURE 6. Near-field photodiode trace of ignition in air at 1 atmosphere (q" = 670 W/cm^2)

photodiode trace given in Fig. 6. There is a very abrupt "spike" in the signal which drops and then rises again to a relatively steady state. This ignition delay time was normally within 2-3 milliseconds of the time elapsed until the first light emission observed in the direct image. After ignition, combustion of the samples was sustained until they were totally consumed, regardless of the ambient conditions.

Figure 7 exhibits the ignition delay behavior as a function of ambient pressure and incident heat flux. The two curves indicate the pressure dependence, one for ignition at 1 atmosphere and the other in a relative vacuum of less than 0.1 atmosphere. Most importantly, it can be seen that the ignition delay times are shorter at lower pressure. Also, the profiles of both curves as a function of the radiative heat flux from the CO_2 laser are quite similar. In other words, lowering the pressure merely serves to shift the curve down while maintaining the same flux-dependent profile. However, the reason for the shorter delays at lower pressures can most likely be explained by the results given in Fig. 8.

Figure 8 illustrates the effect of the ambient oxygen concentration on the ignition delay behavior at a constant heat flux. The curve shows that the ignition delay time increases with an increase in the ambient oxygen concentration. Although more data is needed to confirm this behavior, the trend is believed to be correct, and it represents a significant point in understanding the ignition mechanisms. This behavior is also the most likely justification for the lower ignition delays at lower pressures observed in Fig. 7, since as pressure decreases, the concentration of ambient oxygen also decreases.

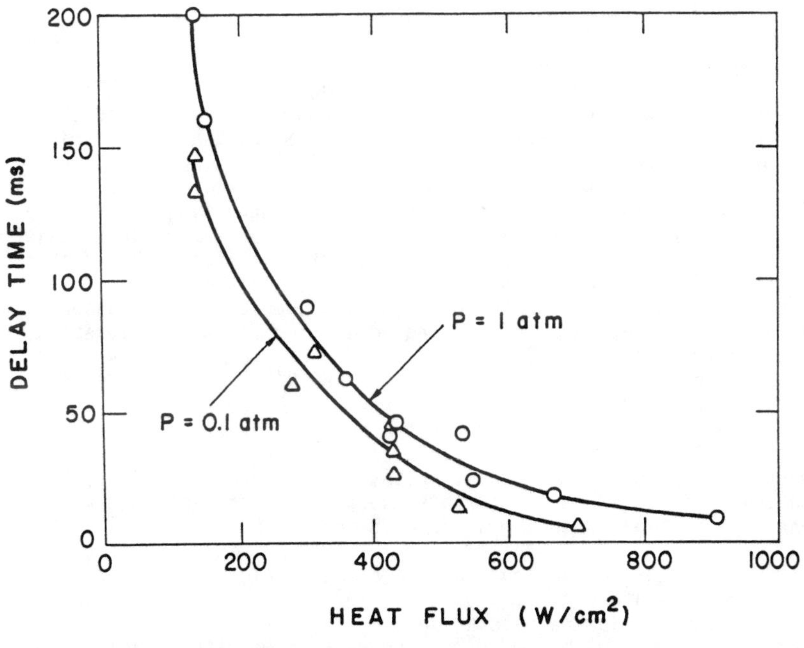

FIGURE 7. Effect of pressure and incident heat flux on ignition delay time of MTV igniter material in air

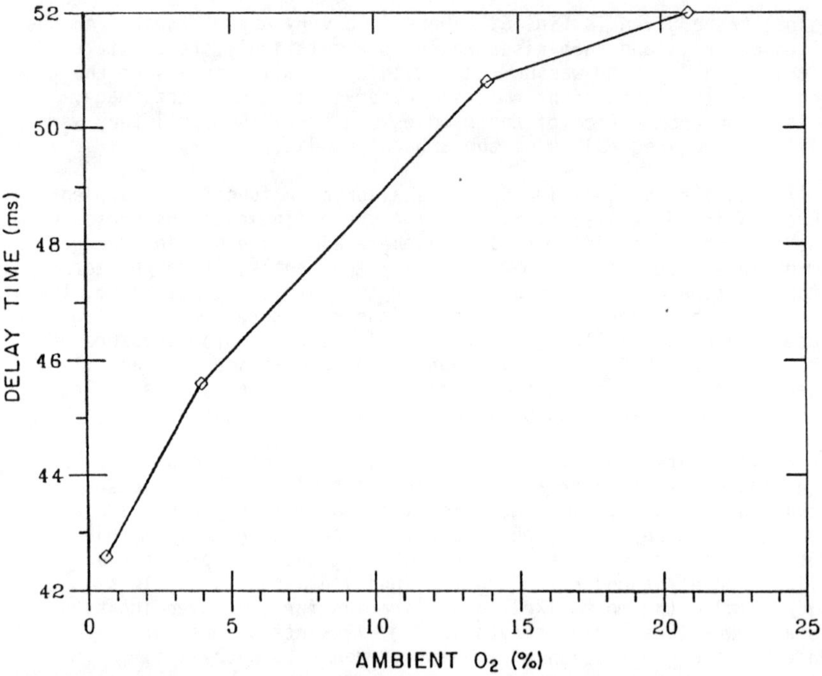

FIGURE 8. Effect of oxygen concentration on ignition delay time of MTV igniter material at 1 atmosphere (q" = 400 W/cm^2)

In this work, it is believed that the inhibiting effect of oxygen observed in Fig. 8 was due to the reaction with Mg to form MgO and thus prevent the fluorine from reacting with the Mg. Since the gravimetric heats of oxidation and fluorination of Mg are 24.75 KJ/g and 46.25 KJ/g, respectively,[11] the reaction of oxygen with magnesium would generate less thermal energy than the reaction of fluorine with magnesium, producing a longer ignition delay in air. Ignition delay tests performed on a pellet composed of a 1:1 mixture of PTFE and Mg without any Viton A showed that the ignition delay was approximately twice as much as that of the MTV igniter material under the same conditions, leading to the conclusion that Viton A is the rate-controlling ingredient for the MTV formulation. Properties of these polymers that would seem to substantiate this behavior are the decomposition of Viton A at 589 K[16] versus 803 K[13] for the PTFE, and activation energies of 210 KJ/mol and 350 KJ/mol, respectively. The effect of oxygen on the ignition delay of the MTV igniter material should then be a result of its effect on Viton A, in addition to its effect on magnesium. Also, the appearance of first light emission always occurred on the surface, suggesting that heterogeneous reactions may be involved in the ignition process.

In an attempt to improve the ignition and combustion of the MTV igniter material, boron was added to several different formulations of Mg and PTFE. The ignition delay results are presented in Fig. 9. The results indicate that increasing the amount of boron in an Mg/PTFE formulation significantly reduces the ignition delay time, with the shortest delay observed for a 50/50 formulation of boron/PTFE without any magnesium.

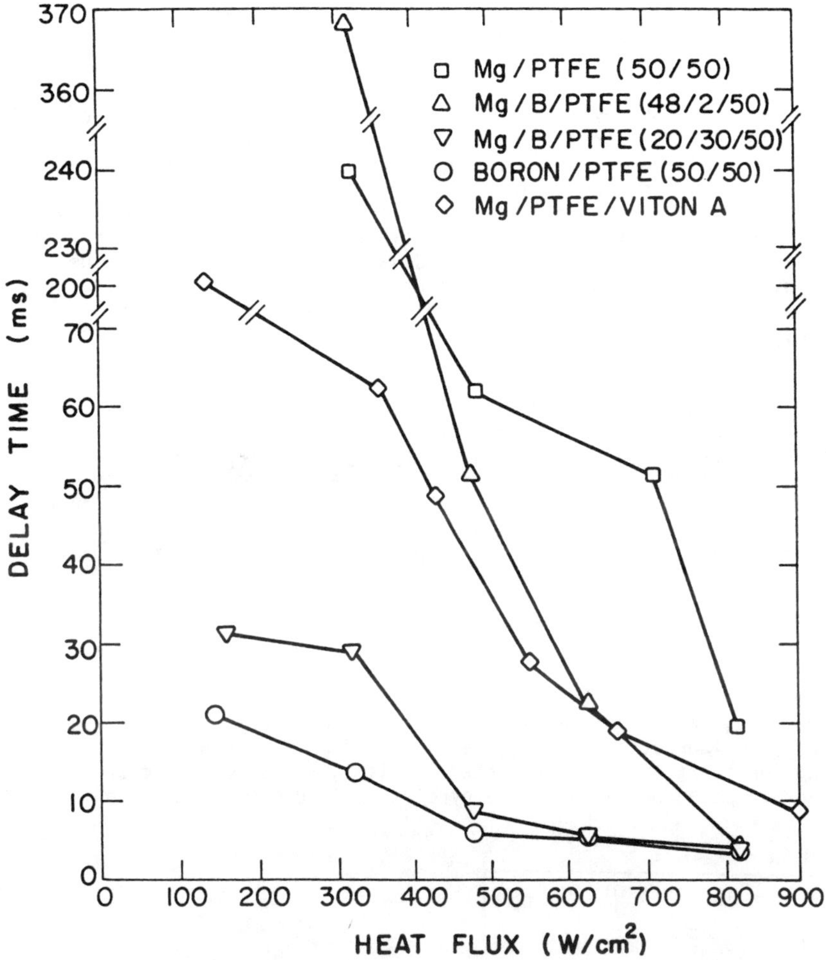

FIGURE 9. Effect on ignition delay of the addition of boron and Viton A to a Mg/PTFE material (ignition in air at 1 atmosphere)

Two factors considered in explaining this trend were radiation reflectivity at the 10.6 µm wavelength of the CO_2 laser and thermal diffusivity. The reflectivity of magnesium is about 0.75,[17] which is much greater than that of boron. Therefore, a large part of the laser energy may be reflected away from the surface of samples containing significant amounts of magnesium. It was considered that the ignition enhancement by boron observed in Fig. 9 may be an artifact of the laser and not a practical consideration for the improvement of the ignition of Mg/PTFE igniter material. However, the combustion after laser cutoff of the Mg/B/PTFE formulation with 30% boron was much more vigorous than the formulation without any boron, indicating that the addition of boron does have a significant physicochemical effect. Also, the thermal diffusivity of magnesium is 0.876 cm^2/sec, which is eight times higher than that of boron. Consequently, in a sample containing magnesium, the rate of penetration of the thermal wave is much higher. This phenomenon lowers the rate of increase of the surface temperature, which in turn, increases the delay time to ignition.

Another important observation from Fig. 9 is that the replacement of PTFE in the 50/50 Mg/PTFE formulation with an amount of Viton A significantly reduced the ignition delay time. Kubota and Serizawa[13] reported that the addition of 3% Viton A to their Mg/PTFE formulation did not alter the burning rate. However, the MTV igniter material tested here contained a higher percentage of Viton A, and the effect on the ignition delay time was obviously significant. The lower delay times are believed to be caused by the lower decomposition temperature and lower activation energy of Viton A, in comparison with those of PTFE, which were reported earlier.

5. PYROLYSIS AND IGNITION PRODUCT SPECIES ANALYSIS

The gaseous species evolved during the pyrolysis of PTFE and Viton A, and the ignition of the MTV igniter material, were analyzed with a gas chromatograph/mass spectrometer. Tests were conducted in both an inert atmosphere and in air to investigate the role of oxygen in the ignition processes. Solid ignition products were also analyzed by X-ray diffraction.

The pyrolysis of PTFE in an inert N_2 environment at one atmosphere produced the total ion chromatograph given in Fig. 10. The sample was heated twice by a heat flux of 430 W/cm^2 for two seconds each time. The most abundant species evolved are C_2F_6, C_2F_4 (the monomer), C_3F_6 (two isomers), CF_4, and two isomers of C_4F_8. Interestingly, the most abundant species, perfluoroethane (C_2F_6), was not listed in a depolymerization/degradation chart for PTFE presented by Elias,[18] but was the most abundant compound observed in the pyrolysis of both PTFE and Viton A.

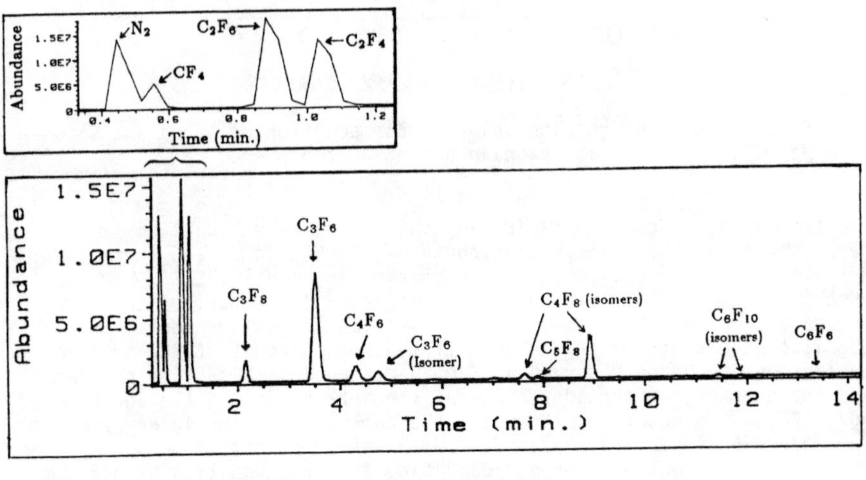

FIGURE 10. Chromatograph of the pyrolysis of PTFE in an inert N_2 environment at 1 atmosphere (q" = 430 W/cm^2 for 2 heatings of 2 seconds)

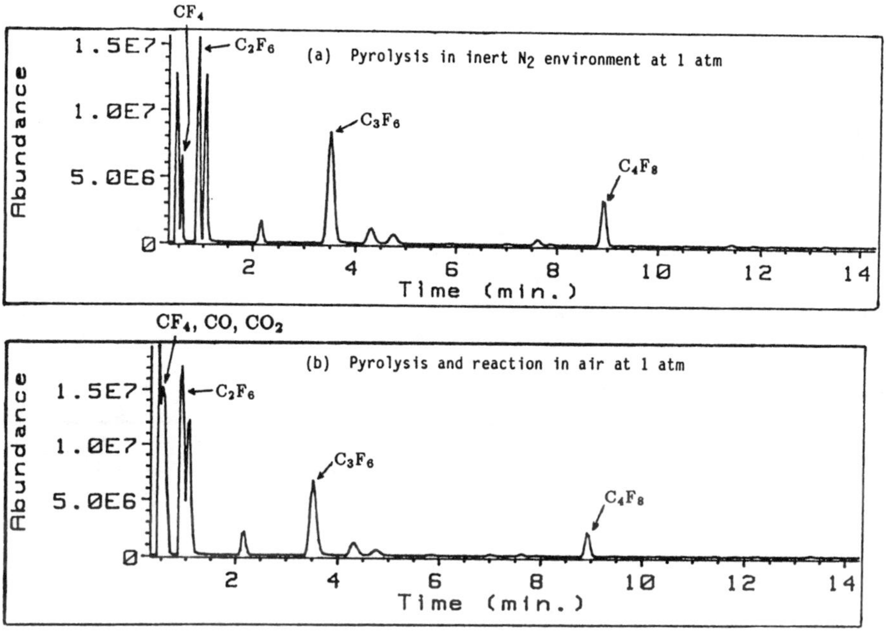

FIGURE 11. Chromatograph of the pyrolysis and reaction of PTFE in both an inert N_2 environment and air at 1 atmosphere (q" = 430 W/cm^2 for 2 heatings of 2 seconds)

Figure 11 compares the inert pyrolysis of PTFE (shown in Figs. 10 and 11a) with the pyrolysis and reaction of PTFE in air at 1 atmosphere. It can be seen that the amounts of C_3F_6 and C_4F_8 decreased slightly for pyrolysis and reaction in air (Fig. 11b), compared to pyrolysis in nitrogen (Fig. 11a). The amount of C_2F_6 appeared to remain the same. The amount of CF_4 appeared to increase in Fig. 11b, but the mass spectrometric breakdown of that peak indicated that CO and CO_2 were also eluted at that time, and the three compounds could not be individually resolved. Griffiths et al.[12] reported that CO_2 and CF_2O had been found by several other researchers for the reaction of PTFE in air. However, the existence of CF_2O could not be confirmed by the present apparatus. Using a microburner to study the premixed burning of several fluorocarbon-oxygen mixtures, Matula[19] reported significant burning of mixtures containing C_3F_6 and C_4F_8. He also reported that C_2F_6-O_2 and CF_4-O_2 mixtures could not be ignited at one atmosphere of air. These results are consistent with the results presented in Fig. 11.

Figure 12 gives the total ion chromatograph for the pyrolysis of Viton A in an inert N_2 atmosphere. The major species evolved were CF_3H, C_2F_6, C_2F_4, C_2F_3H, C_2F_5H, $C_3F_4H_2$, C_3F_6, C_3F_5H, $C_4F_6H_2$, C_5F_{10}, C_5F_7H, and C_4F_6. The two most abundant species evolved were C_2F_6 and C_3F_6, which were also the most abundant species, along with the monomer, that were evolved from PTFE. The similarity of products between PTFE and Viton A substantiates the claim that the effect of oxygen on the decomposition products for Viton A and PTFE may be quite similar. Interestingly, no CF_4 was observed for Viton A, but a peak observed at the same time it is normally

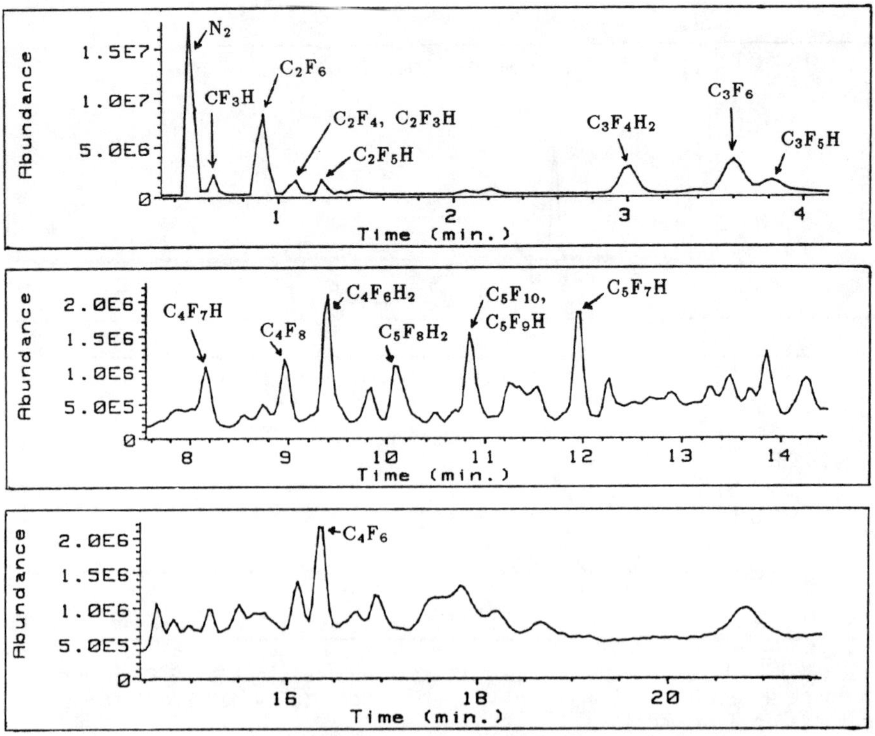

FIGURE 12. Chromatograph of the pyrolysis of Viton A in an inert N_2 environment at 1 atmosphere (q'' = 430 W/cm^2 for 2 heatings of 2 seconds)

eluted contained CF_3H. A trend was observed where fluorine atoms were replaced by hydrogen atoms in some of the more common C_mF_n structures (i.e., $C_3F_6 \Rightarrow C_3F_5H$).

The chromatographs of the gaseous product species evolved for the ignition and combustion of the MTV igniter material are presented in Figs. 13 and 14. Since the 5 millimeter cubes of igniter material were totally consumed after ignition and the gas samples were taken after the test was over, the species analyzed were the products of sustained combustion and not just ignition. Figure 13a displays the results of tests conducted in air, and the tests for Fig. 13b were conducted in an inert nitrogen atmosphere. Magnesium compounds were not observed in the gaseous state in either case. As was seen in Fig. 11, tests run in air also produced CO and CO_2 as well as most of the other products identified in the combustion in an inert atmosphere.

The rest of the peaks observed in these chromatographs establish an important point to be considered in the mechanisms of ignition and subsequent combustion of the igniter material. Almost all of the species observed for combustion in both air and an inert N_2 atmosphere were C_aH_b compounds, and the numerous C_mF_n and $C_xH_yF_z$ observed in the pyrolysis of Teflon and Viton A were not found in the combustion of the MTV igniter

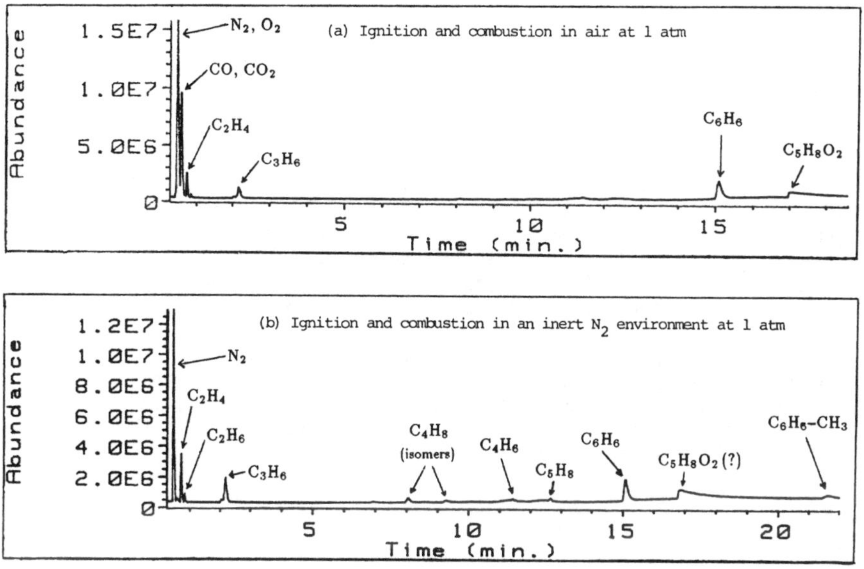

FIGURE 13. Chromatograph using Carbopack B/5% Fluorcol Column to analyze the gaseous products of ignition and combustion of the MTV igniter material (q" = 430 W/cm^2)

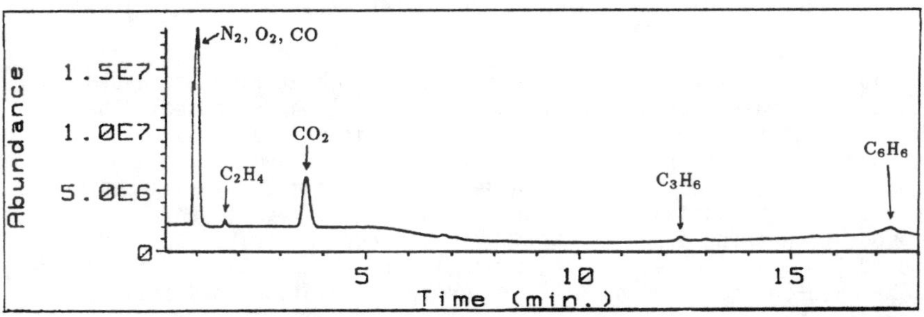

FIGURE 14. Chromatograph using a series of Porapak Q and R columns to analyze the gaseous products of the ignition and combustion of the MTV igniter material in air at 1 atmosphere (q" = 430 W/cm^2)

material. A probable explanation is that the primary pyrolysis species are consumed to form undetected HF or to form the observed MgF_2 solid products. Also, the small peaks from 8-13 minutes observed in the inert combustion tests are not evident in the combustion in air; clearly, the combustion in an inert atmosphere is not as complete as that in air.

In Fig. 14, gas samples of the MTV material ignition and combustion in air were also run through a series of Porapak Q and R columns for species identification. These columns are more applicable to detecting C_aH_b compounds. However, only small amounts of a few compounds of this type were detected, as well as significant amounts of CO and CO_2. These results verify the results given in Fig. 13a for the same experimental conditions using the Carbopack B/5% Fluorcol column for analysis.

X-ray diffraction analysis was also used to investigate the composition of the condensed-phase species evolved during the combustion of the propellant. MgF_2, MgO, and carbon were detected in varying amounts, depending on the test conditions. Of course, MgO was only detected in significant amounts for the combustion in air.

6. SUMMARY AND CONCLUSIONS

The pyrolysis and ignition characteristics of a Mg/PTFE/Viton A igniter material were studied under radiative heating by a high-power CO_2 laser. The ignition sequence was recorded by both direct high-speed video photography of the flame structure and schlieren flow visualization of the gas-phase dynamics. Ignition delay times were obtained as a function of incident heat flux, pressure, and ambient oxygen concentration. A gas chromatograph/mass spectrometer was used to analyze the pyrolysis and ignition products of the test sample and its constituents.

High-speed video photography showed that ignition always occurred on the sample surface. A primary flame zone attached to the surface with a height of 1.2-1.5 millimeters was observed in both air and inert gas environments at one atmosphere. The size of the luminous gas-phase plume above the primary flame decreased with a decrease in ambient oxygen concentration for tests conducted at one atmosphere. However, at a pressure of ~ 0.1 atmosphere, no evident plume evolution and gas-phase dynamics were observed, and the igniter material decomposed in large fragments.

Ignition delay times decreased monotonically as heat flux increased, pressure decreased, or ambient oxygen concentration decreased. The effect on ignition delay of adding boron or Viton A to a basic composition of Mg/PTFE was also investigated. The addition of boron significantly lowered the ignition delay time and greatly enhanced combustion. The addition of Viton A, which has a much lower decomposition temperature and activation energy than PTFE, greatly lowered the ignition delay time, which led to the conclusion that Viton A is the rate-controlling constituent for ignition of the MTV igniter material.

For combustion of the igniter material in both inert N_2 and air environments at one atmosphere, the only gaseous product species detected were several hydrocarbons, with the addition of CO and CO_2 in air. However, for pyrolysis of PTFE and Viton A individually in an inert N_2

environment, numerous fluorocarbon species, and hydrofluorocarbons for Viton A, were detected. It was concluded that these primary pyrolysis species are consumed to form the detected hydrocarbons, CO and CO_2 for combustion in air, undetected HF, and solid MgF_2 and carbon. These condensed phase products were identified by X-ray diffraction. Solid MgO was also observed for combustion in air.

In summary, the MTV igniter material with the addition of boron is highly recommended for base bleed applications, due to its suitable ignition behavior under various gas and low-pressure environments.

ACKNOWLEDGMENT

This work was sponsored by the Office of Naval Research, Contract No. N00014-86-K-0468, under the management of Drs. Richard S. Miller and Gabriel D. Roy. We would like to express our appreciation to them, and also to NWC for providing the igniter material that was used in this research.

REFERENCES

1. Brody, F., Snelling, L., Steinle, M., "A High Volume Cost Effective Manufacturing Process for the Production of Propellant Grains for Base Burning Artillery Projectiles," 1986 JANNAF Propulsion Meeting, CPIA Publication 455, New Orleans, LA, August 26-28, 1986.

2. Sahu, J., "Supersonic Flow Over Cylindrical Afterbodies with Base Bleed," Final Report, BRL-TR-2742, June 1986.

3. Hellgren, R., "Base Bleed Projectiles: Range Calculation Program and Efficiency Comparisons," FOA-C-20516-D2, November 1983.

4. Fox, J. H., "A Generalized Base-Flow Analysis with Initial Boundary-Layer and Chemistry Effects," Final Rept., AEDC-TR-79-46 October 1, 1976-May 18, 1979.

5. Bauer, R. C., Fox J. H., "An Application of the Chapman-Korst Theory of Supersonic Nozzle-Afterbody Flows," Final Rept., AEDC-R-76-158, July 1, 1975-Sept. 30, 1976.

6. Peters, C. E., Phares, W. J., "Analytical Model of Supersonic, Turbulent, Near-Wake Flows," Final Rept., AEDC-TR-76-127, July, 1973-Dec., 1975.

7. Strahle, W. C., Hubbartt, J. E., Walterick, R., "Base Burning Performance at Mach 3," AIAA Paper 81-1469, AIAA/SAE/ASME 17th Joint Propulsion Conference, July 27-29, 1981.

8. Kayser, L. D., "Effects of Base Bleed and Supersonic Nozzle Injection on Base Pressure," Final Rept., BRL-MR-2456, March 1975.

9. Kloehn, R., Rassinfosse, A., "Base Bleed Solid Propellants Containing Thermoplastic Elastomers as Binders," Joint Symposium (3rd) on Compatibility of Plastics/Materials with Explosives, Propellants, and Pyrotechnics Ingredients Processing, Phoenix, AZ, May 31-June 2, 1982.

10. Peretz, A., "Investigation of Pyrotechnic MTV Compositions for Rocket Motor Igniters," Journal of Spacecraft and Rockets, Vol. 21, 1984, pp. 222-224.

11. Peretz, A., "Some Theoretical Considerations of Metal-Flourocarbon Compositions for Ramjet Fuels," 8th International Symposium on Air-Breathing Engines, Cincinnati, OH, June 15-19, 1987.

12. Griffiths, T. T., Robertson, J., Hall, P. G., and Williams, R. T., "Thermal Decomposition of PTFE in the Presence of Magnesium," Proceedings of the 10th International Pyrotechnics Seminar, July 2-5, 1985, Karlsruhe, Federal Republic of Germany, page 19-1.

13. Kubota, N., Serizawa, C., "Combustion of Magnesium/ Polytetrafluoroethylene," Journal of Propulsion, Vol. 3, No. 4, July-Aug. 1987, pp. 303-307.

14. Kubota, N., Serizawa, C., "Combustion Process of Mg/TF Pyrotechnics," Propellants, Explosives, Pyrotechnics, Vol. 12, 1987, pp. 145-148.

15. Chen, D. M., Hsieh, W. H., Snyder, T. S., Yang, V., Kuo, K. K., "Study of the Thermophysical Properties and Combustion Behavior of Metal-Based Solid Fuels," AIAA/ASME/SAE/ASEE 24th Joint Propulsion Conference, Paper No. AIAA-88-3041, Boston, MA, July 11-13, 1988.

16. Haaf, F., "Occasional Exothermic Decompositions During Processing of 'Viton' Fluoroelastomer," Dupont Bulletin VT-1001, April, 1983.

17. Siegel, R., Howell, J. R., Thermal Radiation Heat Transfer, 2nd Edition, McGraw-Hill, New York, 1981, pp. 129-137.

18. Elias, H. G., Organic and Macromolecular Chemistry, 1972.

19. Matula, R. A., Orloff, D. I., Agnew, J. T., "Burning Velocities of Fluorocarbon-Oxygen Mixtures," Combustion and Flame, Vol. 14, 1970, pp. 97-102.

Dr. N. E. Gunners' Question: In your paper you say that "the MTV igniter material with the addition of boron is highly recommended for base bleed applications." In conventional base bleed applications, the ignition takes place under conditions rather different from your experiments. Would you please comment on this? Do you have other information--not reported here--supporting your recommendation?

Answer: We understand that the actual conditions for ignition of the base bleed igniter material in the gun environment are quite different from the well-controlled laboratory conditions under which these experiments were conducted. In the gun environment, ignition is due to thermal energy input from hot gases, while ignition by a high-power laser is due to radiative energy input. However, in both cases, ignition of the igniter material is caused by an external thermal energy input to increase the energy in the thermal layer near the surface of the igniter material to a critical level.

The ignition delay is governed by the rate of this thermal energy input to the test sample. The rate of energy input experienced by the igniter material in the gun can be simulated by the controlled radiative energy flux from a high-power laser. In addition, the fundamental understanding of the ignition processes of the MTV igniter material can be better achieved by the well-controlled experimental setup.

Presently, we do not have additional results to support our recommendation. However, we are conducting a comparative study between convectively-induced ignition of the igniter material in a shock tunnel environment and the CO_2 laser-induced ignition.

Dr. A. Gany's Question: You relate your ignition data to the heat flux to the surface, while the heat balance may be different under practical system conditions. Did you consider a possibility that a certain surface temperature could be an ignition criterion? Is there any indication of the propellant surface temperature at the instant of ignition?

Answer: In reference to your first statement, please refer to our answer to Dr. Gunners' question. Regarding your question about a critical surface temperature, fine-wire surface thermocouples were not employed in our ignition experiments. They cannot survive the vigorous ignition of the MTV igniter material and the subsequent self-sustained flame attached to the surface. However, separate tests were conducted in a strand burner using embedded thermocouples to determine the burning surface temperature, which is believed to be comparable to the critical ignition temperature. This temperature was found to be 1000K ± 100K.

Development and Production of Base Bleed Grain for 155 mm Rounds

STEINAR HAUGEN, KNUT MELBY, and ARTHUR OPPEGÅRD
Defence Products Division
RAUFOSS A/S
2830 Raufoss, Norway

0. ABSTRACT

The continued development and production of a base bleed grain described here is based on initial development performed by the Swedish National Defence Research Institute. The baseline composite propellant which has been used previously for base bleed grains for smaller caliber gun ammunition, was modified with the primary goal of meeting more rigid requirements for mechanical properties. Other goals included improving the aging properties, especially in exposure to humid air, and reducing health hazards in production. The means used besides conventional tailoring by optimizing curative ratio and oxidizer particle size distribution, were introducing a new bonding agent, a new antioxidant, and introducing fiber reinforcement of the propellant. Selected results from the development activities are presented.

Special test methods and equipment for evaluating effects of specific influences on base bleed systems of high linear acceleration and rotation rate, are described.

The RAUFOSS production process for base bleed grains for 155 mm rounds is briefly described.

1. INTRODUCTION

A base bleed system on gun ammunition rounds is a unit which bleeds gas into the space behind the projectile as it travels at high speed through the atmosphere. The effect of the unit is to reduce the pressure drop in the wake of the projectile, thereby increasing the range of the projectile by reducing the base drag.

Raufoss A/S first involvement in the base bleed system was production of base bleed grains for BOFORS in Sweden. The first deliveries were for 75 mm ammunition in 1983. Shortly after production of grains for 120 mm rounds followed.

This paper deals primarily with adaptation of the Swedish baseline propellant to the Raufoss process and improvement of properties to better meet the very severe structural requirements assosiated with base bleed application. The following sections discuss the propellant tailoring activities including critical test methods employed. Finally, a brief description of the Raufoss process for production of the 155 mm base bleed grain is presented.

2. BACKGROUND

A central patent which covers several aspects of the base bleed concept including propellant type, is a Swedish patent held by Gunners, Hellgren and Liljegren (ref. 1). The baseline propellant was a composite type with hydroxylterminated polybutadiene (HTPB) binder system and with 80% ammonium perchlorate (AP). The development of this baseline propellant base bleed grain was performed mainly at the Swedish National Defence Research Institute (known as FOA in Sweden).

The majority of results from the initial development work upon which Raufoss' subsequent activity design in this paper is based, are available in public reports. Some of the main reports which will also serve as a source of reference to other work are: Liljegrens study of the combustion of fuel-rich HTPB-propellants with ammonium perchlorate as oxidizer (ref. 2) reported in 1979, Gunners, Hellgren, Liljegren and Nilssons report dated 1979 on designing and testing of a base bleed system for 155 mm gun (ref. 3), Liljegrens report from 1980 on experience from production of composite propellant base bleed grains for 105 mm gun rounds (ref 4), Nilsson and Sandens report from 1981 on ageing studies of HTPB composite propellants and liners for rubber based inhibitors (ref. 5) and Schwarts' report dated 1982 on testing of propellant mechanical properties and bond strength between inhibitor and propellant (ref. 6).

The initial production of base bleed grains for 75 and 120 mm ammunition was based on a complete data package, which contained some deviations from RAUFOSS' composite propellant formulation and production knowhow. Improvements were proposed to, and accepted by, BOFORS. Changes made to the propellant during that period will be described subsequently.

During development of the 155 mm base bleed unit strain problem was encountered with unchanged propellant and upscaled grain configuration. The problem originated from forces acting on the grain during the acceleration phase in the gun. Inside the gun barrel the grain is exposed to an environment with extreme deformation potential. Deformation of the grain is a result of the combined forces from gas pressure and linear and rotational accelerations. Consequently a main design goal for base bleed systems is to assure that the grain is not strained beyond its failure limits.

A propellant development program was initiated to further inprove the propellant for the 155 mm application. A problem in the propellant development was that the loads acting on the grain due to the deformation rates and pressure, is of a magnitude far outside the range of test equipment which is normally used in propellant development activities. We had access to special test equipment and simulation computer programs, which to some degree gave valid data. But final verification of the quality of the base bleed grains could only be proved by expensive gun testing.

3. DEVELOPMENT OF PROPELLANT GRAIN

3.1 Propellant Improvements.

To adapt the baseline propellant to the Raufoss process and to improve its properties special attention was given to the bonding agent, anti-oxidant, and additive to improve structural characteristics as described in the following sections.

3.1.1 Bonding Agent

The baseline propellant contained a bonding agent for which a production process was developed by Nilsson and Sanden, FOA, for use in HTPB/AP propellants (ref. 7). Bonding agents improve the mechanical properties of composite propellants by increasing the adhesive strength between the AP particles and the polymer, and are recognized as necessary ingredients in HTPB composite propellants.

The specified agent was a condensation product of tartaric acid and methyl-aziridyl-phosphine oxide (MAPO) which is a curing agent for carboxyl terminated polybutadienes. RAUFOSS proposed to use another bonding agent from the amine family of chemicals. The reason for this change was mainly to avoid exposing personnel to carcinogeic agents in the processing of bonding agent and propellant. Based on Raufoss experience the proposed agent also results in propellants with better aging properties in a humid environment.

Tests were performed where propellant samples containing different bonding agents were exposed to air with varied relative humidity. The samples were shaped as standard JANNAF dog bones which have a moisture diffusion web of approximately 3.5 mm. After exposure the samples were dried. The mechanical properties were measured before and after the treatment. Results are given in Fig. 1.

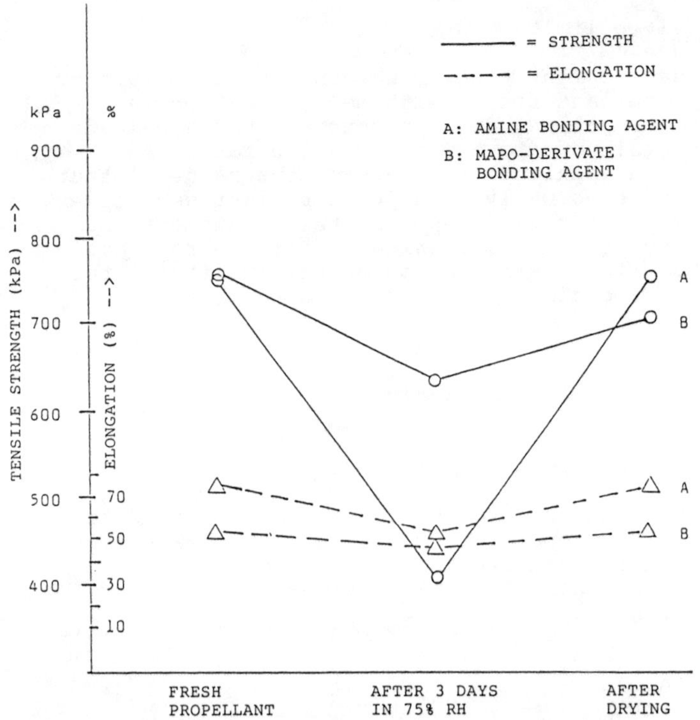

FIGURE 1. Changes in mechanical properties of 2 composite propellants with different bonding after short-time exposure to moist air, after drying.
Sample type: Dogbones, 7 mm thickness.

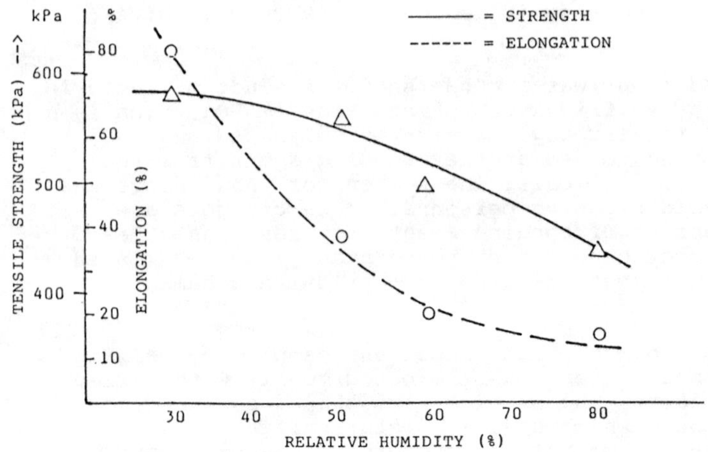

FIGURE 2. Effect of humidity on mechanical properties of a composite propellant.

Even this short duration test indicated differences in moisture sensitivity. When exposed to moist air, the propellant with the amine bonding agent was able to regain its initial properties when dried. The previously used bonding agent became irreversibly deteriorated after exposure to moisture, but is less sensitive to short time exposure. It is important to be aware of these differences when establishing procedures for handling of grains during processing and by the user.

However, the presence of moisture in propellants will always cause problems concerning mechanical properties. This is illustrated by Fig. 2, which is a plot of the relation between mechanical properties of a propellant versus relative humidity in the air with which the propellant is in equilibrium after 1600 hours. The propellant contains the amine bonding agent.

3.1.2 Antioxidant

A number of antioxidants are available for HTPB. RAUFOSS produces a number of propellant formulations and strives to reduce the number of chemicals stored for economical reasons. Accordingly we proposed to change the antioxidant in the base bleed propellant to one already in the Raufoss inventory.

Ageing tests at elevated temperature were performed to verify that no reduction in propellant shelf life was introduced by these changes in the formulation.
The effect of humidity on ageing was included in the tests. Propellant samples were stored in contact with air with a range of different relative humidities up to 75%.

3.1.3 Fiber Reinforcement of Propellant

The main goal for the propellant development was to improve the mechanical properties. The main limitations to be observed was the requirement of fuel rich propellant (max. 80% solids) and stable burning at low pressure which as a general rule determines a lower limit for oxidizer content. The initial studies were mainly variation of oxidizer particle size distribution and cure level. Binder system was not changed. These are conventional techniques, and will not be discussed here. The novel parameter we introduced, was the use of fibers to obtain a tougher propellant. Previous use of fibers is described by L.J. Van de Kieft et.al. (ref 8), who in 1982 published investigations on the effect of fiber reinforcement of gun propellants.

Screening of potential formulations was undertaken employing one gallon mixer to obtain processing data, dogbone samples for tensile testing and strands for burning rate testing. Selected more promising formulations have been tested in the special test equipment for base bleed grains.

Different types of short fibers made from aramide, carbon, glass and silica have been added to HTPB/AP propellants. We have found that just very small additions of fibers to propellants with low solids loading has a significant effect on the mechanical properties of the propellant. As an example effects of carbon fibers added to the baseline propellant for the base bleed system are illustrated in Figs. 3-8. Fig. 3-5 are plots of the effect of carbon fiber concentration on mechanical properties. Fig. 6 - 8 are plots of carbon fiber length vs. mechanical properties.

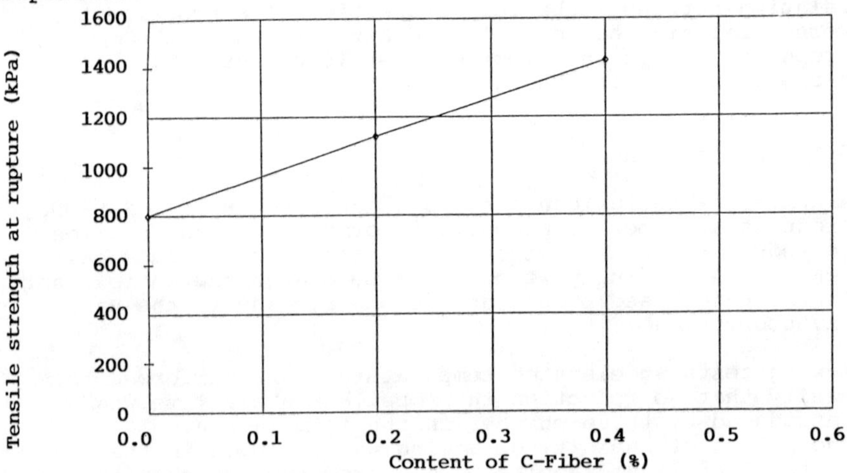

FIGURE 3. Tensile strength (at rupture) versus concentration of C-fibres in composite propellant (6 mm fibres, temp.= 20°C.

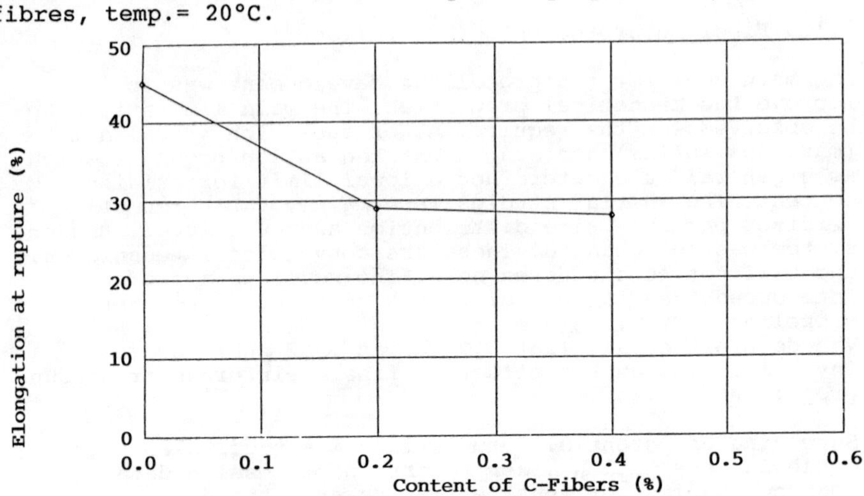

FIGURE 4. Elongation (at rupture) versus concentration of C-fiber in composite propellant (6 mm fibers, temp.= 20°C)

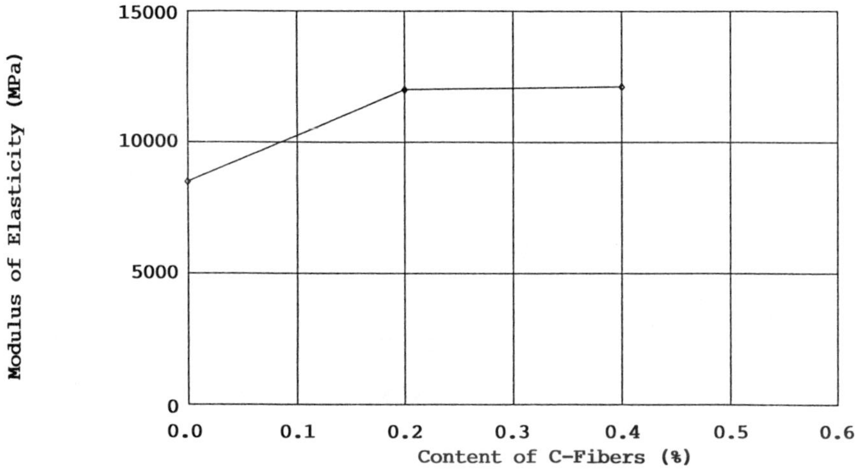

FIGURE 5. Modulus of elasticity (at 5% elongation) versus concentration of C-fiber in composite propellant (6 mm fibers, temp.= 20°C).

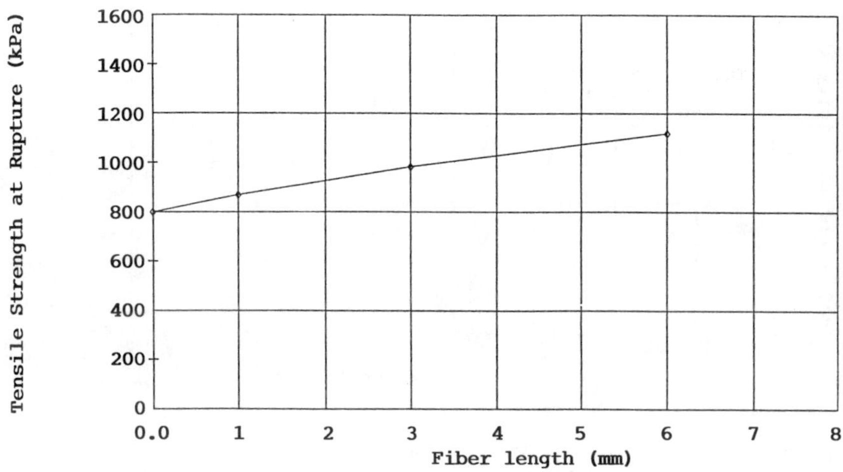

FIGURE 6. Tensile strength at rupture versus length of C-fibers in composite propellant (conc. of fibers = 0.2%, temp.= 20°C).

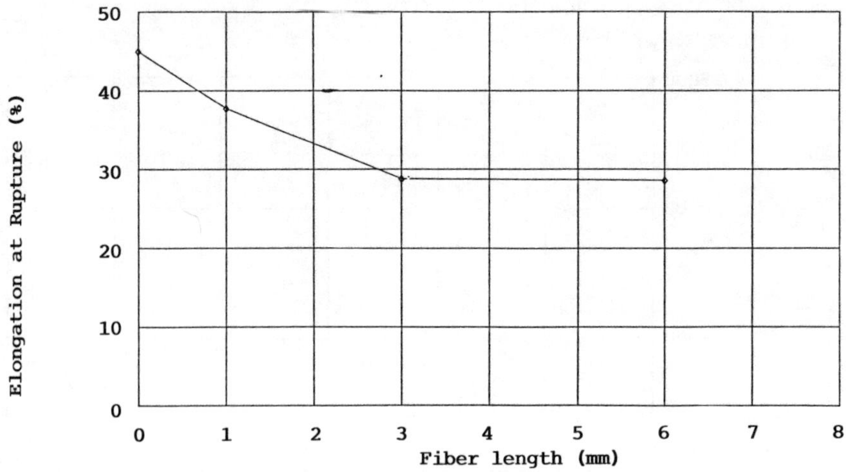

FIGURE 7. Elongation (at rupture) versus length of C-fibres in composite propellant (conc. of fibers = 0.2%, temp.= 20°C).

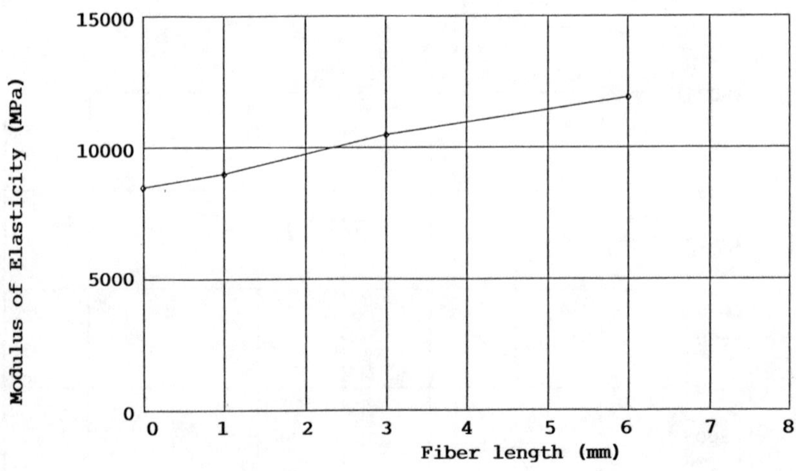

FIGURE 8. Modulus of elasticity versus length of C-fibers in composite propellant (conc. of fibers = 0.2%, temp.= 20°C).

Fig. 9 demonstrates the effect of fibers in high rate compression tests. Four samples are compressed to different degrees. The broken line represents the baseline 80% solids propellant.

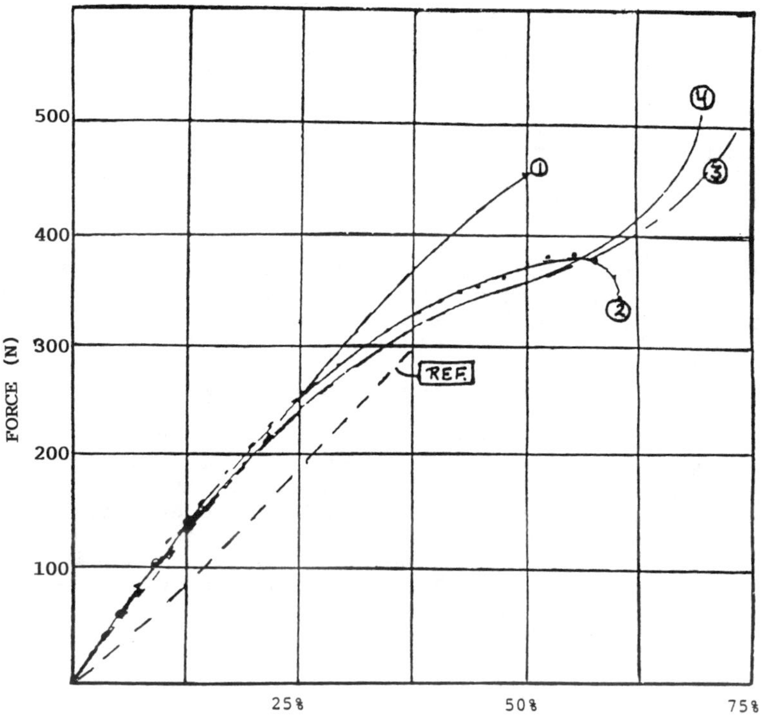

FIGURE 9. High-speed compression (500%/min) reinforced propellant samples. Broken line represents baseline propellant.
1 : 50% compression, no cracks.
2-4: 56-70% compression, cracks.

Base bleed grains with 0.2 % carbon fibers were burned in a spinstand. These initial tests indicated that carbon fibers had no significant effect on burning properties on neither static nor rotating grains.

Concurrent with RAUFOSS' study of fiber to improve propellant mechanical properties, BOFORS was evaluating mechanical design approaches to provide required structural characteristics for the 155 mm application. At the present time one of the BOFORS mechanical design concepts has been implemented in production.

3.2 Description of Test Equipment and Methods.

Special equipment and methods were employed to test propellant samples exposed to high rate of compression to simulate the environment in a gun barrel. A high rate tensile testing machine is run in compression mode. The machine is delivered by MTS Systems GmbH. It has a 100 kN load frame and 80 mm/s deformation rate. A DEC computer system is controlling the machine and performing data registration and reduction. The test specimen are small cylinders (8 mm diameter and 16 mm high) which are compressed at a predetermined rate. A number of samples are compressed to different preset values (% of initial height) with registration of load versus compression. After the test, the samples are visually inspected for cracks. In this way the failure limits for a propellant are established.

An example of compression force versus deformation curves is given in Fig. 10. In this example the line represents test data for a sample which is compressed to 64% of its initial height. The points in squares are calculated from uniaxial tensile test data, and correlate very well with the experimental curve up to about 35% compression.

Fig. 11 is a sketch of equipment which is used for exposing propellant grains to axial accelleration of the same magnitude as in the guns, and subsequent retrieval of the grain for visual inspection. Its main components are a 120 mm gun and a long tube. The projectile, which contains the test object, is fired into the tube, and is braked by compressing the air in front of it. The compressed air is vented out at a rate which ensures that the test object is retarded without being damaged.

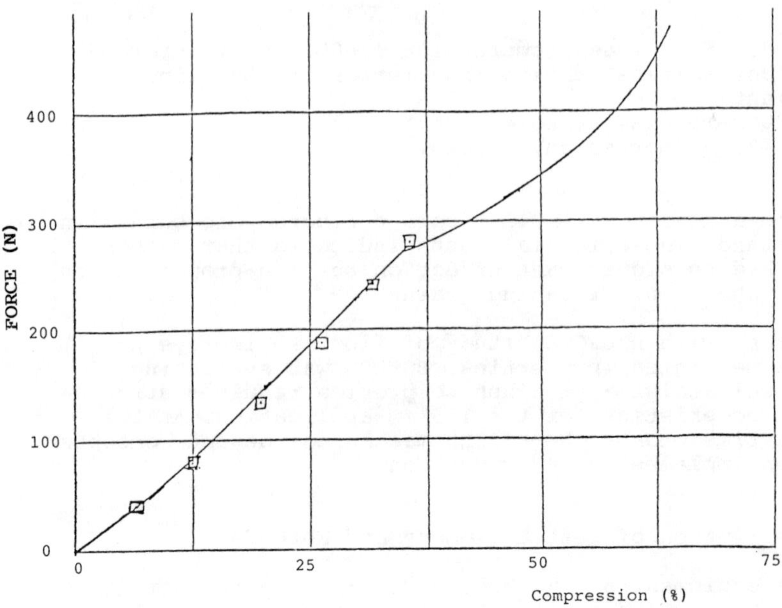

FIGURE 10. Compression force versus compression of a cylinderical propellant sample. (Rate: 500% per second, temp. 65°C).
⊡ Calculated values based on tensile testing of dogbone.

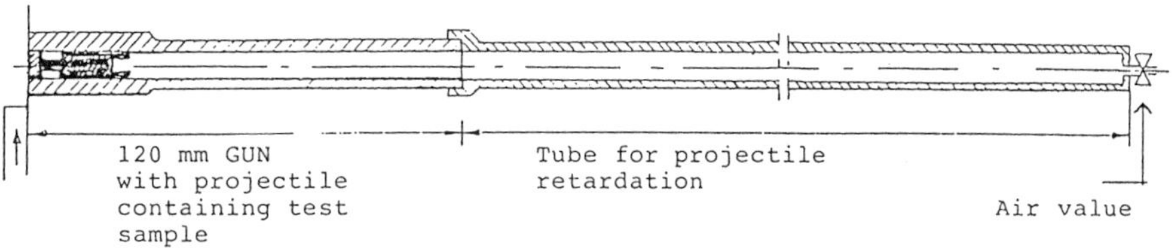

FIGURE 11. Acceleration testing arrangement.

Fig. 12 is a drawing of the test projectile which contains a base bleed grain to be tested. To measure maximum deformation during acceleration, a witness rod is positioned in the grain bore. It is covered with wet paint which will be transfered to the grain where contact has been established during the test. By varying the witness rod diameter, the deformation can be registered. The grain also is visually inspected for cracks after the test. The validity of this simulation of gun firings is somewhat limited due to the lack of rotation and exposure to gas pressure.

The high rotation rate of spin-stabilized projectiles may have an influence on the burning characteristics of a base bleed grain. Possible effects of propellant composition changes were tested on a spinstand. Fig. 13 shows examples of pressure - time traces for baseline propellant grains burned with and without rotation. In this example the rotation increases the pressure in the middle of the burntime.

4. PRODUCTION OF BASE BLEED GRAINS FOR 155 MM GUN

RAUFOSS is presently producing composite propellant base bleed grains for 155 mm gun ammunition employing special production equipment designed and built by RAUFOSS. Fig. 14 is a flowchart for the present base bleed grain production. A blend of coarse and ground AP is mixed with polymer, additives and curing agent in a 300 gallon Baker-Perkins vertical mixer. Propellant is cast and cured according to standard procedures. Grains are machined to correct dimensions and inhibited with curable inhibitor.

Quality control of the propellant and the grains includes the testing of

- Mechanical properties
- Burn rate at ambient pressure and no rotation
- Peel test of propellant/inhibitor bond
- Weight and dimensions.

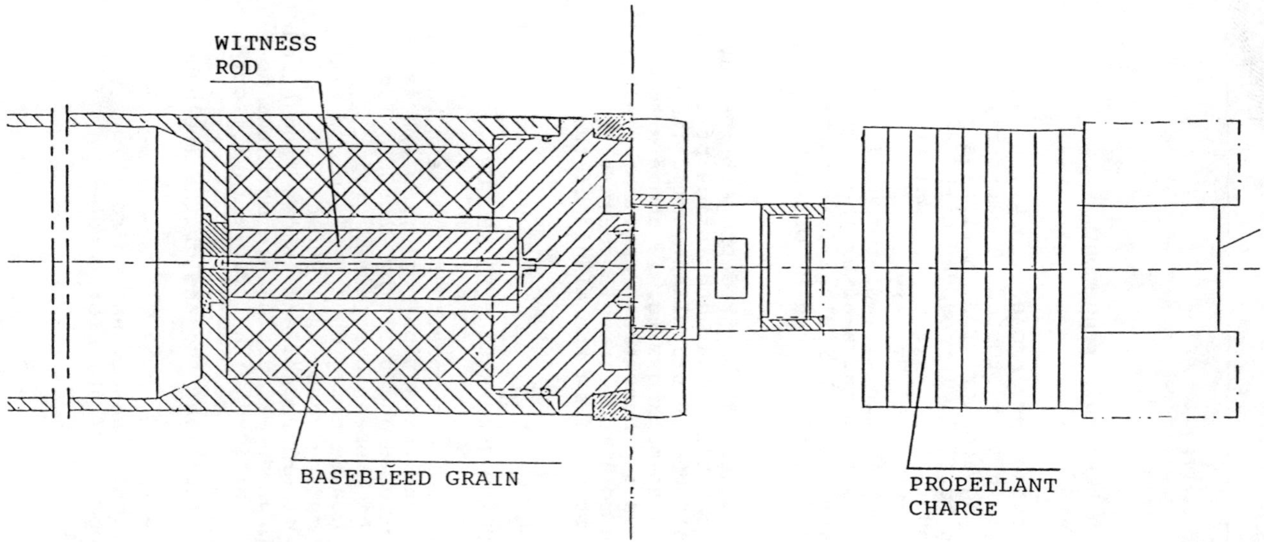

FIGURE 12. Projectile containing sample for acceleration testing. The sample is a basebleed grain.

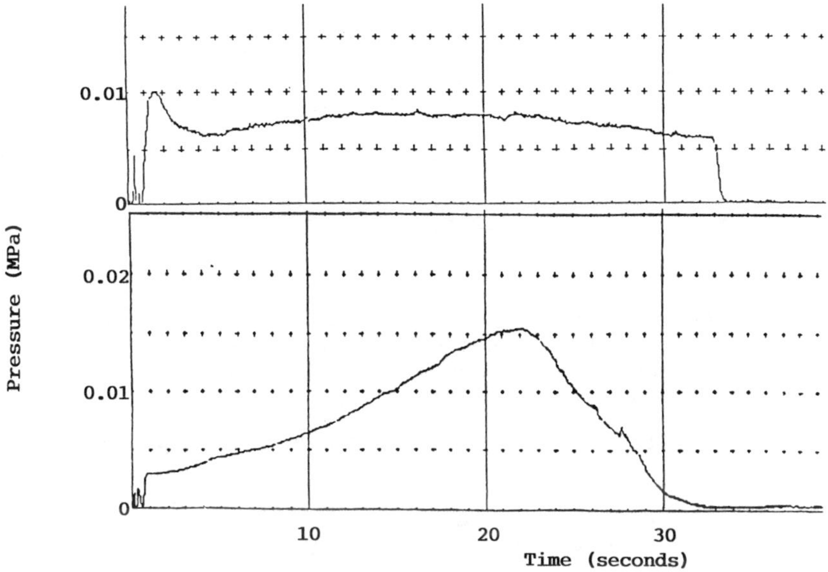

FIGURE 13. Overpressure versus time traces for burning of 155 mm gun basebleed grains at sea level pressure and 20°C
Upper: Static test.
Lower: Rotation test, 16000 rpm.

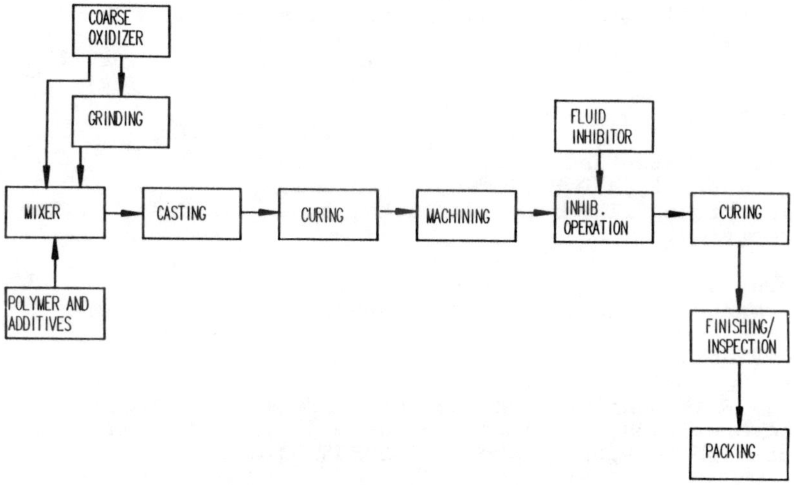

FIGURE 14. Process Flow Chart for BASEBLEED Grains.

5. CONCLUSIONS

RAUFOSS development activities have resulted in improvements to the baseline propellant for base bleed grains. Change in bonding agent has provided more resistance to humid environments and use of a different anti-oxidant has simplified inventory control.

Although more development is required, addition of reinforcing fiber to the propellant composition appears to offer an approach to meet the structural demands of base bleed grain designs.

Although final verification of a base bleed grain design must be obtained from actual gun testing, significant test equipment and methods are avaiable to provide assurance of success in the actual gun environment. These include

* High rate tensile testing machine for compression loading

* Axial acceleration equipment providing for post-test grain examination

* High rotation rate spin stand to evaluate effect of spin on propellant burning rate.

6. ACKNOWLEDGES

The authors wish to acknowledge BOFORS AB for carrying out experiments on their spinstand and high rate deformation apparatus, and Försvarets Forsöksplats, Karlsborg, for carrying out linear acceleration tests. We also thank our consultant William G. Haymes, previously at Hercules Inc., USA, for reviewing the paper.

REFERENCES

1. Swedish patent no. 340.076: N-E Gunners, R.V.Hellgren, T.Liljegren, Kanonprojektil med anordning för ökning av dess skottvidd.

2. Torsten Liljegren, Bert Anderson, Forbränningsegenskaper för bränslerika HTPB-krut med ammoniumperklorat (AP) som oxidator, FOA-rapport C 20299-D1, Mars 1979.

3. Nils-Erik Gunners, Rune Hellgren, Torsten Liljegren, Yngve Nilsson, Basflödesaggregat för 155 mm SGR M/77 for haubits F, FOA-rapport C 20305-D3 Maj 1979.

4. Torsten Liljegren, Erfarenheter från tillverkning av krut for basflödeladdningar till 105 mm STGR M 40/74, FOA-rapport C 20350-D1, April 1980.

5. Bengt Nilsson, Roland Sanden, Åldringsundersökningar av HTPB-baserade kompositkrut och bindeskikts-material, FOA-rapport C 20405-D1, April 1981.

6. Anders Schwartz, Provning av basflödesladdningar, FOA-rapport C 20452-D1, April 1982.

7. Bengt Nilsson, Roland Sanden, Framställning av ett medel som ökar hållfastheten hos kompositkrut med hydroxylterminerad polybutadien (HTPB) som bindemedel, FOA rapport C 20033-D1, November 1975.

8. L.J.Van de Kieft, B.J.Pacanowsky, W.W.Hillstrom, Fiber Reinforcement of Gun Propellants, ADPA Meeting, USA, May-June 1982.

COMMENTS

Mr. Ton Wijker, Muiden Chemie, The Netherlands. Is there a change in viscosity of the mix by adding fibers?

Author's reply. There is a slight increase corresponding to the increase in solids loading.

*

Mr. Ton Wijker, Muiden Chemie, The Netherlands. Did the compressed specimen after 35% compression show any cracks? If not, at what compression level did cracking occur? (Ref. Fig. 3).

Author's reply. The figure shows one sample of series. They were compressed to different levels. This particular one, compressed to 63%, cracked. the sample compressed to 50% had no cracks, and the failure value is somewhere between 50% and 63%.

*

Mr. Ton Wijker, Muiden Chemie, The Netherlands. What was the storeage time at the various humidity levels of the specimen, mentioned/showed in Fig. 2?

Author's reply. All were stored for 168 hours.

*

Mr. Ton Wijker, Muiden Chemie, The Netherlands. Can you say something about the changed mechanical design by BOFORS, that solved your overstrain problems?

Author's reply. The answer is no. BOFORS has applied for a patent.

*

Mr. Ton Wijker, Muiden Chemie, The Netherlands. What would you consider to be the most optimal mechanical properties for a base bleed?

Author's reply. According to our mechanical analysts, no general answer can be given. It depends very much on the design of the base bleed unit. In principle, brittle propellants may crack because of the sudden and very high deformation rates in the gun barrel. Further, propellants with very different characteristics as to modulus of elasticity may be used. For a specific modulus, a corres-ponding minimum elongation must be achieved by tailoring the propellant properties.

SESSION II: MODELS AND PERFORMANCE CALCULATIONS OF BASE BLEED PROJECTILES

A Review of the Fluid Dynamic Aspects of the Effect of Base Bleed

J. SAHU
Computational Aerodynamics Branch
U.S. Army Ballistic Research Laboratory
Aberdeen Proving Ground, Maryland 21105, USA

W. L. CHOW*
Department of Mechanical and Industrial Engineering
University of Illinois at Urbana-Champaign
1206 West Green Street
Urbana, Illinois 61801, USA

ABSTRACT

The fluid dynamic aspect of the effect of base bleed is briefly reviewed. Earlier understanding on the basis of interaction between the viscous and inviscid streams can adequately explain the three different flow regimes resulting from base bleed or mass injection into the base region. The effect of energy addition to the wake has also been ascertained from this approach. Detailed numerical computations of base flow can be made by solving the Navier-Stokes equations. The effect of base bleed can then be illustrated by providing appropriate bleed boundary conditions at the base.

INTRODUCTION

Flow studies on the pressure prevailing behind blunt-based bodies in flight have been prompted by the considerable importance in its practical applications. Since this pressure, the base pressure, is usually much lower than that of the free stream, it constitutes a major portion of the overall drag experienced by the body. Although early studies of low-speed flow around blunt-based bodies tended to be overshadowed by the phenomenon of vortex shedding, research initiated since the advent of high-speed flight resulted in a slow unraveling of the processes and mechanisms which control and establish these flows. The essentially inviscid external stream establishes and determines the overall flow pattern including the low pressure prevailing within the major portion of the wake. On the other hand, the viscous flow processes, such as the jet mixing along the wake boundary, recompression at the end of the wake, and the ensuing process of flow redevelopment, establish and determine the "corresponding inviscid body geometry." Thus, a low base pressure is the result of the strong interaction between the inviscid and viscous flows, the latter being attached to the inviscid flow in the sense of the boundary layer concept. Naturally the method of controlling and reducing the base drag also becomes part of these investigations.

One of the effective ways of reducing the base drag is to increase the base pressure through base bleed. It is known that by bleeding a certain amount of relatively hot fluid in the wake, the base drag can be reduced. It is the intention of this paper to present a brief review on the fluid dynamic aspects of the base bleed. Flow models for such studies and the influence of base bleed from the early crude analyses up to the sophisticated approach through solving the Navier-Stokes equations are presented in the following sections.

---

* Present Address: Department of Mechanical Engineering, Florida Atlantic University, Boca Raton, FL 33431

STUDIES ON THE BASIS OF STRONG INTERACTION

Nearly three decades ago, the second author had the opportunity of examining the effect of base bleed with a small-scale model (Figure 1) simulating a projectile in supersonic flight.[1] For increasing stagnation pressures of the bleeding jet issuing from the center portion of the base into the wake, it was learned that three distinctly different flow regimes exist (Figure 2). Starting from the no-bleed condition, where a low base pressure prevails throughout the base-wake region, the base pressure tends to increase as the stagnation pressure of the base bleed increases. Under this condition, all the mass of bleed is entrained into the mixing region along the wake boundary of the slip stream, and the flow condition (indicated as Regime I) is depicted in Figure 3. This trend of increase of the base pressure persists until a maximum base pressure is reached. Thereafter, as the stagnation pressure increases, the momentum of the bleed is strong enough to allow part of the jet to overcome the high pressure prevailing at the end of the wake as a result of flow-recompression, causing the base pressure to decrease. This type of flow pattern is shown in Figure 4 (identified as Regime II). The base pressure continues to decrease as the stagnation pressure of the base bleed increases until a relative minimum is reached.

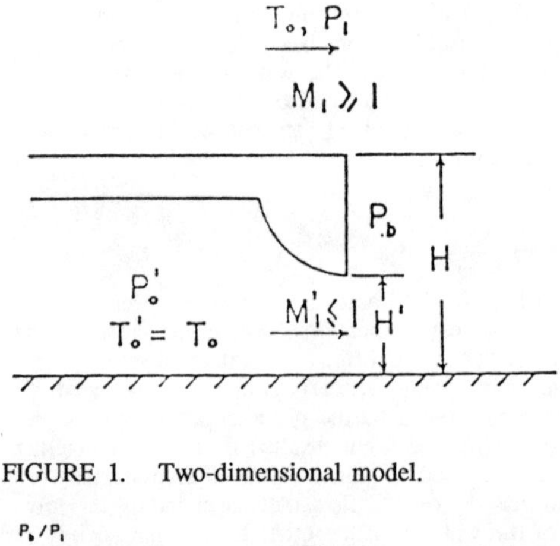

FIGURE 1. Two-dimensional model.

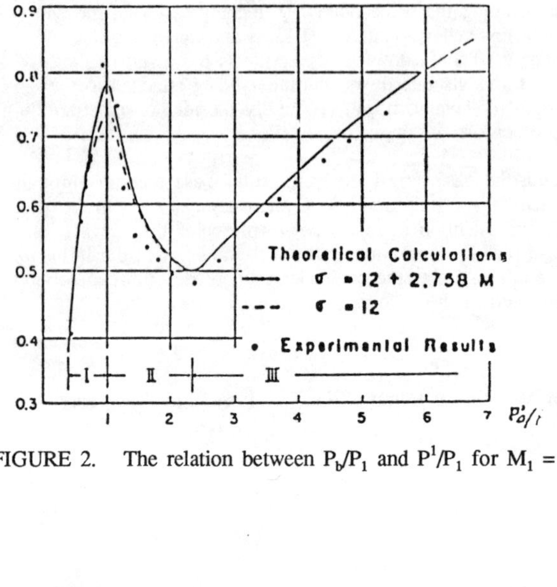

FIGURE 2. The relation between P_b/P_1 and P^1/P_1 for M_1 = 1.92, H/H^Q = 1.94.

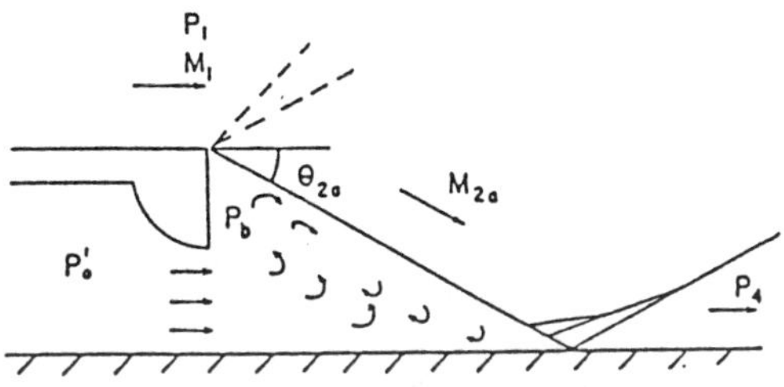

FIGURE 3. Flow model for Regime I.

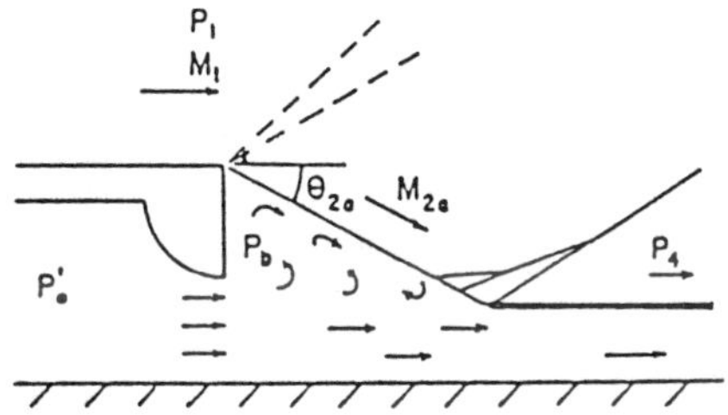

FIGURE 4. Flow model for Regime II.

At higher stagnation pressure ratios, the jet is so strong that it becomes a supersonic stream itself, and the base pressure increases again relative to the pressure of the slip stream. The flow pattern pertaining to this region (Regime III) of the flow is depicted in Figure 5a. The interaction between the two supersonic streams is such that an equilibrium base pressure is reached. In Figure 5b, the detailed flow field at the end of the wake is shown. The slip stream is pumping out a certain amount of fluid from the wake while the jet stream is feeding an equal amount into the wake so that an equilibrium situation can be maintained. For unusually high jet stagnation pressure ratios, the base pressure may even be higher than that of the approaching free stream, and the plume jet may even cause separation of the slip stream away from the wall ahead of the base (plume-induced separation).

The above description on the effect of the base bleed has been observed and supported by many experimental investigations.[2-5] The theoretical description and explanation of the phenomena[1] were based on a simple, crude, yet effective component analysis originally developed by Korst.[6] By studying the constant pressure turbulent mixing process in conjunction with an "escape" criterion at the end of the wake, an equilibrium base pressure solution can be reached by imposing the principle of mass conservation within the wake. Although this study[1] was carried out with a two-dimensional model in a supersonic slip stream, it is equally applicable to an axisymmetric configuration as the separated flow field is governed by the same flow mechanisms of viscous interactions. It may be conjectured that similar phenomena also exist for other flow regimes of the slip stream.

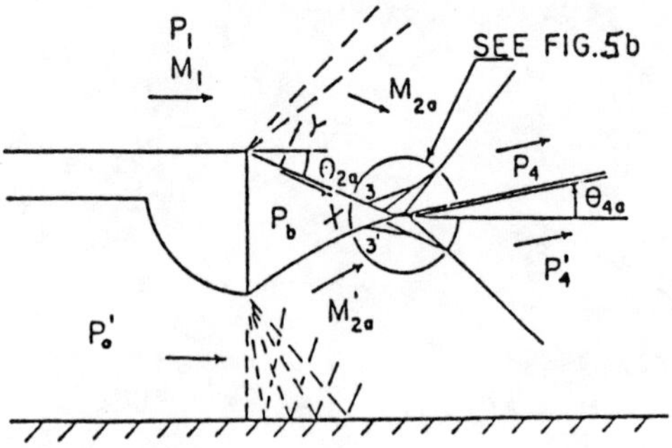

FIGURE 5a. Inviscid flow model for Regime III.

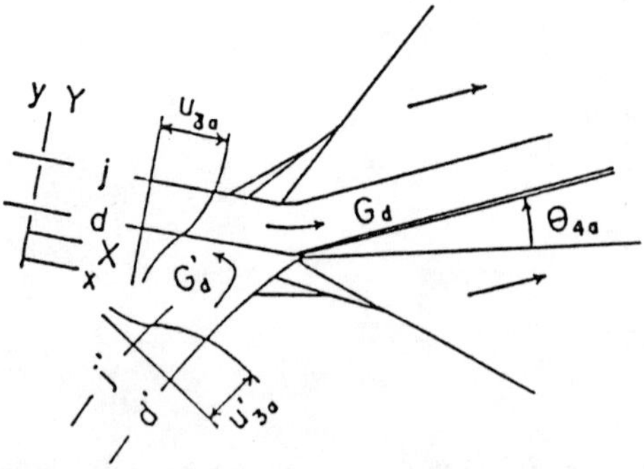

FIGURE 5b. Details in dissipative flow near the end of the wake.

Through similar considerations, the effect of heat addition into the wake through conducting walls on the base pressure has also been examined.[7] Figure 6 presents a set of results for a flow past a backward facing step showing the influence of heat addition (or extraction) on the base pressure ratio without any mass addition, while Figure 7 shows the effect of both mass and heat addition for the same problem. Experimental verification on the effect of heat addition is also presented in Figure 8.[7] Obviously heat addition provides the same favorable influence as the mass bleed.

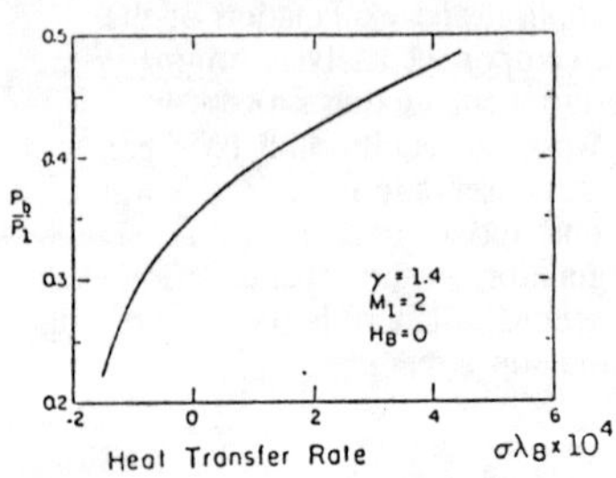

FIGURE 6. The effect of heat addition on base pressure.

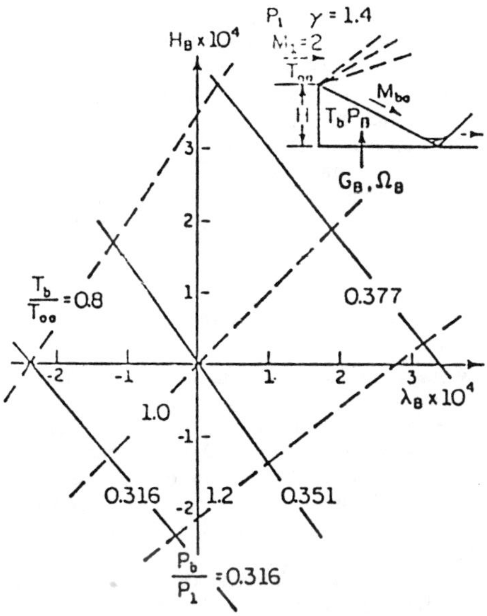

FIGURE 7. The effect of both mass and heat additions.

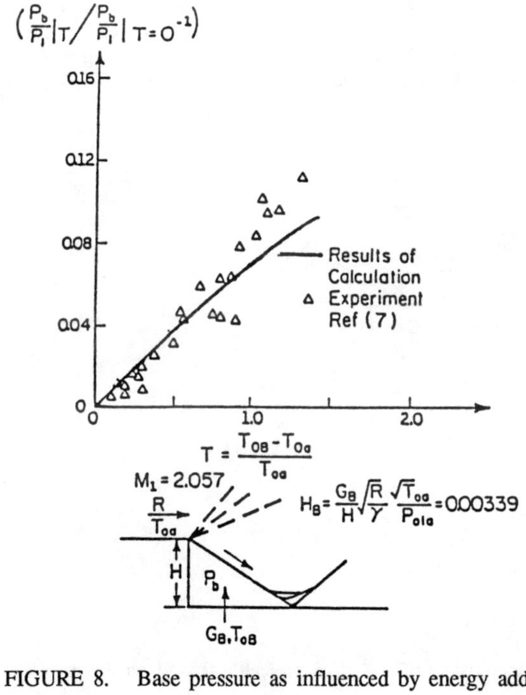

FIGURE 8. Base pressure as influenced by energy addition.

The methods of analysis on viscous flow recompression and redevelopment were developed and improved later.[8-12] These more refined analyses have been employed to study base pressure problems in supersonic, incompressible, and transonic flow regimes. Specifically, the study of transonic flow past a projectile based on an equivalent body concept yielded the detailed breakdown of the fore-body drag, the skin friction, and the base drag. Indeed, the base drag is always a major portion of the total drag experienced by the projectile (see Figure 9). The merit of boattailing has also been substantiated and explained.[12] It is conceivable that the effect of base bleed may be examined with these improved analyses. Since schemes for large-scale numerical computations have been developed, this effort has not been pursued.

STUDIES THROUGH SOLVING THE NAVIER-STOKES EQUATIONS

As more powerful (both in speed and memory space) digital computers become available, it is quite popular to examine flow problems by solving the Navier-Stokes equations with adequate turbulence modeling. Indeed one of the major research activities of the first author has been the study of base pressure problems, with or without bleed, by solving the Navier-Stokes equations in the transformed coordinates. Upon adopting a thin-layer approximation of the transformed Navier-Stokes equation, the phenomena of base bleed in any of the flow regimes described earlier can be completely predicted from this approach by simply providing the appropriate boundary conditions in the base region (conditions of the bleeding jet). Results obtained by the first author and his colleagues[13-17] have supported the observed influences of the base bleed throughout all three flow regimes. Figure 10 shows the detailed streamline pattern of the flow within the Regime I, where all mass of bleed has been entrained into the mixing region. It can be easily observed from this figure that: i) the horizontal bleed core flow persists farther downstream within the wake for a larger mass bleed parameter (I) before it is completely entrained into the mixing region; and ii) judging from the shade (shear layer) simulating the Prandtl-Meyer expansion at the corner, a larger mass bleed parameter corresponds to a higher base pressure ratio. Indeed, this increase in pressure is also observed in the early phase of the bleed for the case shown in Figure 11 and compares well with the experimental data.[4] Figure 11 also shows a drop in the base pressure ratio for a higher stagnation pressure of the bleeding jet (or a larger bleed parameter) which corresponds to the flow conditions within the Regime II.

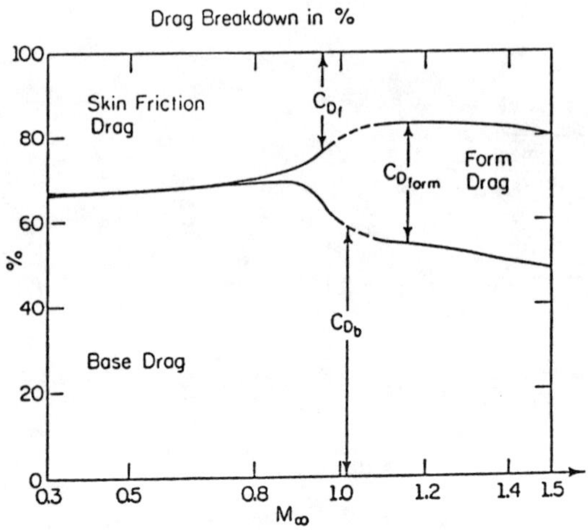

FIGURE 9. Breakdown of drag characteristics of a SOC-6 projectile (based on the results obtained from Reference 11).

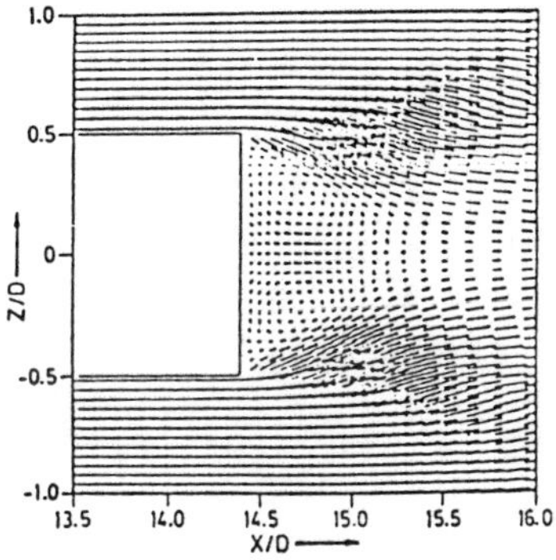

$M_\infty = 2.5$, $\alpha = 0$, $I = 0$

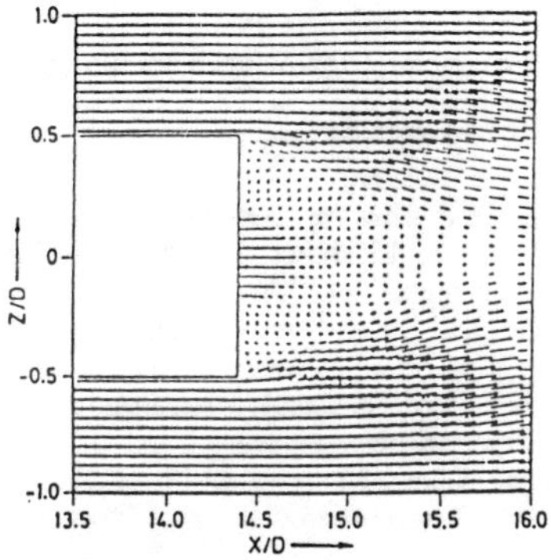

$M_\infty = 2.5$, $\alpha = 0$, $I = 0.01$;

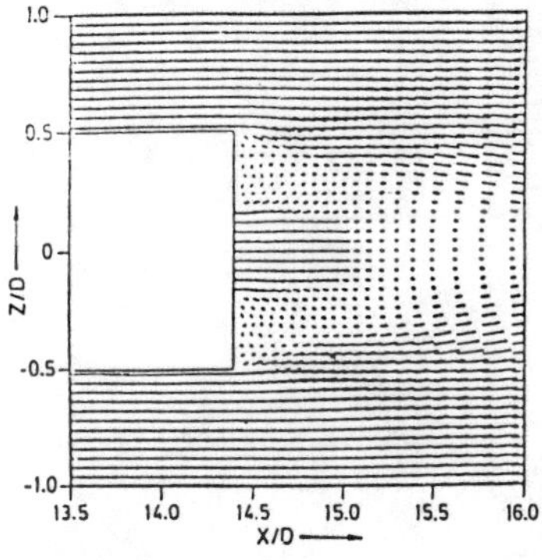

$M_\infty = 2.5$, $\alpha = 0$, $I = 0.02$

FIGURE 10. Velocity vectors in the base region (from Reference 16).

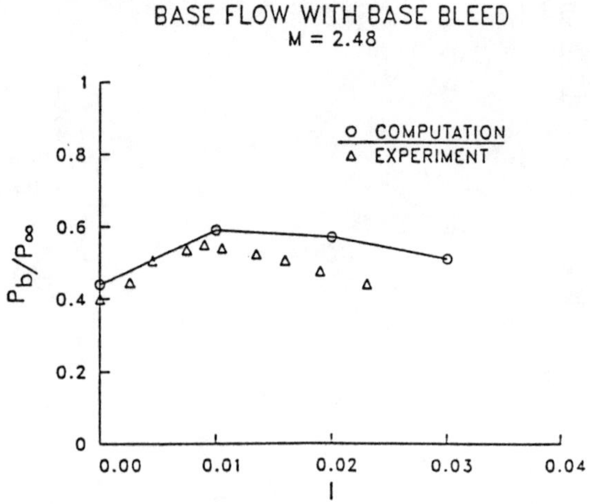

FIGURE 11. Base pressure as influenced by the bleed parameter I (showing both Regimes I and II).

A detailed streamline plot does indicate that some of the bleed flow can penetrate the downstream region of a higher pressure (see Figure 12). Figure 13 shows the case of higher jet stagnation pressure ratios where the bleed stream becomes a plume jet. The computed base pressures have been compared with the experimental data[18] and are generally in good agreement. This is obviously the flow condition in Regime III. The base pressure increases as the stagnation pressure (or the static pressure) of the jet is increased. A detailed streamline flow pattern of the plume jet is shown in Figure 14. The interaction between the jet and the free stream results in a pair of recirculating bubbles in the near wake. Many other investigators have also computed the popular MICOM problem[19-21] within this flow regime.

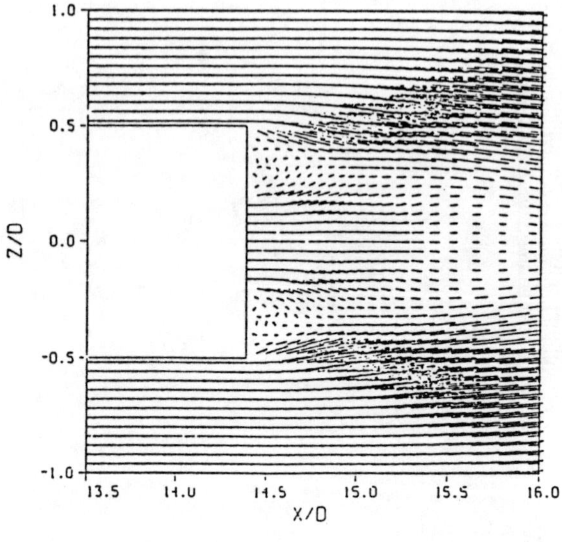

FIGURE 12. Detailed velocity vectors diagram in Regime II.

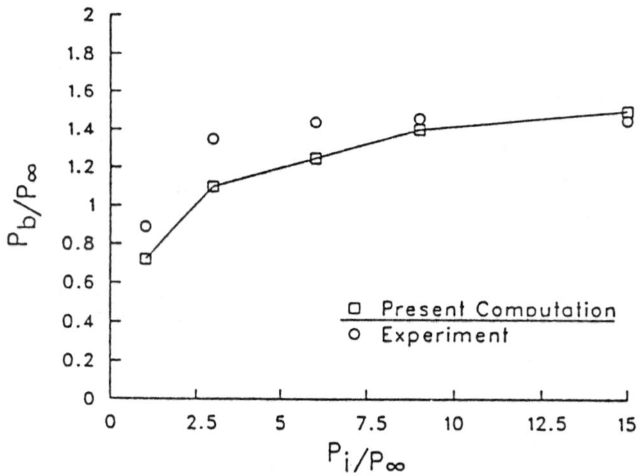

FIGURE 13. Base pressure as influenced by the jet pressure ratio--Regime III.

In all these numerical computations, we have considered only the effect of mass bleed since the stagnation temperature of the jet, T_{oj}, was set equal to that of the free stream, $T_{o\infty}$. Should $T_{oj} > T_{o\infty}$, the results correspond to the combined influences of both the mass and energy additions. Such a simulation can easily be produced by simply changing the temperature boundary condition at the bleed exit.

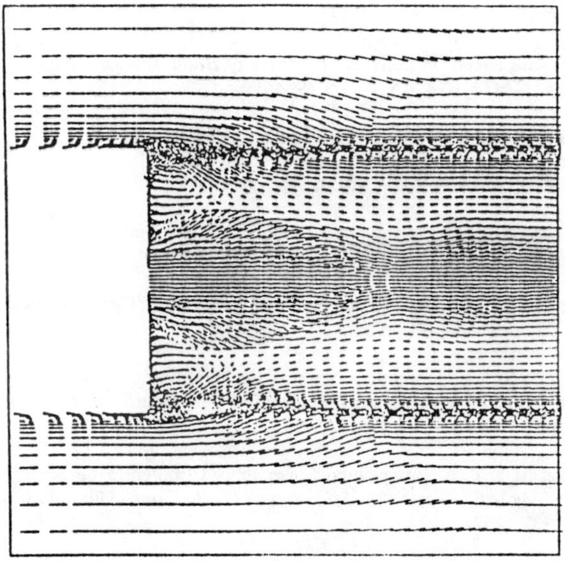

FIGURE 14. Detailed velocity vectors diagram--Regime III.

CONCLUSIONS

A short review of the fluid dynamic aspects of the effect of base bleed is given. From the interactive study, it was observed that, depending upon the relative level of the stagnation pressure of the bleed stream, its influence on the base pressure ratio can be classified into three distinctively different flow regimes. The effect of energy addition to the wake can also be extracted from a simple component analysis. In addition, through large-scale numerical computation of the Navier-Stokes equations, it has been shown that the effect of base bleed can be effectively simulated by providing the appropriate boundary conditions of the bleeding stream at the base. It may be anticipated that the effect of combustion by bleeding fuel into the wake may be interpreted in terms of the effects of mass and energy additions. Again, when a chemical reactive computer code is incorporated into the present Navier-Stokes solver, this phenomenon can be simulated by simply specifying the concentration of various reactive species at the bleed exit.

ACKNOWLEDGMENT

The authors are grateful to Dr. W. B. Sturek and Mr. C. J. Nietubicz at the Ballistic Research Laboratory for their continued interest and guidance on the numerical computations of the base pressure problems. The second author wishes to acknowledge the support of the U.S. Army Grant DAALO3-87-K-0010, and NSF Grant No. MSM-8603985 which made large-scale numerical computations possible.

REFERENCES

1. Chow, W. L., On the Base Pressure Resulting From the Interaction of a Supersonic External Stream With a Sonic or Subsonic Jet, *J. Aero. Sci.*, vol. 26, no. 3, pp. 176-180, 1959.

2. Fuller, L., and Reid J., Experiments on Two-Dimensional Base Flow at M = 2.4, Brit. RAE Report Aero, 2569, 1956.

3. Wimbrow, W. R., *Effect of Base Bleed on the Base Pressure of Blunt-Trailing-Edge Airfoils at Supersonic Speeds*, NACA RM A54A 07, 1954.

4. Bowman, J. E. and Clayden, W. A., Cylindrical Afterbodies in Supersonic Flow With Gas Ejection, *AIAA Journal*, vol. 5, pp. 1524-1525, 1967; Cylindrical Afterbodies at M_∞ = 2 With Hot Gas Ejection, *AIAA Journal*, vol. 6, pp. 2429-2431, 1968.

5. Schilling, H., Experimental Investigation on the Base-Bleed-Effect for Body-Tail-Combinations. *Proc., 8th Intern. Symp. Ballistics, Amsterdam*, 1984.

6. Korst, H. H., A Theory for Base Pressures in Transonic and Supersonic Flow, *Journal of Applied Mechanics*, vol. 23, pp. 593-600, 1956.

7. Page, R. H., and Korst, H. H., Non-Isoenergetic Compressible Jet Mixing With Consideration of its Influence on the Base Pressure Problem, *Proc., Fourth Midwestern Conference on Fluid Mechanics, Purdue Univ.*, pp. 45-68, Sept. 1955.

8. Chow, W. L., Recompression of a Two-Dimensional Supersonic Turbulent Free Shear Layer, *Development of Mech.*, vol. 6, *Proc., 12th Midwestern Mech. Conf.*, University of Notre Dame Press, pp. 319-332, 1971.

9. Chow, W. L., and Spring, D. J., The Viscous Interaction of Flow Redevelopment After Reattachment With Supersonic External Streams, *AIAA Journal*, vol. 13, no. 12, pp. 1576-1584, 1975.

10. Chow, W. L., and Spring, D. J., Viscid-Inviscid Interaction of Two-Dimensional Incompressible Separated Flows, *J. Applied Mech.*, vol. 43, series E, pp. 387-395, Sept. 1976.

11. Chow, W. L., Base Pressure of a Projectile Within the Transonic Flight Regime, AIAA Paper 84-230, Presented in the AIAA 22nd Aerospace Sciences Meeting, Reno, Nevada, Jan. 1984, also *AIAA J.*, vol. 23, no. 3, pp. 388-395, 1985.

12. Chow, W. L., The Effect of Boattailing of a Projectile in Transonic Flow, *Proc., of the Third Symposium on Numerical and Physical Aspects of Aerodynamic Flows, California State University, Long Beach, CA.*, 9-15 to 9-23, Jan. 1985.

13. Sahu, J., Nietubicz, C. J., and Steger, J. L., Numerical Computation of Base Flow for a Projectile at Transonic Speeds, U.S. Army Ballistic Research Laboratory, Aberdeen Proving Ground, MD, ARBRL-TR-02495, June 1983. (AD A130293) (See also, AIAA Paper no. 82-1358, August 1982.)

14. Sahu, J., Nietubicz, C. J., and Steger, J. L., Navier-Stokes Computations of Projectile Base Flow at Transonic Speeds With and Without Mass-Injection, U.S. Army Ballistic Research Laboratory, Aberdeen Proving Ground, MD., ARBRL-TR-02532, Nov. 1983 (see also *AIAA Journal*, vol. 23, no. 9, pp. 1348-1355, Sept. 1985).

15. Sahu, J., and Nietubicz, C. J., Numerical Computation of Base Flow for a Missile in the Presence of a Centered Jet, U.S. Army Ballistic Research Laboratory, Aberdeen Proving Ground, MD, ARBRL-MR-3397, October 1984. (AD A148784) (See also, AIAA Paper no. 84-0527, January 1984.)

16. Sahu, J., Supersonic Flow Over Cylindrical Afterbodies With Base Bleed, U.S. Army Ballistic Research Laboratory, ARRADCOM, Tech. Report BRL-TR-2742, June 1986 (see also *Computational Mechanics*, no. 2, pp. 176-184, 1987).

17. Sahu, J., Computations of Supersonic Flow Over a Missile Afterbody Containing an Exhaust Jet, *J. of Spacecraft and Rockets*, vol. 24, no. 25, pp. 403-410, Sept.-Oct., 1987.

18. Agrell, J. and White, R. A., An Experimental Investigation of Supersonic Axisymmetric Flow Over Boattails Containing a Centered Propulsive Jet, FFA Technical Note AU-913, 1974.

19. Petrie, H. L., and Walter, B. J., Comparisons of Experiment and Computations for a Missile Base Region Flow Field With a Centered-Propulsive Jet, AIAA Paper 85-1618, 1985.

20. Venkatapathy, E., Lombard, C. K. and Luh, C. C., Accurate Numerical Simulation of Supersonic Jet Exhaust Flow With CSCM on Adaptive Overlapping Grids, AIAA Paper 87-0465.

21. Hoffman, J. J., Birch, S. F., Hopcroft, R. G., and Holcomb, J. E., Navier-Stokes Calculations of Rocket Base Flows, AIAA-Paper 87-0466.

COMMENTS

Dr. Hasselaar, Vd. Burchlaan, Netherlands. Do you consider an injection parameter of 0.01 as optimal for base bleed projectiles?

Author's Reply.

The injection parameter of 0.01 is a small enough number that it will have a beneficial effect in reducing the base drag. As the evidence of difference between the flow patterns of Regime I and II, the momentum related to the bleed flow is the other important quantity of concern. Many factors, such as the flow geometry and the flight Mach number of the projectile, will have their direct influences in reaching the optimal condition. Thus, it is difficult to conclude that the bleed parameter of 0.01 corresponds to the optimal bleed flow condition. For atmospheric flight conditions, Reynolds number will be larger than those considered here for wind tunnel conditions. In addition, for actual base bleed projectiles, one has to consider both mass and energy additions in the base region which would allow the same bleed effect to occur at a lower bleed parameter.

Navier-Stokes Computations of Base Bleed Projectiles

CHARLES J. NIETUBICZ and JUBARAJ SAHU
Launch and Flight Division
U.S. Army Ballistic Research Laboratory
Aberdeen Proving Ground, Maryland 21005-5066, USA

ABSTRACT

The M864 is an extended range projectile which includes both a dome base cavity and a base burn capability. A Navier-Stokes computational technique has been applied to the M864 base burn projectile. The Navier-Stokes technique is described and includes a discussion of applicable boundary conditions. Computed results have been obtained for a flat base and dome base configuration both with and without base bleed. A small reduction in base drag was found for the dome base configuration without base bleed. The addition of hot bleed gas in the base region was found to significantly change the recirculating base flow; depending on the bleed parameter has resulted in a large decrease in base drag. The computed results cover a Mach number range of 0.7 thru 2.5.

INTRODUCTION

The ability to compute the base region flow field for projectile configurations using Navier-Stokes computational techniques has been developed over the past few years. This capability is most important for determining aerodynamic coefficient data and in particular the total aerodynamic drag. The total drag as described in this paper consists of the pressure drag (excluding the base), viscous drag and base drag components. At transonic and low supersonic speeds the base drag component is the major contributor to the total aerodynamic drag.

The majority of base flow calculations to date have modeled the base region as a flat solid surface. Many of the actual configurations have some form of base cavity. General opinion has been that the inclusion of a base cavity or modifications to the interior cavity of a projectile base would have little or no effect on the overall flight performance parameters. In addition to base cavities, projectiles are currently being designed which include base bleed and base burning.

The M864 base burn projectile is a configuration of current interest and includes both a dome base cavity and a base burn capability. A schematic is shown in Figure 1. The shape of the M864 is similar to that of the product improved M825 which also contains a base cavity.[1] Very limited data exist for the M864 and the effects of both the dome base and the addition of base bleed are not well understood. Range firings have been

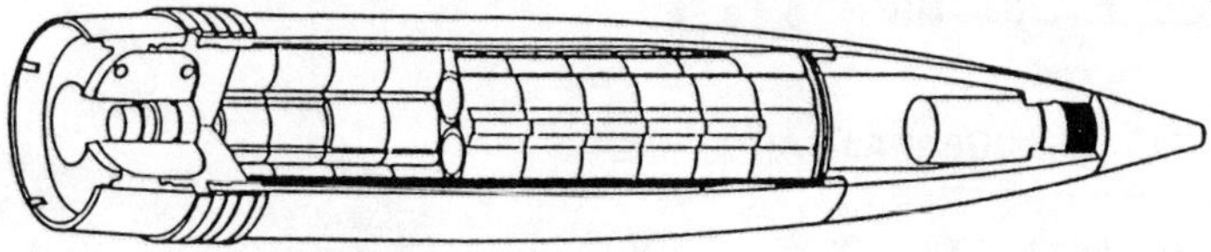

FIGURE 1. Schematic of M864

conducted for the M864 with an inert base burn motor and active motors. Reductions in base drag of between 30-50% have been found during the 30 second motor burn.[2]

A Navier-Stokes computational study is in progress to determine the aerodynamic coefficients for the M864 both with and without base bleed. A flat base M864 calculation was performed to determine the effect of the base cavity. The use of Navier-Stokes codes can provide a detailed description of the flow field associated with these configurations as well as the integrated aerodynamic coefficients. The work to date has been accomplished using an axisymmetric base flow code. Numerical computations have been completed for a Mach number range of $0.7 \leqslant M \leqslant 2.5$. The computed aerodynamic drag show a small reduction in the total aerodynamic drag for the dome configuration. Additional drag reduction was found by modeling cold mass addition in the base region. Further reduction in drag was found with hot mass injection. Qualitative features of the computed flow fields are presented in the form of velocity vectors, and particle paths.

GOVERNING EQUATIONS AND SOLUTION TECHNIQUE

Governing Equations

The complete set of time-dependent, thin-layer, Navier-Stokes equations is solved numerically to obtain a solution to this problem. The numerical technique used is an implicit finite difference scheme. Although time-dependent calculations are made, the transient flow is not of primary interest at the present time. The steady flow, which is the desired result, is obtained in a time asymptotic fashion.

The time-dependent, thin-layer, Navier-Stokes equations written in strong conservation law form for the axisymmetric formulation[3] are:

$$\frac{\delta \hat{q}}{\delta \tau} + \frac{\delta \hat{E}}{\delta \xi} + \frac{\delta \hat{G}}{\delta \zeta} + \hat{H} = \frac{1}{Re} \frac{\delta \hat{S}}{\delta \zeta} \tag{1}$$

The general coordinate transformations are defined as:

$\xi = \xi(x,y,z,t)$ - is the longitudinal coordinate
$\zeta = \zeta(x,y,z,t)$ - is the near normal coordinate
$\tau = t$ - is the time

The vector $\hat{q}$ contains the dependent variables [ρ, ρu, ρv, ρw, e] and the flux vectors $\hat{E}$, $\hat{G}$ contain terms which arise from the conservation of mass, momentum and energy. The source vector $\hat{H}$, in Equation (1), contains terms which result from an analytic determination of the circumferential flux vector given the assumption of axisymmetric flow and constant angular velocity.[3] The viscous terms are contained in the vector $\hat{S}$ which is seen to have variation in the ζ direction only. This is representative of the thin-layer approximation.

Solution Technique

The numerical algorithm used for the solution of Equation (1) is the Beam and Warming[4] implicit, approximately factored, finite-difference scheme which uses central differencing in both ξ and ζ directions. Code improvements have been made to include a variable time step, numerical smoothing based on local solution gradients and code vectorization.[5] The Beam-Warming implicit algorithm has been used in various applications for the equations in general curvilinear coordinates. The algorithm is first-order accurate in time and second- or fourth-order accurate in space. Central difference operators are employed; the algorithm produces a block tridiagonal system for each space coordinate. The main computational work is contained in the solution of these block tridiagonal systems of equations.

To suppress high frequency components that appear in regions containing severe pressure gradients e.g., shocks or stagnation points, artificial dissipation terms are added. In the present application, a switching dissipation model is used which is a blend of second- and fourth-order dissipation terms. This is similar to the model used by Pulliam[6] which uses a fourth-order dissipation in smooth regions and switches to a second-order dissipation in regions containing high pressure or density gradients. Incorporation of this dissipation model has resulted in an improvement in the quality of the results and has made the code more robust.

The axisymmetric code that solves Equation (1) uses a unique flow field segmentation procedure to compute the full flow field over a projectile or a missile including the base region. Although the thin-layer approximation is not valid in the wake region, this segmentation procedure allows for turbulent quantities to be determined across the shear layer. This has provided reasonable accuracy for supersonic flows. The details of this can be found in Reference 7. For the computation of turbulent flows, the two-layer algebraic, Baldwin-Lomax, turbulence model[8] is used over the projectile body.

BOUNDARY CONDITIONS

The outer computational boundary was kept at approximately 12 body diameters from the body surface. This allows free stream conditions to be imposed at the outer boundary. At the downstream boundary extrapolated outflow conditions were used.

Viscous boundary conditions were imposed on the surface using no slip at the wall. The normal momentum equation is solved at the wall for p_n, the normal pressure gradient. The free flight wall temperature is specified. Using the time lagged pressure, the density at the wall is determined.

The base bleed boundary conditions initially used were as described in Reference 9, where the bleed conditions are iteratively set. The current boundary conditions do not require iteration and follow from knowing the computed pressure and flow field variables at each time step. The bleed boundary conditions are thus lagged in time. The bleed conditions $(\rho u)_j$, and T_{o_j} are specified while $p_j = p_1$ is the pressure at the interface boundary at time step n. The bleed stagnation pressure, p_{o_j}, is defined and the bleed properties of p_j and T_j are known. By specifying the mass injection rate as:

$$I = \frac{\rho_j u_j A_j}{\rho_\infty u_\infty A_b} \tag{2}$$

the bleed Mach number (M_j) can be directly solved by:

$$M_j^2 = \frac{1}{\gamma-1} \quad -1 + \sqrt{1 + 2(\gamma-1)\frac{T_{o_j}}{T_\infty} \quad \frac{IM_\infty \rho_\infty A_b}{p_j A_j}^{\,2}} \tag{3}$$

where A_b and A_j are the area of the base and bleed opening respectively. Using isentropic relations the bleed stagnation pressure p_{o_j} and static temperature T_j can be calculated. Given the temperature and pressure the density is computed and the boundary conditions for the next time step are obtained.

MODEL GEOMETRY AND COMPUTATIONAL GRID

The computational model of the M864 is shown in Figure 2. The features of this projectile which have not been modeled exactly are the meplat on the fuze and the rotating band near the base. The rotating band was eliminated. The meplat was modeled as a hemisphere cap. The model nose is a combination of ogive and conical sections. The cylindrical section is slightly undercut, and the boattail length is 0.524 calibers with a 3° angle. The general shape of the internal base cavity is shown as a dotted line on the boattail. The base burn motor is located internally with the exhaust port centered on the model axis.

The solution technique requires the discretization of the entire flow region of interest into a suitable computational grid. The grid outer boundary has been placed at 1.5 body lengths upstream and surrounding the projectile. The downstream boundary was placed at 2 body lengths. Since the calculations are in the subsonic/transonic regime, the computational boundaries must extend out beyond the influence of the body. This ensures that the specified boundary conditions are satisfied.

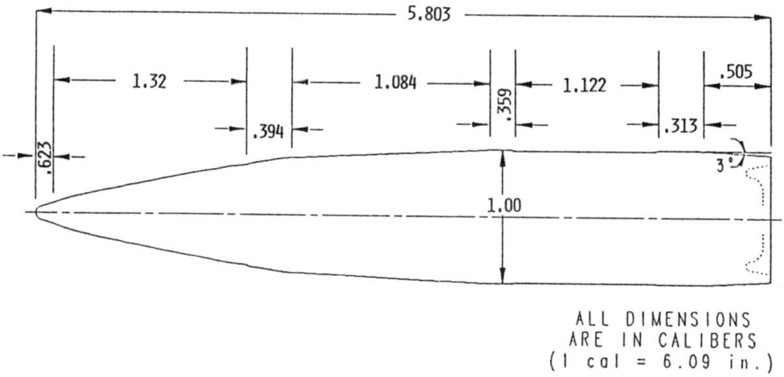

FIGURE 2. Computational model, M864

Figure 3 shows the grid generated over the projectile body including the base. It consists of 265 longitudinal points and 60 points in the normal direction. This is broken down into two sections: a body region, and a base region. The surface points for each region are selected using an interactive design program. Each grid section is then computed separately using a hyperbolic grid generation program.[10] Figures 4a and 4b show an expanded view for both the flat and dome base configurations. The flat base grid was routinely generated. However, due to the extreme concavity, the grid for the dome base required an increase in the smoothing values used by the hyperbolic grid generator as well as the addition of a grid cell averaging technique.

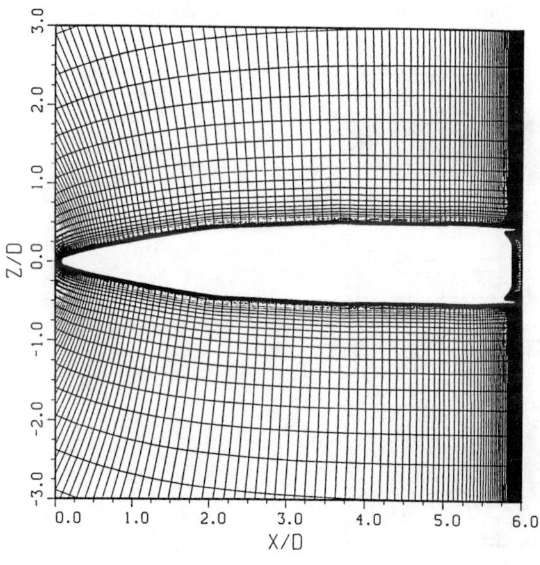

FIGURE 3. Computational grid expanded near the model, M864

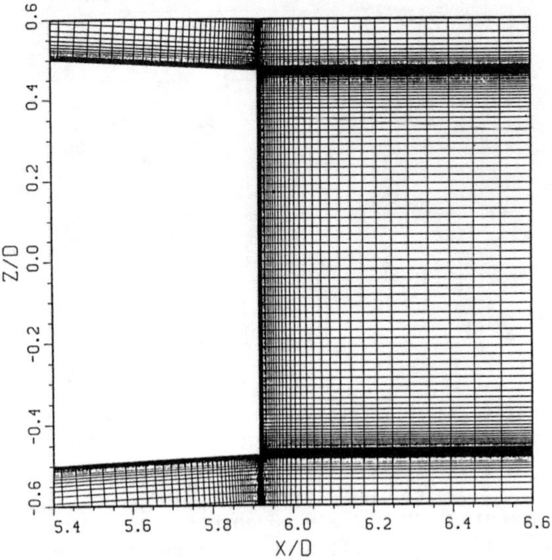

FIGURE 4a. Base region grid, flat base, M864

RESULTS

A series of calculations have been run for both the flat base and the actual domed M864 with and without base bleed. The Mach number range computed was 0.7 < M < 2.5 with $\alpha = 0°$. Figures 5a and 5b show the computed velocity vectors in the wake region for the flat base and domed base, respectively. The effect of the dome base is to generate a second,

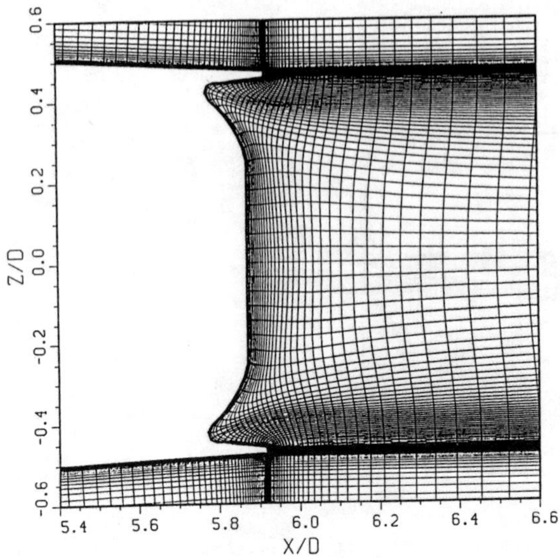

FIGURE 4b. Base region grid, dome base, M864

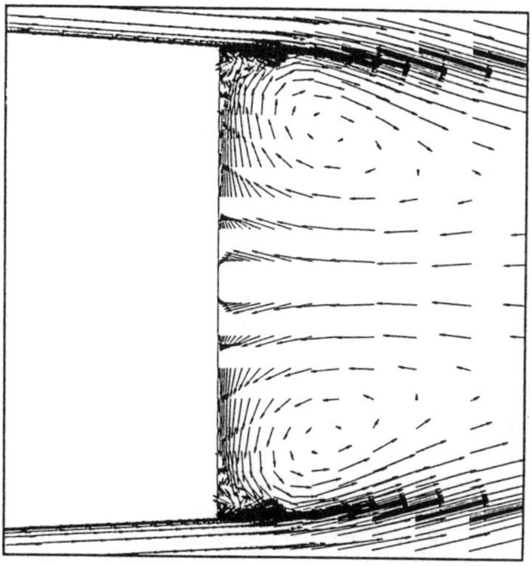

FIGURE 5a. Velocity vectors in base region for flat base M864, M = 1.4, $\alpha = 0°$, I = 0.0

low speed recirculation in the recessed cavity. This recirculation bubble is in the opposite direction of the primary recirculation pattern. This is significantly different than the recirculation pattern found for the flat base and other dome base configurations.

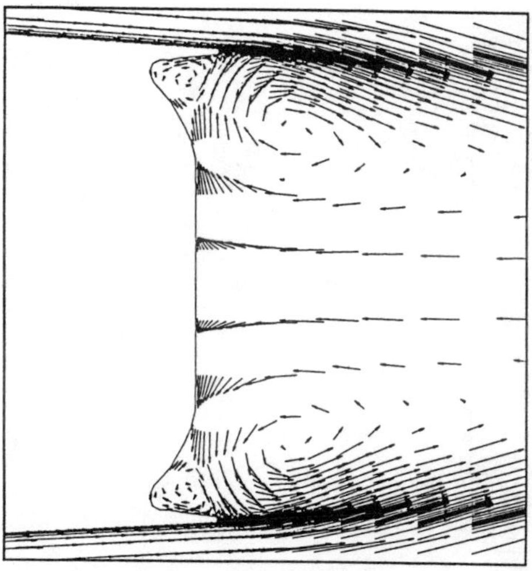

FIGURE 5b. Velocity vectors in base region for dome base M864, M = 1.4, $\alpha = 0°$, I = 0.0

Computations have also been performed for the dome base with the inclusion of base bleed. The mass injection parameter, I_j, defined as $I_j = \dot{m}_j/\rho_\infty u_\infty A_b$; was used to vary the mass flow rate. The computations to date have been for cold and hot mass flow, with mass injection values of up to 0.04. Using the propellant mass and a burn time of 30 seconds, the estimated $I_j \simeq 0.004$. Although the actual mass flow rate varies over the course of the trajectory, the present calculations were run to see the effect on drag across a Mach number range. Figure 6 shows the computed velocity vectors for the dome M864 with I = 0.01. The addition of base bleed has increased both the size and strength of the secondary recirculation region.

Recent ground tests have been completed for the spinning M864 base burn motor assembly. Pressure taps and a thermocouple were contained within the base cavity. The temperature varied as a function of spin and burn time. A maximum temperature level of 1200K was obtained. A second set of computations were run to see the effects of hot mass addition. The results shown in Figures 7a and 7b show particle traces in the vicinity of the base region for the bleed gas temperature of 300K and 1200K cases, respectively. The dark traces are particles which originate from the bleed ejection hole. The lighter traces were released in various parts of the flow field. The effect of temperature is rather dramatic in changing the recirculation pattern within the base flow field. At the lower temperature, the bleed gas is shown to be captured by the recirculation region near the base. As the temperature is increased, all of the bleed gas moves downstream. The higher temperature gives rise to an increased velocity and thus an effective increase in the mass injection parameter. The resultant effect of temperature on total drag is shown in Figure 8. The increased temperature shows a net reduction in drag across a range of mass injection rates.

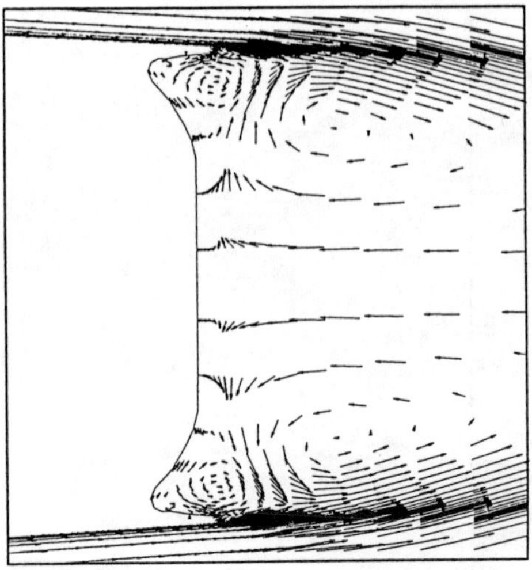

FIGURE 6. Velocity vectors in base region for dome base M864, M = 1.4, $\alpha = 0°$, I = 0.01, T_{o_j} = 300K

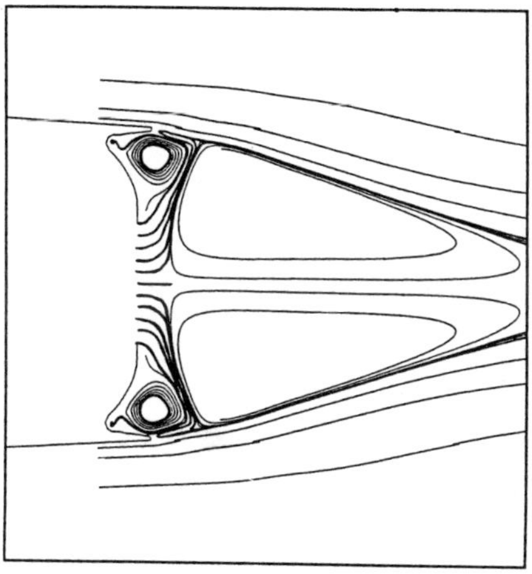

FIGURE 7a. Particle traces in base region, $\alpha = 0°$, M = 2.0, T = 300K

Figure 9 is a plot of the computed drag coefficient, C_{D_0}, for the flat base and dome base with and without base bleed. The computed drag coefficient is the summation of the pressure, viscous and base drag components. The results show a decrease in total drag for the dome base configuration for all Mach numbers. When the base bleed is initiated a further reduction in drag is evident.

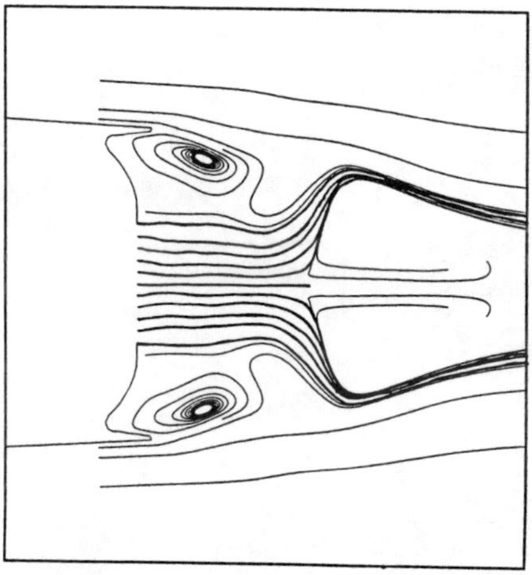

FIGURE 7b. Particle traces in base region, $\alpha = 0°$, M = 2.0, T = 1200K

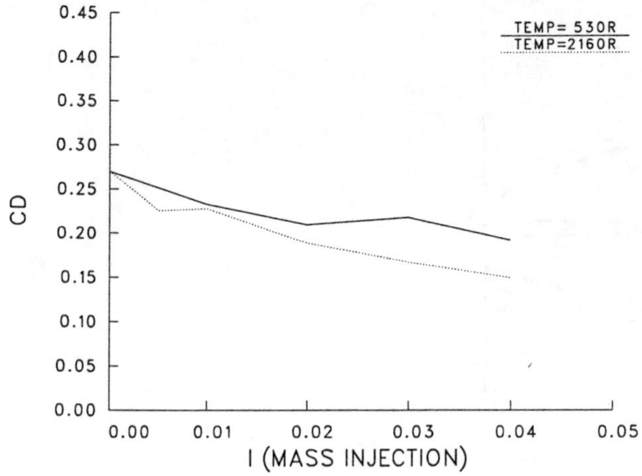

FIGURE 8. Total drag vs mass injection rate at T = 300K and 1200K, M = 2.0

Recent tests performed in the BRL Transonic Aerodynamics Range have provided some limited drag data[11] for the M864. This experimental data is also shown in Figure 9. The experimental data is for the M864 without base burn using an inert burn motor. The computed results are in general agreement at the supersonic velocities, however, in the transonic and subsonic regions, the comparison is poor. The underprediction of the drag in this region has been noticed before and at present lacks a full explanation. Grid refinement studies, turbulence modeling and alternate solution techniques are currently in progress.

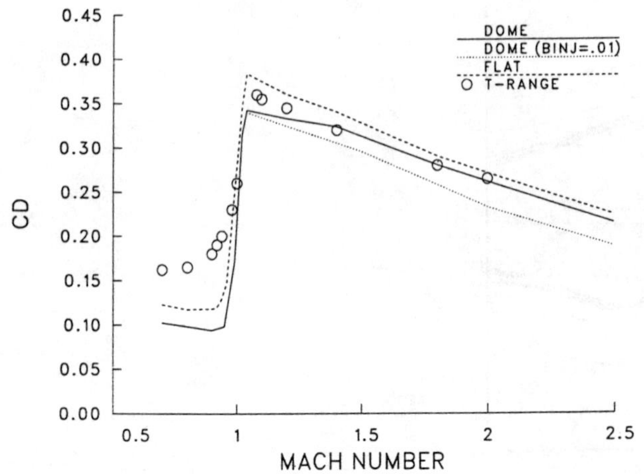

FIGURE 9. Comparison of computed and experimental drag, M864, $\alpha = 0°$

CONCLUDING REMARKS

Numerical computations have been performed for the M864 projectile at supersonic and transonic speeds. The addition of base bleed with a dome base cavity has also been computed. Additional results were also obtained for hot mass addition.

The computed drag coefficients showed an overall decrease in drag between the flat and dome base configuration. An additional decrease in drag was found for cold mass flow addition at the base. Further reductions were found for hot mass addition. A comparison of the computed drag and limited experimental data, for a case with no base bleed, indicated a fair agreement at the supersonic speeds. As the Mach number approaches M = 1 and into the subsonic regime, the comparison becomes poor. Grid refinement work and alternate solution techniques are currently being employed in an attempt to provide more accurate solutions in the subsonic and transonic area.

NOMENCLATURE

a	speed of sound
A	area
C_{D_o}	zero yaw drag coefficient
D	body diameter
$\hat{E},\hat{G}$	flux vectors of transformed Navier-Stokes equations
$\hat{H}$	η-invariant source vector
I	mass injection parameter, $\rho_j u_j A_j/\rho_\infty u_\infty A_b$
M	Mach number
p	pressure
$\hat{q}$	solution vector containing dependent variables
Re	Reynolds number, $\rho_\infty D a_\infty/\mu_\infty$
$\hat{S}$	viscous flux vector
t	physical time
T	temperature
u	Cartesian velocity component

Greek Symbols:

γ	ratio of specific heats
ξ,ζ	transformation coordinates in axial, and radial directions
ρ	density
τ	transformed time

Subscript:

b	base
j	bleed gas conditions
o	stagnation conditions
∞	free stream conditions

REFERENCES

1. D'Amico, W. P., "Transonic Range Testing of the 155mm M825 Projectile with Standard and Dome Steel Bases," U.S. Army Ballistic Research Laboratory, Aberdeen Proving Ground, Maryland, report is forthcoming.

2. Lieske, R. F., private communications, Firing Tables Branch, Launch and Flight Division, U.S. Army Ballistic Research Laboratory, September 1987.

3. Nietubicz, C. J., Pulliam, T. H., and Steger, J. L., "Numerical Solution of the Azimuthal-Invariant Navier-Stokes Equations," U.S. Army Ballistic Research Laboratory, Aberdeen Proving Ground, Maryland, ARBRL-TR-02227, March 1980. (AD A085716) (Also see AIAA Journal, vol. 18, no. 12, pp. 1411-1412, December 1980.)

4. Beam, R., and Warming, R. F., "An Implicit Factored Scheme for the Compressible Navier-Stokes Equations," AIAA Paper No. 77-645, June 1977.

5. Sahu, J., and Nietubicz, C. J., "Improved Numerical Prediction of Transonic Flow," Proceedings of the 4th Army Conference on Applied Mathematics and Computing, Cornell University, May 1986.

6. Pulliam, T. H., and Steger, J. L., "On Implicit Finite-Difference Simulations of Three-Dimensional Flow," AIAA Journal, vol. 18, no. 2, pp. 159-167, February 1980.

7. Sahu, J., Nietubicz, C. J., and Steger, J. L., "Navier-Stokes Computations of Projectiles Base Flow with and without Base Injection," U.S. Army Ballistic Research Laboratory, Aberdeen Proving Ground, Maryland, ARBRL-TR-02532, November 1983. (AD A135738) (Also see AIAA Journal, vol. 23, no. 9, pp. 1348-1355, September 1985.)

8. Baldwin, B. S., and Lomax, H., "Thin-Layer Approximation and Algebraic Model for Separated Turbulent Flows," AIAA Paper No. 78-257, 1978.

9. Sahu, J., "Supersonic Flow Over Cylindrical Afterbodies with Base Bleed," U.S. Army Ballistic Research Laboratory, Aberdeen Proving Ground, Maryland, BRL-TR-2742, June 1986.

10. Nietubicz, C. J., Heavey, K. R., and Steger, J. L., "Grid Generation Techniques for Projectile Configurations," Proceedings of the 1982 Army Numerical Analysis and Computers Conference, ARO Report 82-3, February 1982.

11. Oskay, V., private communications, Free Flight Branch, Launch and Flight Division, U.S. Army Ballistic Research, September 1987.

COMMENTS

D.A. Barents, Ministry of Defense, The Netherlands. In relation to the graph of the measured pressures in the chamber, my question is looking at measured pressure, is this graph independent of trajectory in other words independent of elevation, or is this a special case? If in a special case can you give more data at more cases? At fixed Mach numbers?

Author's Reply. The graph you are indicating was a plot of the computed radial distribution of the base pressure. The experimental points were from measurements made at two locations on the base. The calculations were made at one Mach number at a specific elevation and compared with the measured pressure at that point. The measured base pressure varied through the supersonic portion of the trajectory and became approximately constant over the subsonic portion of the flight. Additional tests are planned at higher launch velocities.

K.K. Kuo, Pennsylvania State Univ., USA. It is a very interesting paper. The CFD model and results presented by you do not contain any combustion consideration. If you were interested in extending the model to a more complete one to include the effect of chemical reactions, how would you treat the problem of chemical reacting flows? Do you believe that it is reasonable to assume the flow field inside the base bleed unit as if it is at chemical equilibrium and on the other hand treat the external flow field by finite-rate chemistry.

Author's Reply. The attempt to include the effect of chemical reactions has been started. The use of hot mass injection is meant to simulate the air heated due to combustion. As you can see from some of my results the base pressure has changed significantly due to the increased temperature of the injected mass. Additional work will follow using energy release in the wake and possibly the use of a variable ratio of specific heats in the wake. The assumption of chemical equilibrium in the combustion chamber is a good place to start. Ideally one likes to treat the entire problem with finite-rate chemistry calculations.

J. Fleming, FCI, England. You mention the re-design of the rear base of the base bleed projectile was changed to a dome base shape. Although it is understood this modification was affected mainly for weight reasons, please comment on how the change in shape has influenced the aerodynamics. What was the change in range observed?

Author's Reply. The base change I was referring to was actually on the M825 projectile. The dome base of the M864 is similar to that of the M825. The major changes in aerodynamics of the M825 for two different base configurations was approximately a two percent change in drag and a seven percent change in lift. The change in range was found to vary with muzzle velocity. The percent change in range (difference between standard base and dome base round) varied from approximately one percent a 200 m/sec to 0.2 percent at 800 m/sec.

M. Hasselaar, DMKL/MBA1, The Netherlands. Could you show your viewgraph of the temperature measurements in free flight. Are you able to determine the burning time of the grain in free flight. If so, do you find the burning time is consistent for several different situations? Please comment.

Author's Reply. An estimate of the burn time can be seen in the temperature trace shown on the viewgraph. However a better estimate was made from looking at the in flight combustion chamber pressure measurement. Since there was only one instrumented round fired, I cannot comment on a constant burn time for different situations. Additional flight tests are being planned.

Computer Codes for Ballistic Performance Calculations of Base Bleed Propellant Grain

YVES FABIGNON
S.N.P.E., C.R.B.
91710 Vert le Petit
France

DOMINIQUE COUPEZ
S.N.P.E., St Médard
33160 St Médard en Jalles
France

ABSTRACT

Concerning shell external ballistics, different technical solutions are studied in order to improve aerodynamic performance of a shell. The solution named base bleed consists of injecting a hot gas at a low velocity into the depressional flow area. Preliminary design of a base bleed propellant grain has to satisfy requirements generally expressed as a shell range increase against available grain volume. In order to adapt the propellant grain ballistic performance to the shell trajectory conditions, two different investigations have been developed. The first investigation is essentially theoretical and consists of computer simulations of the flowfield near the shell base with and without effect of subsonic injection. The second investigation is more empirical and consists of analysing the shell velocity evolution by means of radar measurements in order to determine drag reduction factor due to the base bleed effect. The organization of the developed ballistic computer codes and some results are described in this paper. The results of the two studies give ballistic data necessary for simulations of base bleed shell trajectories.

NOMENCLATURE

Theoretical approach	:
C_D, C_1, C_2	: turbulence model parameters
C_p	: specific heat at constant pressure
D	: diameter of projectile
D_b	: diameter of exhaust port
f	: mixing ratio

Theoretical approach (continuing)

H	: total enthalpy
k	: turbulent energy
M_∞	: projectile Mach number
M_K	: molar mass of specie K
p_C	: base pressure
p	: local pressure
p_∞	: ambiant pressure
Q_b	: mass flow rate
I	: injection parameter $I = \dfrac{Q_b}{\rho_\infty\ u_\infty\ S_{REF}}$
R	: constant of perfect gases
S_ϕ	: source term
T	: temperature of gas
T_b	: stagnation temperature of base bleed gas
u,v	: cartesien velocity components
x,r	: cylindrical coordinates
x_K	: molar concentration of specie K
Y	: altitude
Y_K	: mass fraction of specie K
ε	: dissipation rate
ρ	: density of gas
ϕ	: dependent variable
Γ_ϕ	: turbulent exchange coefficient
μ_t	: eddy viscosity
μ_l	: laminar viscosity
$\sigma_k, \sigma_\varepsilon$	: turbulent Prantl numbers

Empirical approach :

C_D	: total drag coefficient
C_{DTOT}	: total drag coefficient without base bleed effect
C_{DBASE}	: base drag coefficient without base bleed effect
C_{RED}	: base drag reduction factor
g	: acceleration of gravity
I	: injection parameter $I = \dfrac{Q_b}{\rho_\infty\ u_\infty\ S_{REF}}$
m	: projectile mass
M_∞	: Mach number of projectile
r_b	: burning rate

Empirical approach (continuing)

S_b	: propellant burning area
S_{REF}, S_{BASE}	: reference area and base area of projectile
V	: projectile velocity
x,y	: range, altitude
ρ_∞	: air density
ρ_p	: propellant density
θ	: trajectory angle

1. INTRODUCTION

When a shell manufacturer designs a new projectile with a base bleed device, the requirements are expresssed as a shell range increase with the available grain volume. The first step in the preliminary design of a base bleed grain is to derive a propellant grain configuration from shell performance. The designer has first to select a grain internal shape and a propellant. Figure 1 shows general bloc diagram of a new base bleed grain analysis.

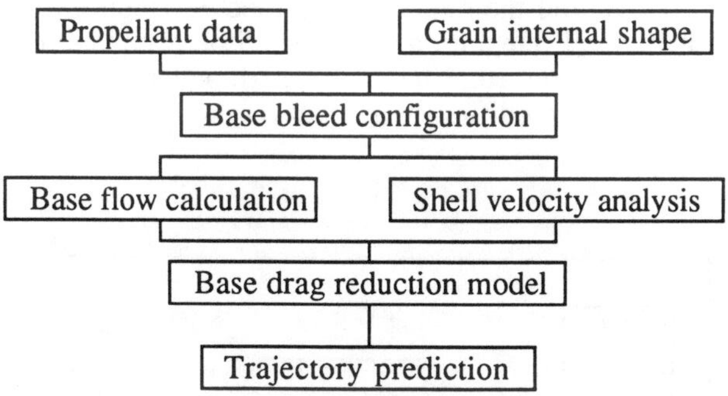

FIGURE 1. Main flow chart

This paper presents the work performed in our company on different methods to establish and to complete a base drag reduction model. The theoretical approach starts with calculation of propellant thermo-chemical properties [1]. These calculated data are input data to simulate afterwards base flows with base bleed effect.

The computer code used to simulate base flow solves Navier-Stockes equations in cylindrical coordinates including a (k-ε) turbulence model. A combustion model has also been developed in order to analyse the effect of propellant gas afterburning in ambient air. To simulate operating conditions of a base bleed gas generator, calculations are performed with and without a reactive injecting flow. Results show the effect of propellant gas on the shell base pressure and the dead air area location. This approach is described in more details in section 2. The empirical approach is based on gun firing tests results and consists in analysing the shell velocity evolution during base bleed operating time. The computer code uses a simple external ballistic model. After some gun firing tests, radar velocity measurements are performed in order to evaluate total aerodynamic drag coefficient with and without the base bleed effect. Results allow the determining of base bleed grain efficiency as it varies with shell configuration. This approach is described in section 3.

Computed results of these two investigations are used to compare ballistic performances of different base bleed configurations, an additional objective of this project is to establish data base for base bleed systems by means of results obtained with these computer codes.

2. THEORETICAL APPROACH OF THE PROJECTILE BASE FLOW

The complexity of base pressure calculation for an axisymmetric body even without stabilizing fins or spinning is well known. In the case of a projectile with base injection of chemically reactive gas, afterburning leads to additional problems. The objective of the present work is to compute the steady reactive flow behind the rear end of projectile (Figure 2). To solve this problem, a numerical integration of Navier-Stokes equations has been used. In fact, this computer code initially developed to simulate Rocket Motor Plumes has been modified to introduce the projectile base flow concept.

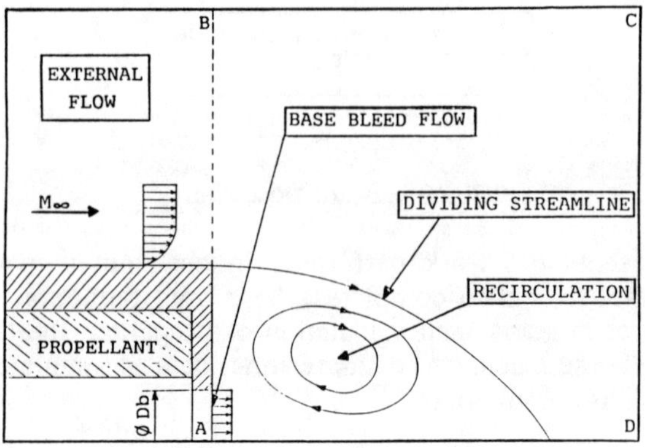

FIGURE 2. Shematic of base area flowfield

2.1. Governing equations

The original Navier-Stokes equations are averaged and completed by a (k-ε) turbulence model and a simplified combustion model. For a two dimensional, axisymmetric, steady, turbulent flow the governing equations may be written in the general form:

$$\frac{\partial\, \rho u \phi}{\partial x} + \frac{1}{r}\frac{\partial}{\partial r}(\rho r v \phi) - \frac{1}{r}\left[\frac{\partial}{\partial r}(\Gamma_\phi\, r \frac{\partial \phi}{\partial r}) + \frac{\partial}{dx}(\Gamma_\phi\, r \frac{\partial \phi}{\partial x})\right] = S_\phi \qquad (1)$$

Where ϕ represents the dependent variable, Γ_ϕ is the appropriate effective exchange coefficient for turbulent flow and S_ϕ is the corresponding source term. The definitions of Γ_ϕ and S_ϕ for the different dependent variables (ϕ) are listed in table 1.
The local pressure is determined using the equation of state:

$$p = \rho R T \sum_{K=1}^{K=7} \frac{Y_K}{M_K} \qquad (2)$$

Where R is perfect gas constant and Y_K the mass fraction of specie K and M_K its molar mass.
The relation between enthalpy (H) and temperature (T) is given by

$$H - \frac{1}{2}(u^2 + v^2) - k = h^* + \int_{T^*}^{T} C_p \cdot dT \qquad (3)$$

Cp, h* and $\int_{T^*}^{T} Cp \cdot dT$ are given by tables from Ref. [2], k is kinetic turbulent energy.

The gas contains seven species (CO , CO_2 , H_2 , H_2O , HCl , N_2 ,O_2) and the combustion model is based on the following reaction:

$$CO + H_2 + O_2 \rightarrow CO_2 + H_2O \qquad (4)$$

Conserved property	ϕ	Γ_ϕ	S_ϕ
Mass	1	0	0
X momentum	u	μ_t	$-\frac{\partial p}{\partial x}+\frac{\partial}{\partial x}(\mu_t.\frac{\partial u}{\partial x})+\frac{1}{r}\frac{\partial}{\partial r}(\mu_t.r.\frac{\partial v}{\partial x})$
r momentum	v	μ_t	$-\frac{\partial p}{\partial r}+\frac{\partial}{\partial x}(\mu_t\frac{\partial u}{\partial r})+\frac{1}{r}\frac{\partial}{\partial r}(\mu_t\, r\frac{\partial v}{\partial r})-\frac{2\,\mu_t}{r}$
total enthalpy	H	$\frac{\mu_t}{\sigma_H}$	0
turbulent energy	k	$\frac{\mu_t}{\sigma_k}$	$\mu_t\, G - \rho\, \varepsilon$
dissipation rate	ε	$\frac{\mu_t}{\sigma_\varepsilon}$	$(C_1\, \mu_t\, G - C_2\, \rho\, \varepsilon)\ \frac{\varepsilon}{k}$

$G \equiv 2\,[\,(\frac{\partial u}{\partial x})^2+(\frac{\partial v}{\partial r})^2\,]+(\frac{\partial u}{\partial r}+\frac{\partial v}{\partial x})^2+\frac{2\,V^2}{r}$
$\mu t \equiv \mu_\rho + \rho\,\frac{C_D\, k^2}{\varepsilon}$
$C_D = 0{,}09$ $C_1 = 1{,}14$ $C_2=1{,}12$ $\sigma_H = 1$ $\sigma_K = 1$ $\sigma_\varepsilon = 1{,}3$

TABLE 1 - Definition of terms

The atomic conservation equations and two linear relations between a passive scalar f and the mass fraction Y_{O_2} and $Y_{CO} + Y_{H_2}$ provide the gas composition f can be defined as the "mixing ratio" and it is convenient [3] to assume the values zero and unity as boundary conditions. f is given by:

$$f = \frac{H - H_\infty}{H_b - H_\infty} \tag{5}$$

Where the subscripts ∞ denote values in the external flow and b the flow through exhaust port.

2.2. Solution procedure

A numerical code is used to solve the elliptic partial differential system of equations given by (1) and table (1) with the appropriate boundary conditions. The external flow is assumed to be supersonic. In this case, velocity, pressure and temperature are fixed during the solution procedure. For the viscous effect, we fix the boundary layer thickness in the radial direction and k-ε are calculated with a characteristic length equals to base bleed diameter.

This code uses a numerical scheme due to Patankar [4] with a modification to include gas compressibility effects. In the numerical code, the flow area ABCD (Figure 2) is divided in a rectangular grid.

2.3. Predictions

Different data are necessary to calculate the flowfield:

- Two geometrical parameters, the diameter of the projectile (D) and the diameter of the exhaust port (D_b) (Figure 2).
- The altitude (Y) and the Mach Number (M_∞) of the projectile.
- Gas composition (CO, CO_2, H_2, H_2O, HCl, N_2), mass flow rate (Q_b) and stagnation temperature (T_b).

A series of computations have been carried out for 105 mm projectile at Mach number 1,66 and 1,28 for altitude Y = 1210 m and Y = 3000 m, respectively. The calculations simulate operating conditions of an experimental base bleed gas generator. The injection of mass flow is provided by the combustion of composite propellant, and a chemical equilibrium calculation near the ambient pressure gives stagnation temperature (flame temperature) and combustion product composition.

For the present case the results are:

Tb = 2290 K
$x(CO) = 0{,}248$ $x(CO_2) = 0{,}0534$ $x(H_2) = 0{,}217$
$x(H_2O) = 0{,}256$ $x(HCL) = 0{,}147$ $x(N_2) = 0{,}075$
(molar concentration for one mol of gas)

To estimate the base bleed effect, we have computed the base pressure (p_C) with the following assumption:

$$p_C = \frac{1}{S} \int_0^R 2\pi r p \, dr \tag{6}$$

Where p is the pressure given by predictions near the projectile, S its section.
First calculation has been performed without injection for $M_\infty = 1{,}28$ and $M_\infty = 1{,}66$. In this case, Figure 3 shows the stream function contours and Figure 4 the pressure function contours in MPa. The results are presented in an adimensionalized coordinate system (x/D , y/D).

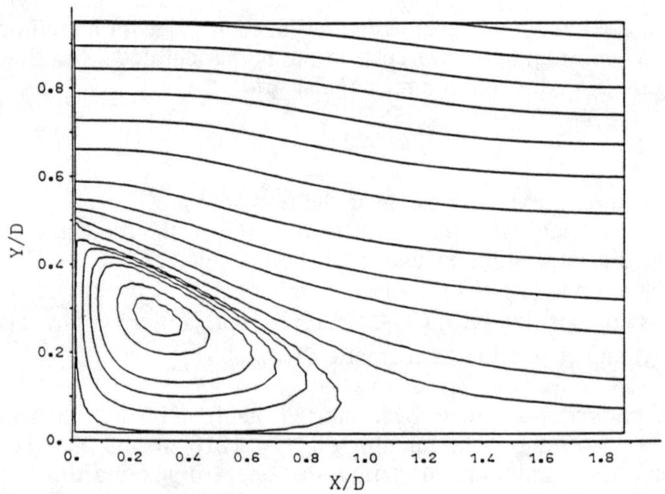

FIGURE 3 - Stream function contours without injection. Mach = 1,66

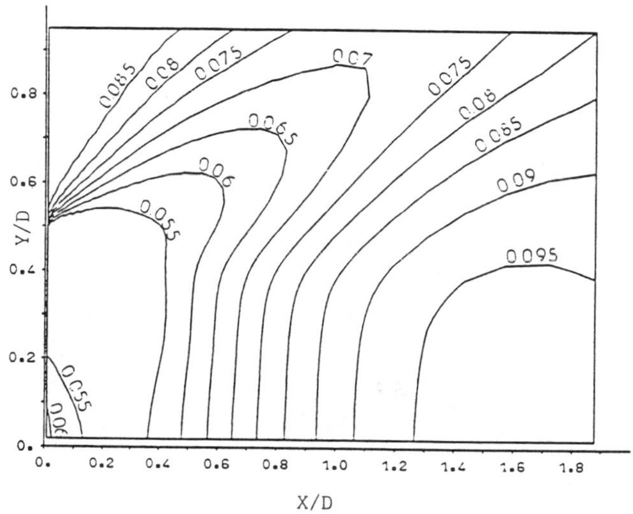

FIGURE 4 - Pressure function contours without injection. Mach = 1,66

Now, an injection is specified (D_b = 30 mm) with Q_b = 18,5 g/s for M_∞ = 1,28 and Q_b = 25,8 g/s for M_∞ = 1,66. To examine the afterburning effect on the base pressure, we have considered at first an inert flow (without reaction (4)) and then a reactive flow. For the case M_∞ = 1,66 and Q_b = 25,8 g/s Figure 5 shows the stream functiun contours taking into account an inert flow . Figure 6 and Figure 7 show the pressure function contours in MPa and the temperature function contours in Kelvin for an inert flow. Figure 8 and Figure 9 show the same variables for a reactive flow.

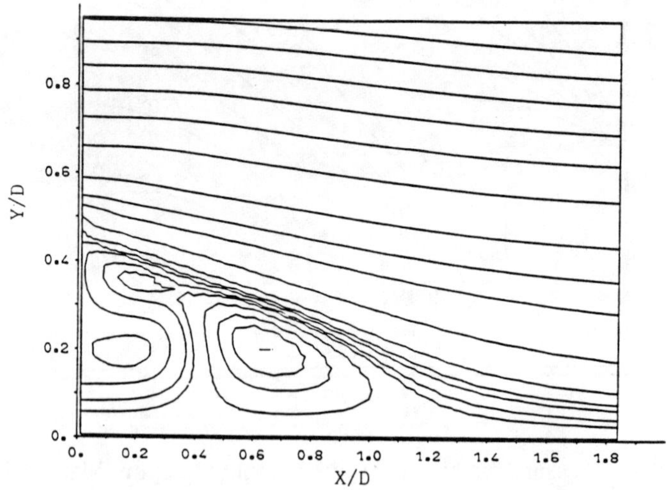

FIGURE 5 - Stream function contours with inert flow
Mach = 1,66 Q_b = 25,8 g/s

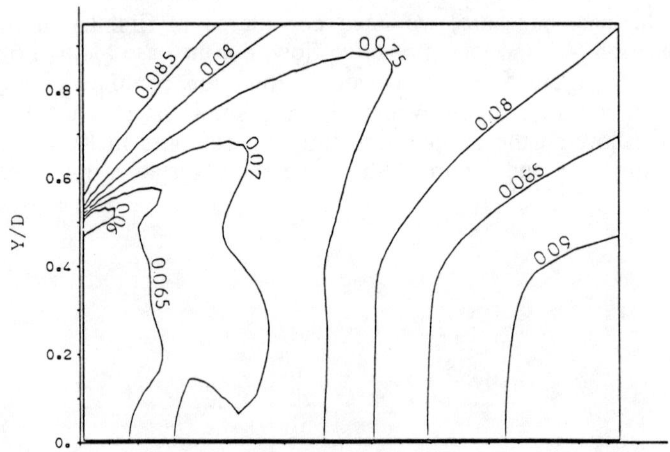

FIGURE 6 - Pressure function contours with inert flow
Mach = 1,66 Q_b = 25,8 g/s

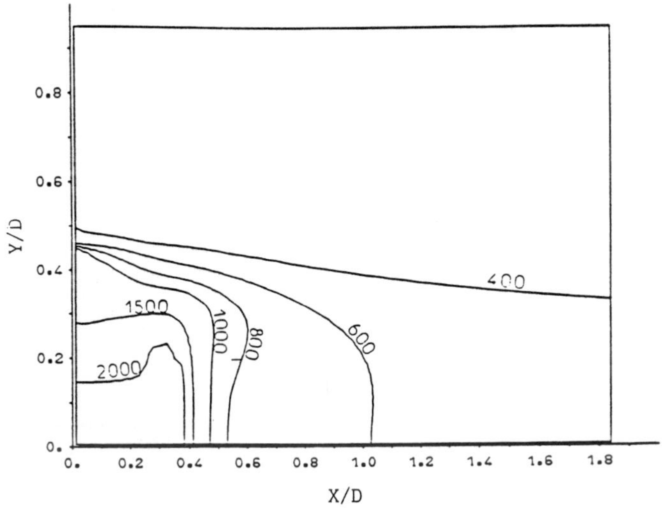

FIGURE 7 - Temperature function contours with inert flow
Mach = 1,66 Q_b = 25,8 g/s

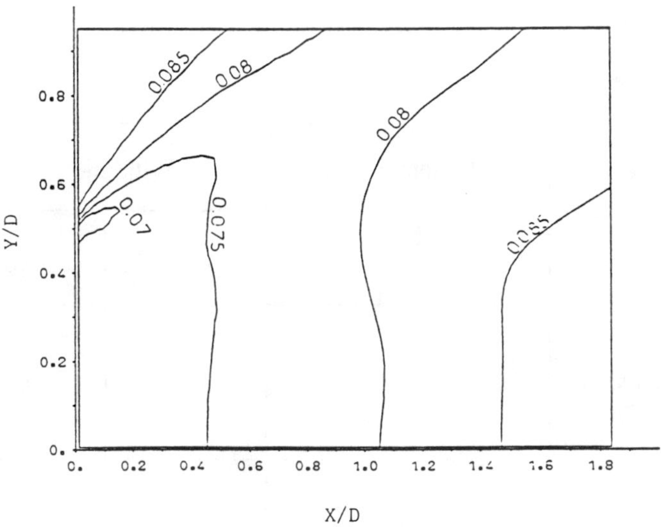

FIGURE 8 - Pressure function contours with reactive flow
Mach = 1,66 Q_b = 25,8 g/s

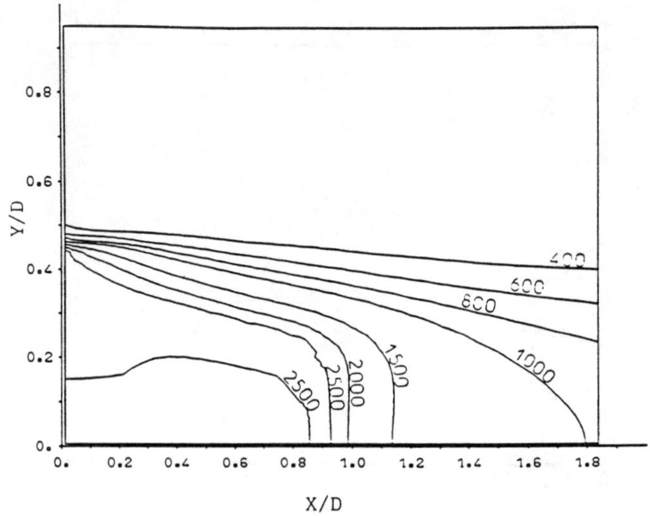

FIGURE 9 - Temperature function contours with reactive flow
Mach = 1,66 Q_b = 25,8 g/s

Table 2 gives the pressure ratio (p_c/p_∞) for the different cases studied with the computer code:

	Without injection $Q_b = 0$	With injection	
		inert fluid	reactive fluid
M_∞ = 1,28 Y = 3000 m Q_b =18,5g/s	0,68	0,77	0,85
M_∞ = 1,66 Y = 1210 m Q_b = 25,8 g/s	0,57	0,69	0,80

TABLE 2 - Pressure ratio (p_c/p_∞) for different calculated cases.

3. EXPERIMENTAL INVESTIGATION: GUN FIRING TEST ANALYSIS

3.1. Description of the method

The method used for the gun firing test analysis is divided in two main steps.

Step 1. We determine total drag coefficient vs Mach number from projectile velocity on its trajectory.

Step 2. Comparing total drag coefficient with and without base bleed effect enables to calculate base bleed reduction efficiency.
For each total drag coefficient calculation the block diagram of modules is the same as the one presented in Figure 10.

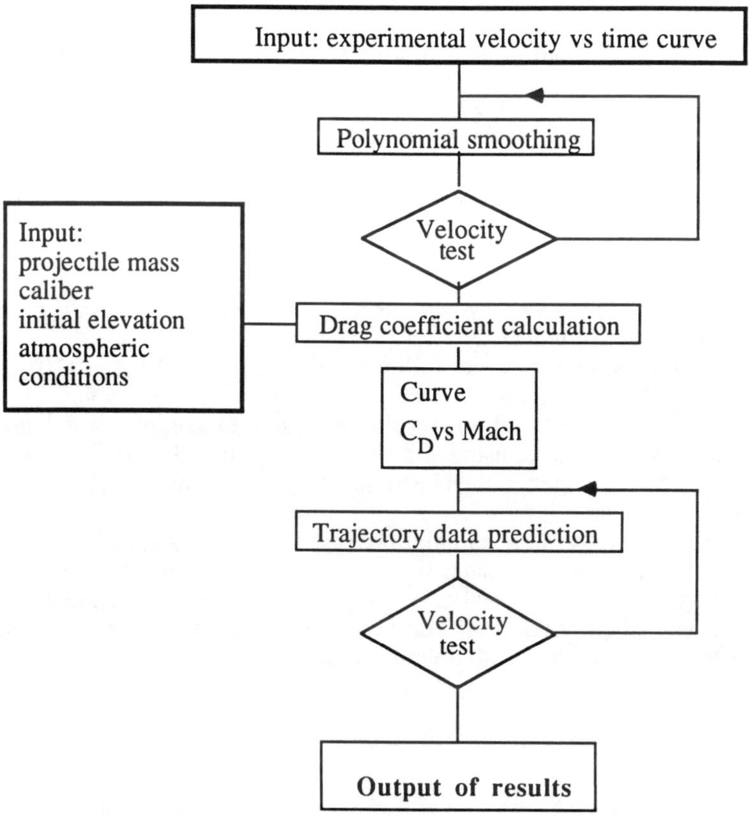

FIGURE 10 - Block diagram of modules

The function of each module is briefly described.

Polynomial smoothing module: Experimental velocity curve is smoothed by means of polynomial functions. At each time step in the curve, we calculate smoothed velocity and its derivative versus time (t). Changing the degree of polynomial functions allows to respect a given criterion between calculated velocity and experimental velocity.

Module of total drag coefficient calculation: total drag coefficient may be calculated from the equations of an external balistic model. The projectile motion is confined to the x y -plane and is given by the equation of motion:

Drag equation : $m\dot{V} = - m\, g \sin\theta - \frac{1}{2}\rho_\infty \, . \, S_{REF} . V^2 . C_D(M)$ (7)

Lift equation : $mV\dot{\theta} = - m\, g \cos\theta$ (8)

Kinematical relations and an equation of projectile mass balance for base bleed configuration complete this model.

$$\dot{x} = V\cos\theta \tag{9}$$

$$\dot{y} = V\sin\theta \tag{10}$$

$$\dot{m} = \rho_p \, . \, S_B \, . \, r_b \tag{11}$$

Initial conditions are given by gun firing test conditions (elevation and muzzle velocity measurements).
The calculation procedure is as follows. Equations (8) (9) and (10) are solved using a Runge-Kutta scheme and determine x,y and θ variables. These data and polynomial smoothing results are used to compute total drag coefficient vs Mach number using equation (7). Note that the projectile mass with a base bleed gas generator is estimated by means of propellant mass flow calculation (11).
Module of trajectory data calculation: then the calculated total drag coefficient curve is an input date. Equation (7) (8) (9) and (10) are solved using a Runge-Kutta scheme and give V,x,y and θ variables. By an interative procedure we adapt total drag coefficient in order to respect a given criterion between predicted and observed velocity.

The comparison of total drag coefficient with and without base bleed effect is based on the Swedish approach of Ref. [5] . The drag coefficient of the projectile during base bleed operating time is:

$$C_D = C_{DTOT} - C_{RED} \cdot C_{DBASE} \tag{12}$$

Where:

C_D	= total drag coefficient with base bleed effect.
C_{DTOT}	= total drag coefficient without base bleed effect.
C_{RED}	= base drag reduction factor.
C_{DBASE}	= base drag coefficient without base bleed effect.

C_{RED} is considered as a polynomial function of Mach number and of an injection parameter I defined as the ratio of the propellant mass flow through exhaust port to the mass flow of air through the section of the shell base.
Then, base pressure (p_C) may be calculated from the expression given by Ref. [5].

$$p_C = p_\infty \left[1 - \frac{0.7\,(1 - C_{RED}) \,.\, C_{DBASE} \cdot M^2}{\frac{S_{BASE}}{S_{REF}}} \right] \tag{13}$$

Where:

p	= free stream pressure
S_{REF}, S_{BASE}	= reference area and base area of the projectile
C_{RED}, C_{DBASE}, M	= ballistic data from the range calculation

3.2. Some examples

These results are proposed to illsutrate code capabilities.
Table 3 shows an example of velocity comparison for a 155 mm projectile.

Time of flight (s)	Experimental velocity	Smoothed velocity	Calculated C_D (M)	Predicted velocity (m/s)
0	783,1	783,0	0,241 (2,30)	783,1
5	610,2	610,2	0,274 (1,82)	609,3
10	499,3	499,4	0,290 (1,51)	498,2
15	423,1	423,2	0,311 (1,29)	421,7

TABLE 3 - Velocity comparison

Figure 11 and Figure 12 show experimental velocity and calculated total drag coefficient for a 105 mm projectile with and without base bleed effect.

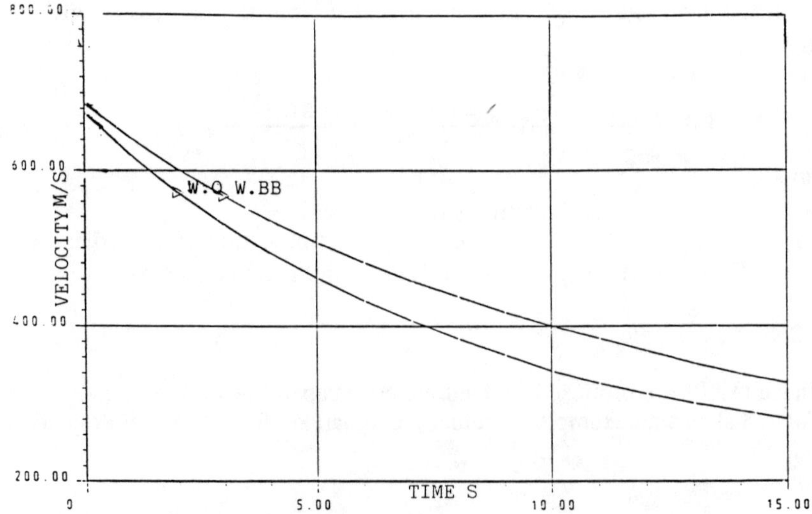

FIGURE 11 - Experimental velocity versus time

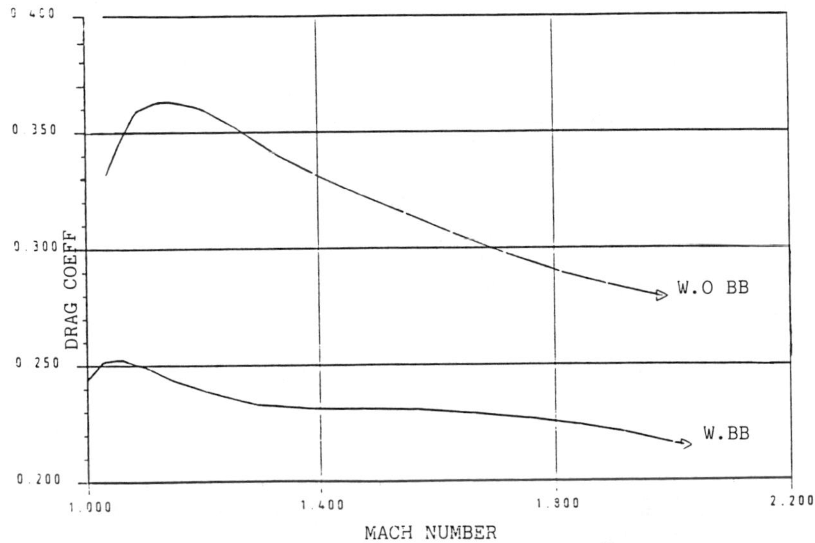

FIGURE 12 - Calculated total drag coefficient versus Mach number

Figure 13 shows pressure ratio (p_C/p_∞) and base drag reduction factor (C_{RED}) as a function of injection parameter (I) and Mach number (M). The method to determine (C_{RED}) is to suppose that base drag coefficient without base bleed effect (C_{DBASE}) is 40% of (C_{DTOT}).

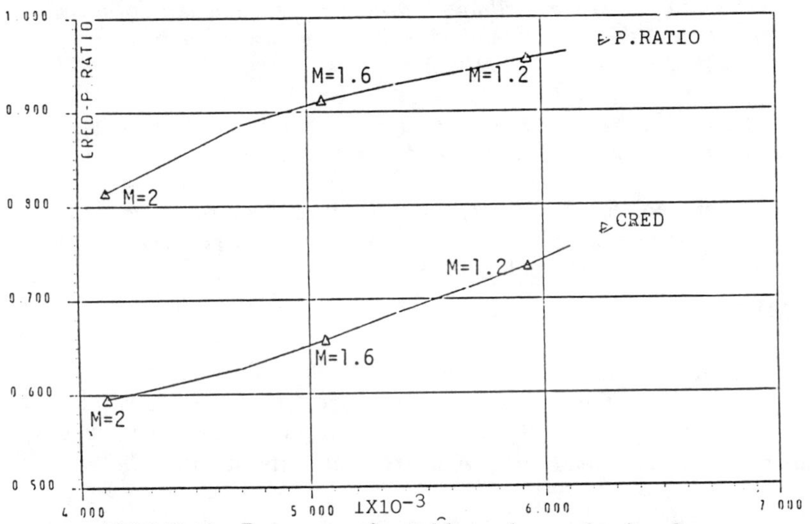

FIGURE 13 - Pressure ratio and base drag reduction factor versus injection parameter and Mach number

4. CONCLUSION

Different computer codes have been developed in order to study the effectiveness of base bleed systems. An intensive use of these programs will allow to quickly perform preliminary design of base bleed grains.
Calculations are computed to analyse tendencies or to compare results of different base bleed configurations by acting on injected mass flow rate and propellant gas composition. Pressure ratio results (p_c/p_∞) show a good comparison between the two presented methods.
Further studies will consist of improving conditions of the simulation.
1 - Theoretical approach : boat - tail cone effect, comparison with wind tunnel tests.
2 - Empirical approach : good knowledge of aerodynamical data such as base drag coefficient.

REFERENCES

[1] Gordon S., Mac Bride B.G., *Computer programm for calculation of complex chemical equilibrium composition - rocket performance - incident and reflecting shocks and CHAPMAN JOUGET detonation,* NASA Lewis research center (1971)

[2] Stull D. R., Prophet H., JANAF thermochemical tables (second edition), NSDRS - NBS 37

[3] Lockwood F.C., *The modelling of turbulent premixed and diffusion combustion in the computation of engineering flows,* Combustion and Flame 29, p. 111 - 122 (1977)

[4] Patankar S.V., *Numerical heat transfer and fluid flow,* Hemisphere publishing corporation, MC Graw Hill book company

[5] Rune Hellgren, *Basfloedesprojektiler Banberaekningsprogram och Effecktivitetsjaemfoerelser (French translation),* Foersvarest Forskningsanstalt Stockholm (1983)

COMMENTS

C. J. Nietubicz - The upstream outer flow field, that is, the boundary layer profile coming into the base corner is important. you show the computation in regions A, B, C, D. How did you determine the upstream boundary condition along long A-B (for the external flow)?

Reply

The external flow is assumed to be supersonic. In this case, velocity, pressure and temperature are fixed during the solution procedure. For the viscous effect, we have fixed the boundary layer thickness in the radial direction and k - ε are calculated with a characteristic length.

A Model of the Axial Drag Coefficient for Base Bleed Projectiles

NILS BARTELSON
Furusundsgatan 12
S-11537 Stockholm, Sweden

SVEN LINDE
Studsvik AB, Studsvik Data
S-61182 Nyköping, Sweden

ABSTRACT

The axial drag on a projectile is usually expressed in terms of the axial drag coefficient C_D. For a conventional projectile C_D is often regarded as a function of the Mach number only. For base bleed projectiles, in their bleed phase, ballistic range data indicate that some factor beside the Mach number has a significant influence on C_D. It is reasonable to expect that the ambient atmospheric pressure is one such factor.

We attempt to relate, in an empirical way, the drag of a base bleed projectile to the geometry of its trajectory and the state of the atmosphere. The relations are not derived from physical principles - our approach is descriptive and statistical. The purpose is to summarize observations of flight performance in a way that permits us to predict a trajectory from its initial conditions and from a picture of the atmosphere.

We propose a functional relation between the Mach number, the ambient atmospheric pressure and the drag coefficient. Its parameters are determined by a regression procedure based on time-of-flight recordings from a Doppler radar instrument. The model proves capable of reproducing observed C_D's along trajectories of widely differing shapes. As a side effect some light is thrown upon the extinction of the bleed charge. It turns out to be a more drawn-out process than is often assumed. We report on a verification of the model and also touch upon the problems of modelling the transition from active to inert base bleed.

1. INTRODUCTION

Artillery ammunition has been in a state of rapid change in Scandinavia for several years. New ideas have emerged and enough funding has been received to result in significant improvements in range and precision.

Ballistic measuring equipment has been developed by a Danish company, OPOS Electronics AS; new methods for analysing data have been formulated by the Royal Swedish Navy together with the National Defence Research Institute and Studsvik AB.

This paper is concerned with long range ammunition. For the designer of a long range projectile the main task is to keep the drag as low as possible, thus increasing the range or decreasing the time of flight.

The axial drag on a projectile depends on the shape, diameter and velocity of the projectile and on the physical state of the ambient atmosphere. The shape dependency is usually expressed in terms of a few dimensionless aerodynamic coefficients, in particular the axial drag coefficient C_D. For a conventional projectile C_D is found to depend mainly on the projectile shape and the ratio of the projectile velocity v to the local sound velocity in air, a. The ratio v/a is referred to as the Mach number.

In the early stages of the development of what is now called base bleed ammunition it was found that the simple relationships which are good approximations for a conventional projectile, were no longer valid. It was concluded that experimental resources and manpower for modelling were required to study this phenomenon. A reliable tool was needed to predict trajectories for the design of fire control equipment and for the calculation of range tables.

The objective of this paper is to formulate the problem and to propose a solution, given experimental data.

The tools used are of two kinds, namely:

- Doppler radar records from the firing in Sweden of slender-bodied 120 mm and 155 mm ammunition with inert and active base bleed,
- computer programs written at Studsvik AB identifying the drag function from recorded radar data, atmospheric conditions, gun elevation, the mass of the projectile and its loss of mass during time of flight etc.

We attempt to relate, in a simple, empirical fashion, the drag of a gun-fired base bleed projectile to the geometry of its trajectory and the state of the atmosphere. The relations are not derived from physical principles - our approach is descriptive and statistical. Our purpose is to summarize observations of flight performance in a way that permits us to predict a trajectory from its initial conditions and from a picture of the atmosphere.

The proposed ballistic model is the aerodynamic point mass or modified point mass model extended to account for base bleed effects. It should be possible to identify the parameters from observations at the proving ground using ground-based instruments. The model should be accurate enough to serve as the basis for the design of range tables and data to the fire control equipment for a specific projectile design. We do not aim to model the burning of the base bleed charge or the dynamics of the injection of mass into the base flow region.

The essential results of the study were published in a Studsvik report [1].

2. FORMULATION OF THE PROBLEM

The theoretical basis is the aerodynamic law and its proportionality factor, called the drag coefficient.

The drag of a projectile is usually written

$$D = \tfrac{1}{2} A \rho C_D v^2 \tag{1}$$

where

- A is a reference area
- ρ the air density
- C_D the drag coefficient, in German literature often denoted C_W
- v the projectile velocity

In the aerodynamic point mass model the drag coefficient of a conventional projectile is usually regarded as a function of the Mach number only. Experience from the proving ground indicates, however, that this simple model does not generally apply to base bleed projectiles in their bleed phase. The C_D curves reproduced in Fig 1 testify to this. They are estimated from two sets of trajectories in quadrant elevations 20° and 50°. The average distance between the curves is large compared to the standard deviations of the estimated C_D's, at most 0.001 for $1.3 \leq \text{Mach} \leq 1.9$.

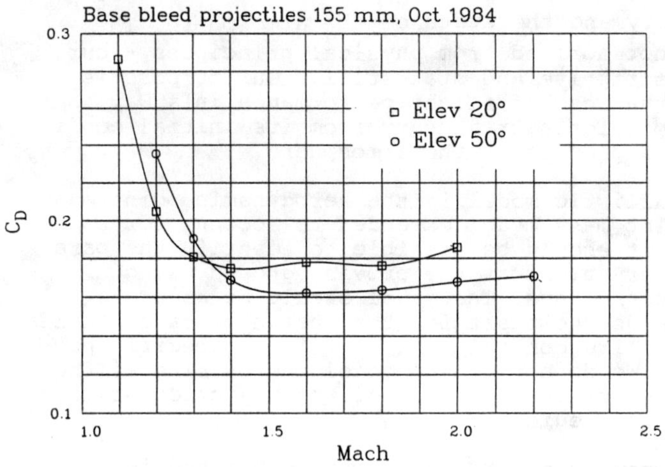

FIGURE 1. Mean C_D estimates from trajectories in two quadrant elevations.

Some factor beside the Mach number has evidently had a significant influence on the drag coefficient. It is reasonable to expect that the ambient atmospheric pressure may be one such factor.

Thus we are led to consider a functional relation of the form

$$C_D(M,p) = u(M) + \lambda(M) \frac{p-\tilde{p}}{\tilde{p}} \tag{2}$$

where

M is the Mach number

p is the local atmospheric pressure

$\tilde{p}$ a reference pressure

$\left.\begin{matrix} u \\ \lambda \end{matrix}\right\}$ are functions to be identified

The reference pressure is introduced to improve the numerical properties of the estimation procedure. The main reason for using the relative deviation from the reference pressure as a predictor is to become independent of pressure units. The form chosen is essentially a first order Taylor expansion of C_D in terms of p.

Next we have to describe the unknown functions u and λ with a reasonably small set of parameters. Since we have no physical theory to indicate a suitable functional form, we have chosen u(M) and λ(M) as cubic spline functions with knots to be prescribed by the analyst. The knots will be denoted

$$x_i,\ i=0..n \quad \text{for} \quad u(M), \tag{3}$$

$$z_i,\ i=0..\ell \quad \text{for} \quad \lambda(M). \tag{4}$$

The end-point conditions for the splines are chosen to be

$$u''(x_0) = u''(x_n) = 0; \tag{5}$$

$$\lambda''(z_0) = \lambda''(z_\ell) = 0. \tag{6}$$

Now let

$$\begin{aligned} b_0 &= u(x_0) & b_{n+1} &= \lambda(z_0) \\ &\ \ \vdots & &\ \ \vdots \\ b_n &= u(x_n) & b_{n+\ell+1} &= \lambda(z_\ell) \end{aligned} \tag{7}$$

be the regression parameters. By the theory of spline functions we construct a linear estimator

$$C_D(M,p) = \sum_{j=0}^{n+\ell+1} a_j(M,p)b_j \tag{8}$$

The a_j's are functions of the local Mach number and atmospheric pressure.

At this point it is appropriate to remark that our use of spline functions to express u(M) and λ(M) is one of several alternatives. We could as well have chosen a piecewise linear function or a Lagrange interpolator of higher order. The advantage of the spline functions is their inherent smoothness.

In practice the same (time, space) data will be used to estimate both M and C_D. So M is a stochastic variable with a certain dispersion, and p is of course not known exactly. Moreover, M and C_D are not independent. The deviations of M and p propagate to the carriers a_j. However, we expect M, which is essentially a first derivative of space data, to be less contaminated with noise than C_D, which is essentially a second derivative. Also, small errors in p will not influence the carriers too much. Thus it seems reasonable to choose the common least squares criterion to find the regression parameters b_j, i.e. to minimize the sum of squared deviations of the observed C_D's from the estimated ones.

3. APPLICATION TO EXPERIMENTAL DATA

Our first approach is to apply the assumed model on tests conducted at Torhamn and in the isle of Öland, Sweden, in October 1984. 36 rounds of a 155 mm projectile were fired, 11 of which were inert to provide baseline drag data. The quadrant elevations ranged from 18° to 70°. Occasionally data will be taken from a second set of experiments conducted with a 120 mm projectile at the Torhamn experimental range in April 1985.

The projectiles were tracked by a Doppler radar system. Recordings of (time, space) values were processed by a drag analyser program providing tables of M, p and C_D for 35-65 time of flight points dispersed along each trajectory.

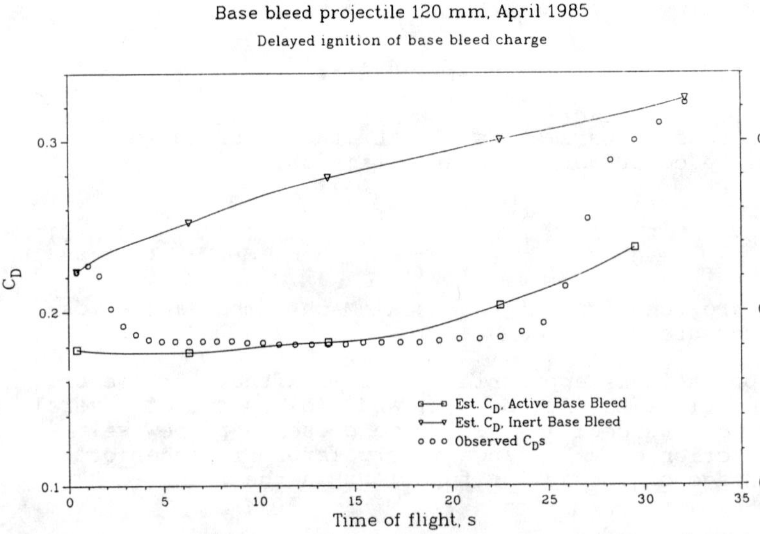

FIGURE 2. Trajectory excluded because of suspected ignition delay.

A few rounds have been omitted from the data set. Observed C_D values for one such projectile are shown in Figure 2. For illustrative purpose, although prematurely, curves of estimated C_D's for inert and active base bleed are displayed in the graph. It is seen that the ignition of the base bleed charge was delayed and that the bleed effect was reduced during the first five seconds. The range to point of impact was 18.0 km for this round compared to an average range of 18.5 km for similar projectiles fired in the same elevation. No omission from the experimental data has been accepted unless a suspected malfunctioning of the bleed charge was confirmed by a deviation in range.

The regression analysis for inert projectiles is based on data from 11 trajectories comprising 321 observations. The C_D estimator was a spline function with 12 knots in the range $0.7 \leq M \leq 2.3$. The result is summarized in Table 1.

Table 1
Estimated C_D function for inert projectile

M	C_D	s dev
.700	.1619	.0059
.850	.1689	.0023
.900	.1713	.0030
.950	.1946	.0031
1.000	.2910	.0013
1.050	.3068	.0009
1.100	.3026	.0006
1.200	.2937	.0005
1.400	.2764	.0004
1.700	.2528	.0005
2.000	.2338	.0006
2.300	.2145	.0025

The standard error of estimate is 0.0036.

The regression analysis for active base bleed is founded on flight performance data from 22 trajectories for which (M,p,C_D) triplets have been calculated at 771 time of flight points. Three trajectories were set aside for later verification of the parameter estimation. For $u(M)$ a spline function with 6 knots was used, for $\lambda(M)$ a spline function with 3 knots.

The first question to be asked is whether the model is appropriate, and, in particular, if the addition of a pressure term improves the fit. Its significance can be assessed by an analysis of variance (Table 2), in which we test the hypothesis

$$H_0: \lambda = 0. \qquad (9)$$

Table 2
C_D regression
Analysis of variance

Source	Sum of squares	Degrees of freedom	Mean square	Mean square ratio
Regression				
mean	21.76161	1		
u(M)	0.00415	5	0.00083	64
λ(M)	0.00923	3	0.00308	239
Residual	0.00982	762	0.000013	
Total	21.78481	771		

The last column is the ratio of the mean square to the residual mean square.

The mean square due to regression on λ should be compared to the residual mean square, and its size is found to be quite large. If the hypothesis is true and under some additional assumptions on the normality of the observations, the test variable for the component λ, i.e. the mean square ratio, will have an F distribution with (3,762) degrees of freedom. The probability of a value $\geq$ 239 is small, indeed. However, we do not wish to stress this formal aspect too much, since little is known about the distribution of the original data.

Anyhow, this exploratory data analysis indicates that the hypothesis H_0: $\lambda = 0$ should be rejected and that the pressure term is significant.

The estimates of the regression parameters are shown in Table 3.

Table 3.
C_D components, active base bleed

M	u(M)	s dev of u(M)	λ(M)	s dev of λ(M)
1.1	0.1745	0.0027	0.0096	0.0038
1.3	0.1797	0.0006	0.0151*	0.0015
1.5	0.1707	0.0003	0.0192	0.0009
1.7	0.1663	0.0003	0.0209*	0.0010
1.9	0.1650	0.0004	0.0207*	0.0020
2.2	0.1612	0.0022	0.0184	0.0056

The values marked by asterisk are interpolations, not actual regression parameters.

The standard error of estimate is 0.0038, close to the value obtained for projectiles with inert base bleed. This may be regarded as another indication that the model is appropriate.

The C_D, u and λ functions are shown in Figure 3.

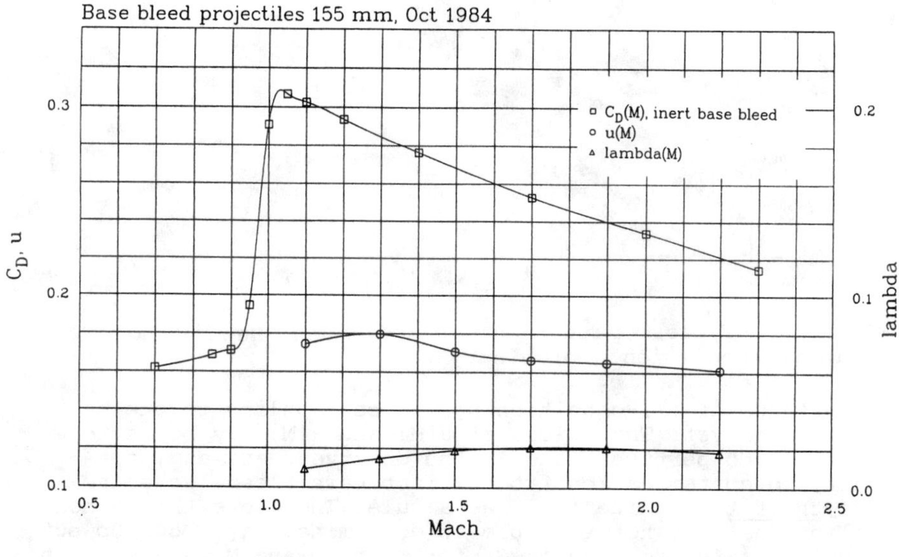

FIGURE 3. Estimated C_D function for a projectile with inert Base Bleed and C_D components for a projectile with active Base Bleed.

Standard deviations of the parameter estimates are low, as a rule. They do not, however, tell the whole story. A glance at the correlation matrix reveals that the parameters defining u(M) are loosely coupled, as are those defining λ(M). But correlation between the two parameter classes may be considerable, indicating that the problem is not so well conditioned as one might wish.

The degree of ill-conditioning depends on the experimental design. To some extent ill-conditioning is inherent in the physical situation on the proving ground. All projectiles set out at a high Mach number M, close to the muzzle, and a high atmospheric pressure p, at ground level. As they proceed to altitudes where p is lower, the Mach number will also decrease, broadly following a common pattern often tempered only by differences in elevation. Information from other regions of the (M,p)-plane, e.g. combinations of large M and low p values, is missing.

A somewhat better design is achieved by reducing the muzzle velocity for some rounds. The pattern of observations from such an experiment is illustrated in fig 4.

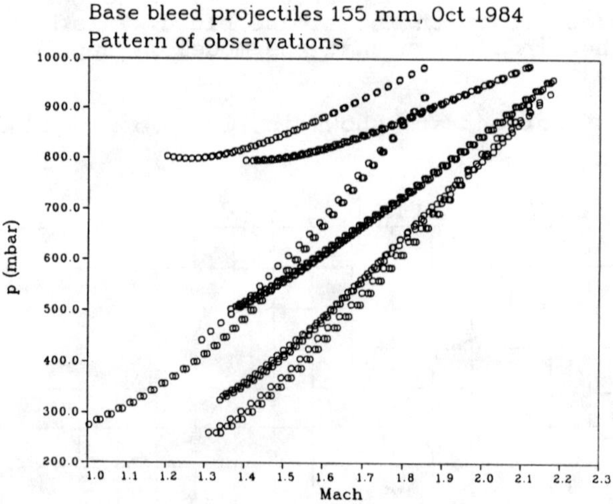

FIGURE 4. Typical pattern of observations, resulting in an ill-conditioned problem.

The effects of ill-conditioning may be puzzling to the beginning analyst. The values of u(M) and λ(M) may be rather sensitive to perturbations in the observed data and, not least, to contamination from extinct base bleed. Nevertheless the fit is quite good, as a rule. Therefore the method outlined here is not reliable to determine, say, $\delta C_D/\delta p$ but is quite useful to predict C_D for those p and M values which occur in ballistic applications.

Three trajectories were excluded from the set used for regression analysis; thus they may be regarded as new and independent experiments. These projectiles were fired in quadrant elevations 20°, 35° and 50°. We will compare the C_D's obtained from these trajectories with the predictions of the model. As a by-product this provides an opportunity to study the extinction of the base bleed charge.

In Figures 5-7 are shown, as functions of time,

- the predicted C_D for inert base bleed,
- the predicted C_D for active base bleed,
- the observed C_D values,
- a confidence band within which the observed C_D's are expected to fall with 95% probability.(1)

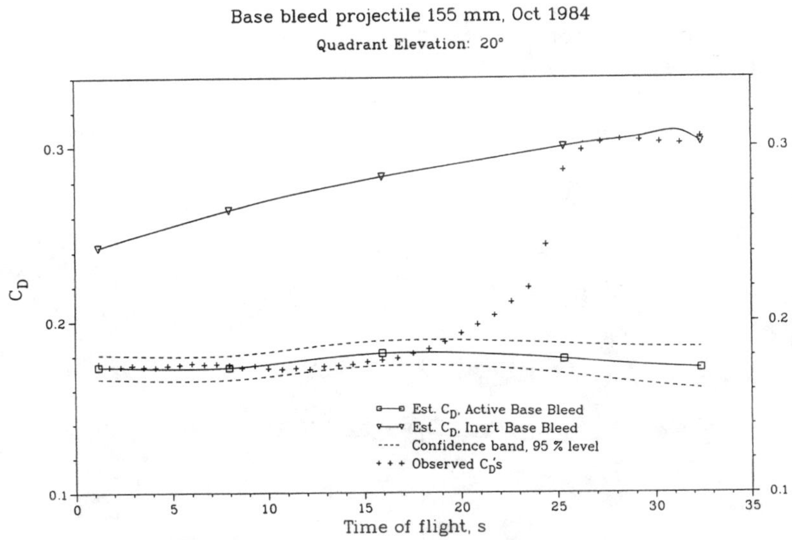

FIGURE 5. Verification of the C_D model, quadrant elevation 20°.

---

(1) The width of the band must not be interpreted as a measure of the uncertainty of the estimated C_D's. The confidence interval for the outcome of a new experiment is in this case much wider than that for the "true C_D", at the same confidence level. This is just an instance of the familiar rule that the outcome of any one experiment is more variable than the mean outcome of, say, ten experiments.

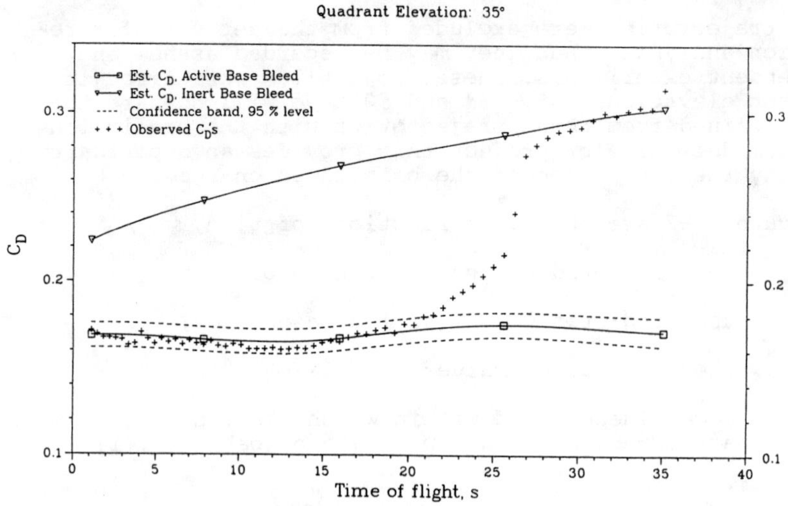

FIGURE 6. Verification of the C_D model, quadrant elevation 35°.

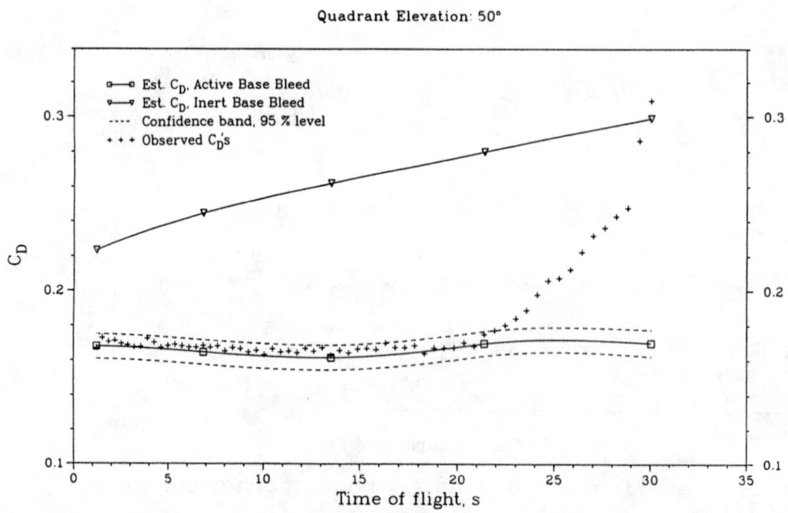

FIGURE 7. Verification of the C_D model, quadrant elevation 50°.

It is seen that up to a certain point in time the observed C_D's fall well within the confidence band. This is true for all the three trajectories, which span a wide range of elevations. So the new experiments verify the model and confirm the significance of the pressure term.

4. THE EXTINCTION OF THE BASE BLEED

The point where the observed C_D's depart from the predicted values coincides with a rough a priori estimate of the burning time of the base bleed charge. The graphs show, however, that the extinction process is slow and continues for 6-7 seconds. During this time interval the actual C_D values will climb gradually from the low active base bleed curve to the higher curve for the inert projectile. We deem it necessary to take the gradual change-over into account in trajectory calculation programs. However, no such program is available to us, and the best we can do is to define an extinction point somewhere in the middle of the extinction interval and use a trajectory program that assumes the active C_D curve to be applicable up to the extinction point, after which a sudden rise to the inert curve will take place.

5. VERIFICATION OF CALCULATED RANGES

Using such a C_D model we can calculate trajectories and find, for instance, the points of impact. Two questions arise:

- How sensitive is the calculated point of impact to a perturbation of the assumed extinction point?
- How does the calculated range compare with the observed one?

Since range observations for the previously treated trajectories are not available, we have to introduce another projectile design including also a new type of base bleed charge. The material as a whole comprises data from more than 20 rounds.

Figure 8 shows the change-over from active to inert base bleed for a 120 mm projectile,fired in a quadrant elevation of 30°.

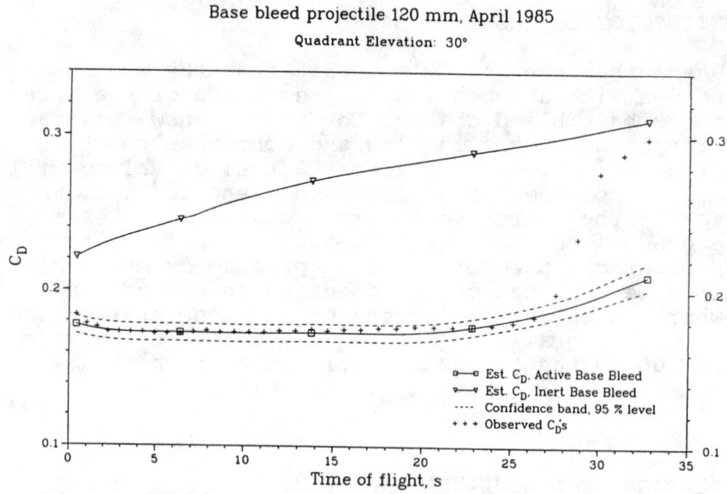

FIGURE 8. Estimated C_D function and observed C_D's for the shell used for range verification.

The bleed is seen to be more sustained and the rise of the C_D curve steeper than in the previous cases. Nevertheless the extinction process is extended over 5-6 seconds.The data for the trajectory shown in the graph was not used for the regression analysis; the round can thus be regarded as an independent experiment.

From the graph the midpoint of the extinction interval can be roughly estimated at 29 seconds. In order to study the sensitivity of the calculated range to this estimate we have performed trajectory calculations with the extinction point set to 27, 29 and 31 seconds.

The ranges obtained are shown in table 4.

Table 4
Range sensitivity to postulated extinction time

Extinction time,s	Calculated range, m
27	25437
29	25580
31	25740

The mean observed range for five rounds was 25539 m. It is evident, that the positioning of the assumed extinction point has an influence on the calculated ranges which is far from negligible. The fact that it also depends on the trajectory geometry further complicates the task of trajectory prediction.

6. INFLUENCE OF BASE BLEED CHARGE TEMPERATURE

The influence of the bleed charge temperature on the C_D function has been studied for 155 mm ammunition. Typical results are shown in figures 9 and 10.

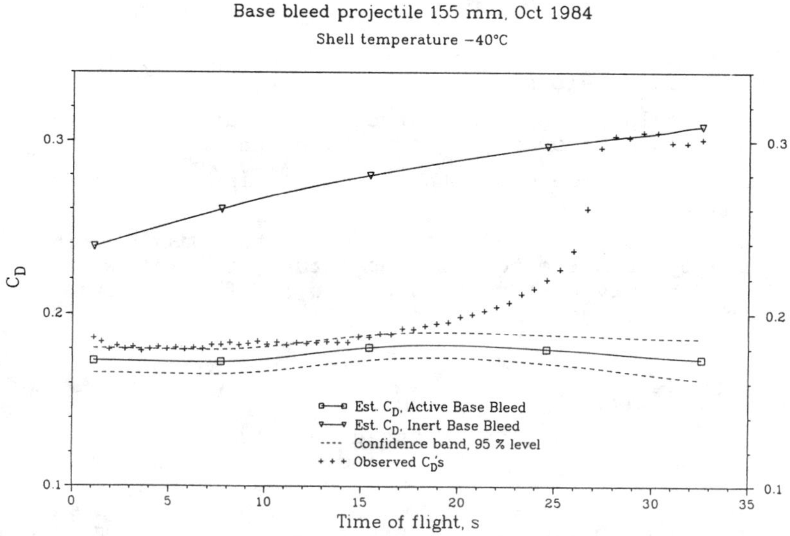

FIGURE 9. Estimated C_D function and observed C_D's for a shell fired with the bleed charge at -40°C.

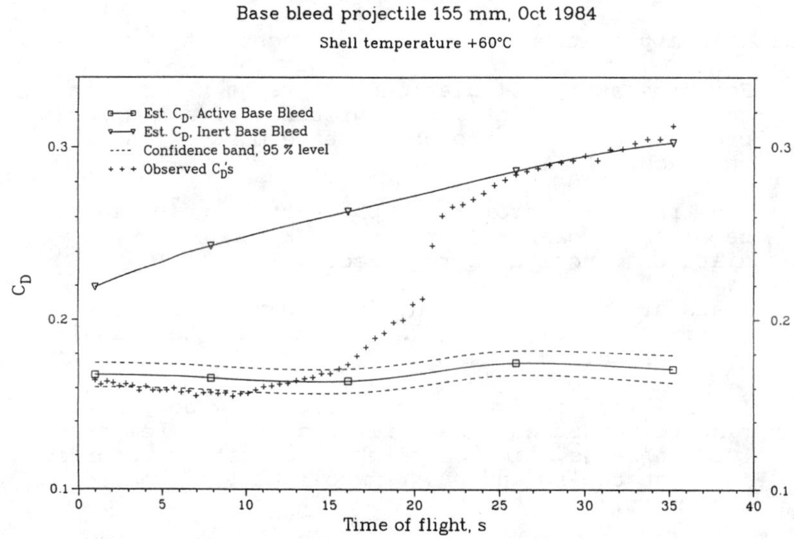

FIGURE 10. Estimated C_D function and observed C_D's for a shell fired with the bleed charge at +60°C.

Figure 9 illustrates the C_D function for a projectile fired with the bleed charge at -40°C. It is seen that the C_D values lie on the upper edge of the 95% confidence band; the low temperature seems to retard the burning. Figure 10 shows the opposite effect. At 60°C the bleed charge burns rapidly and the actual C_D turns out to be low. The burning is not sustained, however.

Since this point is outside the scope of this paper, we summarize our conclusions:

- the influence of the charge temperature depends on the composition of the base bleed charge,
- in the studied cases the influence is far from negligible,
- if the actual bleed charge temperature may differ widely from the normal temperature, the influence must be given as corrections in the fire control and the range table.

7. CONCLUSIONS

The experiments and analyses reported in this document indicate strongly that for slender bodied base bleed ammunition the drag coefficient C_D cannot be represented as a function of the Mach number only. It has been demonstrated that an expression which is essentially a first order Taylor expansion in the ambient pressure, provides a means of predicting the C_D value in a trajectory point where the Mach number and the atmospheric pressure are known.

For ballistic applications a few consequences are obvious.

- For trajectory calculations for range tables, in the fire control etc, the influence of atmospheric pressure on C_D in the bleed phase must be taken into account.

- The transition from bleed phase to inert phase may be of considerable length; some kind of intermediate C_D function is required.

- Means are needed to find, for a particular trajectory, the beginning and the end of the transition period.

- If the bleed charge is to be used at widely differing temperatures, the influence of this temperature on range, time of flight etc must be recorded in range tables and programmed into fire control equipment.

REFERENCE

1. Nils Bartelson, Sven Linde:
A Model of the Axial Drag Coefficient for Base Bleed Projectiles.
Report STUDSVIK/SD-87/67 (1987).

COMMENTS

K. K. Kuo, Pennsylvania State Univ., USA. Several of your figures (fig 5-10) show that the experimental data follow the active base bleed curve initially and then move upward to the inert base bleed curve of C_D versus time. Does this observation imply that we could increase the amount of propellant mass in the base bleed unit so that the drag coefficient could be reduced for longer period of flight.

Author's Reply: The effect would reasonably be the one you mention. However, propellant mass was not among the factors varied in the experiments on which our study is based.

N-Y Wang, East China Institute of Technology, CHINA. Base pressure increased by base bleed $(\Delta C_D)_b$, is a function of mass flow ratio I which is defined by

$$I = \frac{\dot{m}}{\rho_\infty V_m A_b}$$, where ρ_∞ is the atmospheric density.

$$C_D = (C_D)_{\text{without b.b.}} + (\Delta C_D)_b$$

Thus, $C_D = f(M_\infty, I)$.

In the conslusions of this paper, it is true that atmospheric pressure will influence the C_D. But how about the atmospheric temperature T_∞, since $I = f_1(\rho_\infty) = f_2(p_\infty, T_\infty)$.

Author's Reply. Our purpose was too find out if the C_D variations seen, e.g. in Fig. 1, could be predicted by a small set of variables easily observed on the firing range. We found that they could. Adding temperature to the regressor set would probably improve the fit very little and make the problem even more ill-conditioned.

But that is the statistical aspect. From the ballistician's point of view it would of course be nice to connect the model to the theory of Hellgren et al; then the mass flow ratio would be the natural second regressor.

M. D. Chargelegue, Technique de Bourges. In figure one, the drag obtained for two rounds at 20° elevation and 50° elevation are different. Could you explain the physical phenomenon associated with results obtained?

Author's Reply. We can't, but we believe that the mass flow ratio is involved.

For an explanation, see K. Andersson, N-E Gunners and R. Hellgren: Swedish base-bleed - increase the range of artillery projectiles through baseflow.
In
Propellants and explosives, no 1, Oct. 1976.

Effect of Grain Characteristics on Range of Artillery Projectiles with Base Bleed

SLOBODAN JARAMAZ and MILOJKO INJAC
Military Technical Institute
Belgrade, Yugoslavia

Abstract

In this paper the basic points of mathematical model for determining range of artillery projectiles with base bleed are presented. After verification of the model by comparison with experimental data, a parametric study of the effect of some characteristics of base bleed on range was carried out. In was found that the range of a given projectile depends on the number of grain segments, burning rate, internal diameter and grain length. The model can be used for selecting characteristics of grain that will give the maximum range of projectile and for decreasing the number of experimental firings for determining those characteristics.

1. INTRODUCTION

Increasing of range of artillery projectiles by base bleed unit has been the subject of many theoretical and experimental investigations in past few years (1-8). Projectiles with base bleed incease their range as a consequence of decreasing base bleed drag. Base bleed unit discharges comubustion products of grain into the recirculation zone behind the projectile and increases pressure in it. Effect of increasing projectile range is in so mush as expressive as the fraction of base drag in total drag is bigger, that is, as the projectile is of better aerodynamical shape. Selecting of geometrical and physico-chemical characteristics of grain of base bleed is the basic problem which should be solved during the design of such projectiles. For that purpose in this paper the mathematical model which enables selections of these characteristics is developed. Special attention is dedicated to the effect of some grain characteristics (shape, dimensions, burning rate etc.) on the expected range.

2. MATHEMATICAL MODEL

The mathematical model devepoled in this paper is based on the engineering point of view on selecting of grain characteristics. In continuation of this paper, the basic points of mathematical model conected with the general considerations about problem, determining of pressure in base bleed unit, changing of burning surface as a function of burnt web and procedure for range calculation are given.

2.1 GENERAL CONSIDERATION ABOUT PROBLEM

For the projectile trajectory calculation differential equations of projectile motion according the modified point mass trajectory model are used [9]. As a consequence of existing base bleed unit, in the equation of motion projectile drag given by equation (1) is changed.

$$D = C_D \frac{\rho_\infty V_\infty^2}{2} Aref \qquad (1)$$

During the working time of base bleed unit (tg), drag coefficient is changed by ΔC_D i.e.

$$\Delta C_D = C_{DB0} - C_{DBB} \qquad (2)$$

This way the total drag coefficient during the time tg is given by the relation

$$C_D = C_{D0} - \Delta C_D \qquad (3)$$

In general case base drag coefficient for the projectile with base bleed is function several parameters:

$$C_{DBB} = f(M_\infty, I, T_g, Ai, \beta, M, \ldots) \qquad (4)$$

From equations (2),(3) and (4) it is obvious that for determining value ef C_D it is necessary to know values for C_{D0}, C_{DB0} and C_{DBB}. Assuminq that values for C_{D0} and C_{DB0} are known, from theoretical calculations, windtunel results or firing testings. the relation for ΔC_D is [7]

$$\Delta C_D = C_{DB0}(1 - e^{jI})C_K \qquad (5)$$

Impulse I is given by equation:

$$I = \frac{\dot{m}_g}{\dot{m}_B} \qquad (6)$$

Here mass fluxes $\dot{m}_g$ and $\dot{m}_B$ are

$$\dot{m}_g = r A_g \rho_g \qquad (7)$$

$$\dot{m}_B = V_\infty A_B \rho_\infty \qquad (8)$$

In the expression (7) burning rate r is a function given by:

$$r = a + b\, p_g^n \qquad (9)$$

In relation (5) j is funciton of Mach number, i.e., $j = f(M_\infty)$ and its value is determined from firing testings while for first approximaton data for j can be taken from [7] and [10]. From experimental results it could be seen that when impulse I is bigger than some limited value, effect of base bleed is decreasing. Based on that fact in expression (5) exists parameter C_K which can be defined as [7]:

$$C_k = 1.0 \qquad I \leq I_{LIM}$$
$$C_k = 1 - \left(\frac{I}{I_{LIM}} - 1\right)^2 C_{LIM} \qquad I > I_{LIM} \tag{10}$$

Constants I_{LIM} and C_{LIM} should de defined for every projectile. For instance, for 155 mm ERFB/BB projectile these values are: I_{LIM} =0.007 and C_{LIM} = 0.25 [7].
From equations (2) through (10) it is obvious that decreasing of base drag ΔC_D, in general case, is complex function and for given projectile it can be expressed by relation [3] :

$$\Delta C_D = f(I, p_g, M_\infty) \tag{11}$$

For defining ΔC_D it is useful to consider theoretical calculations instead of using only time consuming and expensive experimental testings. We'll first consider the way for determining impulse i.e, the pressure p_g in base bleed chamber.

2.2 Pressure in base bleed chamber

Pressure in base bleed chamber is defined using the equality of mass flux of burning products through the orifice and mass flux of gas formed due to the grain burning. Mass flux of gas due to the grain burning is given by expression (7) .
Mass flux of burning products through the base bleed orifice is a function of ratio of pressure at projectile base durig the flight and pressure in base bleed chamber. For the case when the following inequality is statisfied,

$$p_d / p_g > \left(\frac{2k}{k+1}\right)^{\frac{k}{k-1}} \tag{12}$$

flow of gas from the chamber is subsonic. Then the mass flux of burning products is given by equation.

$$\dot{m}_i = \rho A_i \left(\frac{p_d}{p_g}\right) \sqrt{\frac{2k}{k-1} R_g T_g \left[1 - \left(\frac{p_d}{p_g}\right)^{\frac{k-1}{k}}\right]} \tag{13}$$

When the inequality is not satisfied flow of gas from the chamber is supersonic and mass flux is defined by expression

$$\dot{m}_i = \frac{B\, p_g A_i}{\sqrt{R_g T_g}} \tag{14}$$

In the expression constant B is defined as:

$$B = \sqrt{k} \left(\frac{2}{k+1}\right)^{\frac{k-1}{2(k-1)}} \tag{15}$$

It should be emphasized that occuring of supersonic flow in the orfice is undesirable phenomenon for increasing the range. The

supersonic flow make impossible recirculation of burning products behind projectile and increasing of projectile base pressure. This is the fact which should be given special care during the selection of grain characteristics.
Equalizing mass fluxes $\dot{m}_q$ and $\dot{m}_i$ we get equation for the pressure in the base bleed chamber which is solved iterativly.

2.3 Change of grain burning surface as a function of burnt web

Existing base bleed grains have a cilindrical shape with certian number of segments. External and end grain surfaces are inhibited so that burning take place on internal surfaces and slots. Wiew of one grain wi h three segments is presented on figure 1.

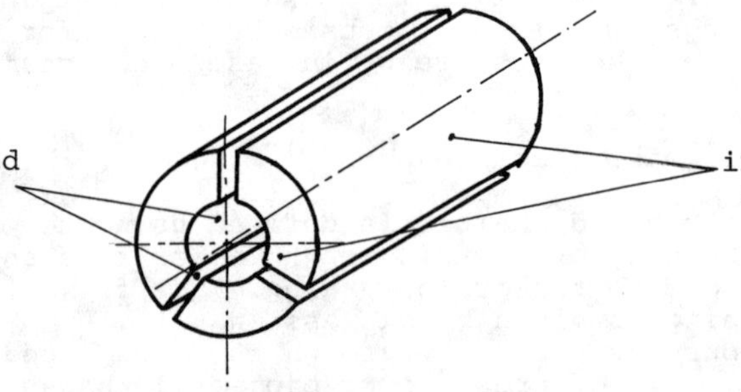

FIGURE 1. View of base bleed grain

Number of grain segments can be different and change of burnt surface as a function of burning web will be given for the case of grain with three segments. Given relations are valid also for grain with N segments but in the next relations value for angle 60° should be replaced with value of $360^\circ/(2N)$ and number of individual elements of burning surface is also function of N.
Cross section of grain with three segments is given on figure 2.

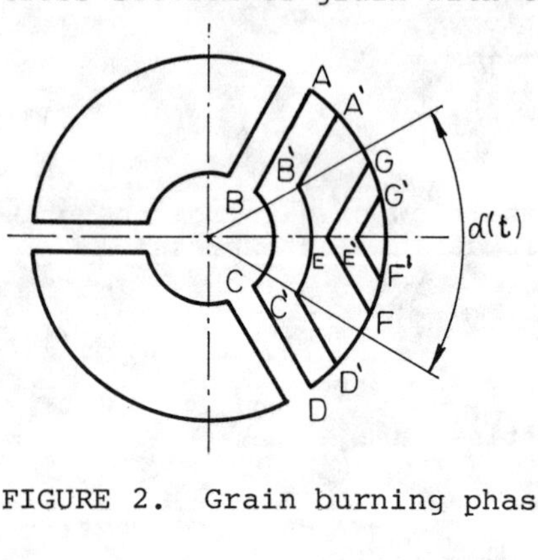

FIGURE 2. Grain burning phases

In general case grain burning is performing in two phases. Each phase has its own shape of grain cross section. At the beginning of the first phase grain cross section is composed from three

egual segments with contours ABCD. Cirsumference of cross section of burning surface for one segment at some moment of first phase is composed from two straight lines A´B´ and C´D´ and arc B´C´. At the end of the first phase arc B´C´ and angle $\alpha(t)$ disappear. At the begining of the second phase circumference of cross section of burning surface is composed from two straight lines GE and EF. At some moment of the second phase circumference of cross section of burning surface is composed from two straight lines G´E´ and E´F´.
Burnt web thickness KA(e) at the some moment at the first phase should be equal to the thickness AG (Figure 3). Then it follows:

$$x_1 + e = (r_o + e)\sin(60^\circ - \frac{\alpha}{2}) \qquad (16)$$

From equation (16) it follows that the burnt web thickness in the first phase is:

$$e = \frac{r_o \sin(60^\circ - \frac{\alpha}{2}) - x_1}{1 - \sin(60^\circ - \frac{\alpha}{2})} \qquad (17)$$

The first phase end occur when $\alpha = 0$. The total burnt web thickness in the first phase is:

$$e_1 = \frac{r_o \sin 60^\circ - x_1}{1 - \sin 60^\circ} \qquad (18)$$

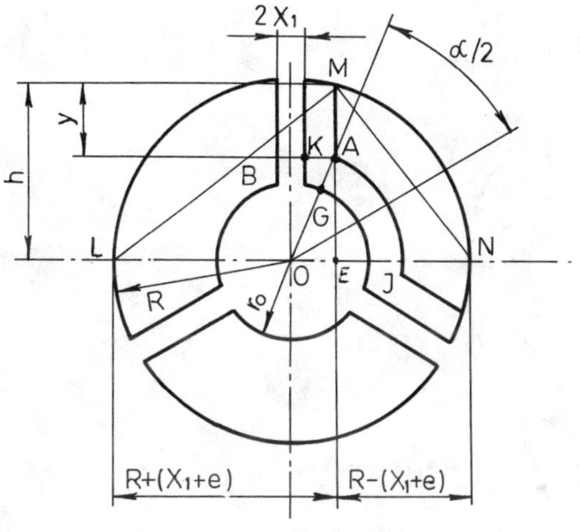

FIGURE 3. First phase of burning

We should calculate the circumference of segment cross section at one moment of the first phase. The height ME at the triangle LMN is the mean geometrical value of LE = $R + (x_1 + e)$ and EN = $R - (x_1 + e)$

$$h = \sqrt{R^2 - (x_1 + e)^2} \qquad (19)$$

From the triangle AOB it follows that $\sin(60 - \frac{\alpha}{2}) = \frac{X_1 + e}{r_o + e}$.
Then $\frac{\alpha}{2} = 60^\circ - \arcsin \frac{X_1 + e}{r_o + e}$. Beside that

$$AE = (r_o + e)\cos(60^\circ - \frac{\alpha}{2}) \tag{20}$$

$$y = ME - AE = \sqrt{R^2 - (r_o + e)^2 \sin^2 (60^\circ - \frac{\alpha}{2})} - (r_o + e)\cos(60^\circ - \frac{\alpha}{2}) \tag{21}$$

The arc length AJ is

$$l = \frac{2\pi}{360^\circ} \alpha (r_o + e) \tag{22}$$

Now, the circumference of grain cross section in the first phase is given by expression:

$$O_I(t) = 6y + 3\,l \tag{23}$$

The area of grain burning surface of length lz is:

$$Ag_I = O_I(t)\,lz \tag{24}$$

From figure 4 it can be seen that the total thickness of burnt web in the second phase is:

$$e_2 = (R - R_1)\sin 60^\circ \tag{25}$$

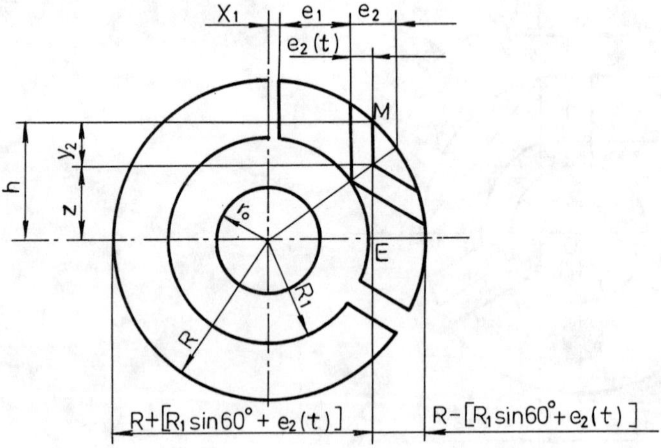

FIGURE 4. Second phase of burning

In the last equation R_1 is radius on which arc disappear from the cross section profile and this radius is equal

$$R_1 = r_o + e_1 \tag{26}$$

Here e_1 is given by the expression (18)
For the second phase one can use the similar procedure.

$$y_2\ (t) = h-z = \sqrt{R^2 - [R_1 \sin 60^\circ + e_2\ (t)]^2} - [R_1 \cos 60^\circ + e_2\ (t)\ tg 60^\circ] \quad (27)$$

$$O_{\underline{II}}\ (t) = 6\ y_2\ (t) \quad (28)$$

$$A_{g\underline{II}}\ (t) = O_{\underline{II}}\ (t)\ lz \quad (29)$$

It should be emphasized that the second phase of grain burning of this geometry need not occur depending on dimensions and number of segments. Then the contour of grain segment cross section for the whole time of burning is composed of straight lines and the arc.

2.4 Procedure for determining range

For determining decreasing of projectile base drag coefficient during the grain burning it is necessary to know several different quantities. First, we should know quantities which are dependant on projectile design characteristics and characteristics of grain i.e. aerodynamic coefficient as a function of Mach number, $j = f(M_\infty)$, and values of I_{LIM}, C_{LIM}, a, b and n. These quantities we get from experimental testings in wind tunel and clossed vessel, firing testings and theoretical calculations. For trajectory calculations it is also necessary to know values for I. For calculations p_g it is necessary to equalize eguation (7) with equations (13) or (14) depending on condition (12). Because of nonlinearity of these equations calculation is performed iterativly with checking condition (12).
The calculation procedure is given on figure 5.

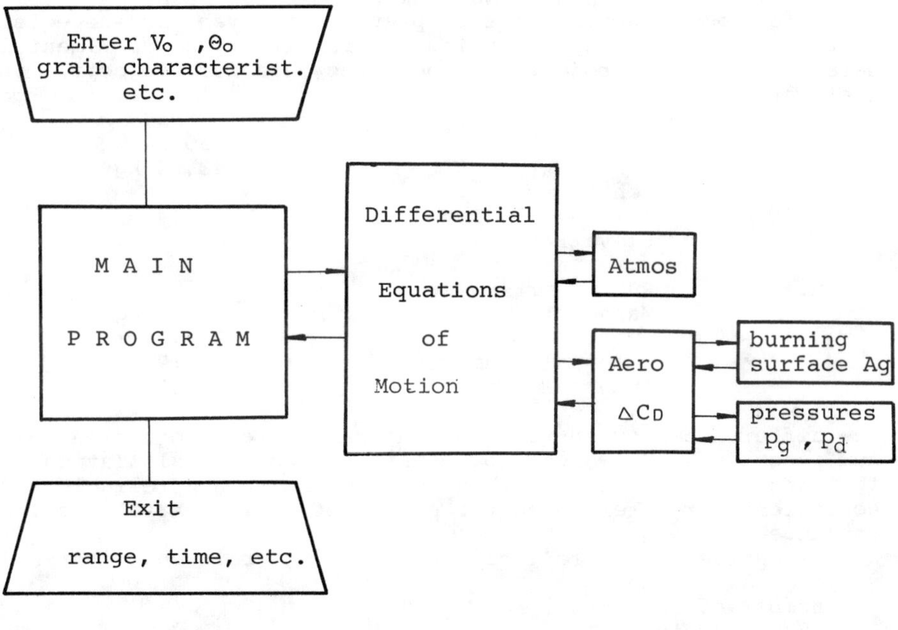

FIGURE 5. Shematic representation of the calculation procedure

Because the calculation procedure is iterative, the end of burning is reached when the web given by the next expression is burnt:

$$e_o = \int_{t=0}^{t_g} r\, dt = R - r_o \qquad (30)$$

The time of grain burning is also calculated from (30).
The iterative procedure can be presented on the following way:

a) assume p_g ($pg_1 = pd + \Delta p$) ;
b) calculation $r = a + b\, pg_i^n$;
c) calculation $e = dr/dt$;
d) calculation $Ag = f(e)$;
e) impuls calculation $I = \dot{m}_g / \dot{m}_B$;
f) calculation of decrease of base drag $\Delta C_D = f(I)$;
g) base pressure calculation $pd = p_\infty \left[1 - kM_\infty^2(C_{DB0} - \Delta C_D)\right]$;
h) ratio calculation pd / pg_i ;
j) calculation pg_i by equations (13) or (14) ;

If $[pg_i - pg_{i-1}] < \varepsilon_{pg}$ the iterative procedure is finished. If this condition is not satisfied we repeat steps from (b) to (j). After the calculation of burnt web we perform the calculation of next step until we don´t satisfy the equation (30).

3. EFFECT OF GRAIN CHARACTERISTICS ON RANGE

For determining the effect of grain characteristics on range of artillery projectiles with base bleed, program based on the mathematical model which basis is previously given, is subjected to the verification by comparing its results with experimental data for one real projectile. The basic data for projectile and grain are:

Projectile:	Caliber	155 mm
	Mass	48.04 kg
	Length	861 mm
	Muzzle velocity	897.4 m/s
	Elevation	50°
Grain:	Number of segments	3
	Mass	1.045 kg
	Length	0.0717 m
	External diameter	0.117 m
	Internal diameter	0.0435 m

The difference between realized and calculated projectile maximum range is 0.09% which shows that this way of calculation of the range of projectile with base bleed is enough reliable and convinient for determining the effect of grain characteristics on range.
We´ll consider the effect of the following grain characteristics:

1. number of segments
2. burning rate
3. internal diameter
4. length

5. delay of grain ignition

For all these considerations shape of grain will be cilindrical with slots, external and end grain surfaces will be inhibited and external diameter will retained constant.

3.1 Number of segments

Number of segments was changed from 2 to 12.The dependance of range on number of segments is presented on figure 6. From the figure it can be seen that maximum ranges are realized with grains which are composed from 6 to 8 segments. For given grain dimensions both burning phases occur when number of segments is greater than 5. Increasing range with grain consisting of 6 segments compareing with common grain number of segments (3) is 750 m or about 2%. Effect of the other grain characteristics on range is defined for grains with 3 and 6 segments.

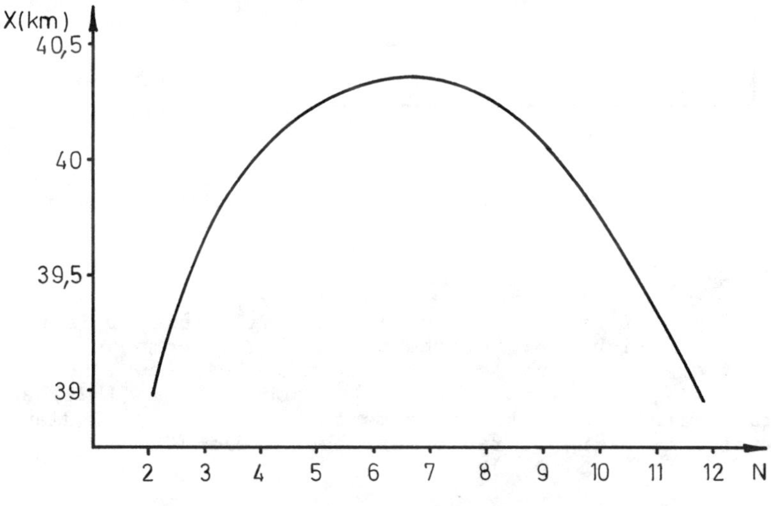

FIGURE 6. Range dependance on number of segments. Standard dimensions ang burning rate.

3.2 Burning rate

Burning rate reflects effect of the grain chemical composition on range. For the grain with common composition constants in the burning law have following values: $a = 0.00072$ m/s, $b = 2.12 \cdot 10^{-7} m/(sPa^n)$, $n = 0.66$. The effect of burning rate on the range is defined by changing value of a which reflects the effect of exotermic reactions which take place beneath burning surface at low pressures. The dependance of calculated range on the burning rate for grains with 3 and 6 segments is given on the figure 7. From the figure it can be seen that the change of a by about 190% increase the range for the grain with 3 segments by 1370m (3.5%) and for the grain with 6 segments by 435 m (1.1%). At given conditions the dependance of the range on a, for the grain with 3 segments has ever increasing character and for the grain with 6 segments this dependance has maximum.

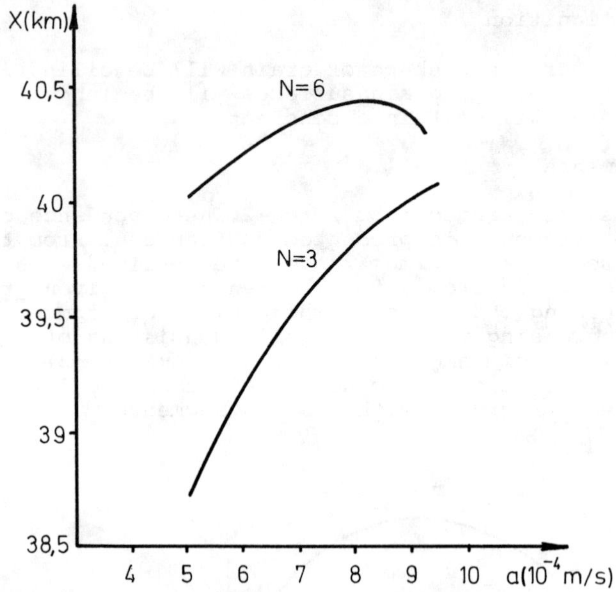

FIGURE 7. Range dependance on burning rate. Standard dimensions and b i n.

3.2 Internal diameter

The depadance of range on the grain internal diameter is presented on figure 8. For the grain with 3 segments the change of grain internal diameter from 0.03045 to 0.05655 m (85.7%) gives decreasing of range by about 490 m (1.2%) and for grain with 6 segments the same change of internal diameter gives decreasing of range for about 1050 m (2.6%).

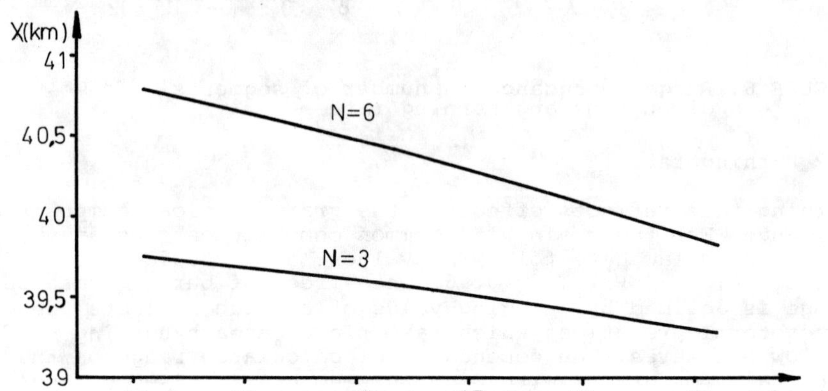

FIGURE 8. Range dependance on grain internal diameter. Standard other dimensions and burning rate.

3.3 Grain length

The dependance of range on grain length for grains with 3 and 6 segments is given on figure 9. The change of grain length from 0.0502 to 0.0932 m (85.7%) gives increasing of range for grain with 3 segments for about 3400 m (8.6%). For the grain with 6 segments the same change of grain length gives increasing of range for 3400 m (8.4%).

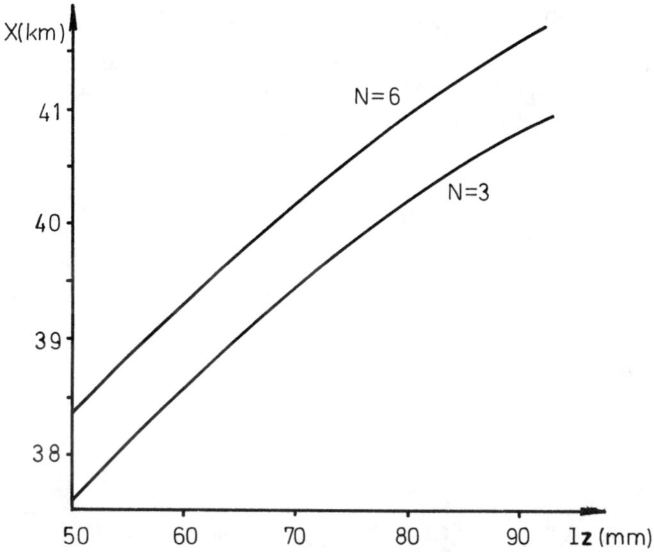

FIGURE 9. Range dependance on grain length. Standard other dimensions and burning rate.

3.4 Diameter of base bleed orfice

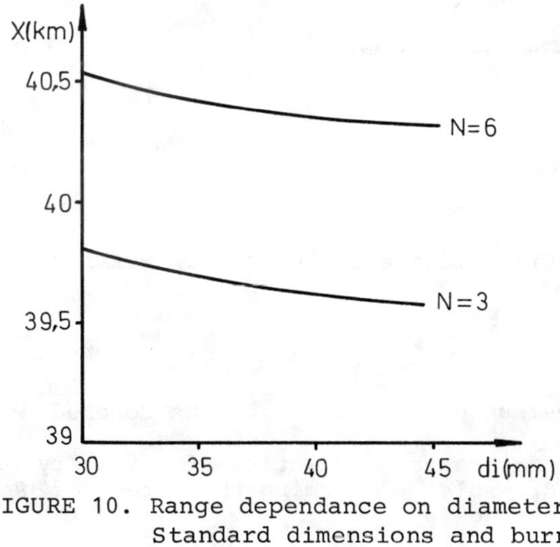

FIGURE 10. Range dependance on diameter of base bleed orfice. Standard dimensions and burning rate.

Efect of the base bleed orifice on range is given on the figure 10. With increasing the diameter of base bleed orifice the projectile range is decreasing. So for the grain with 3 segments the change of diameter of base bleed orifice from 0.0435 m on 0.03115 m (28.5%) gives increasing of range for about 200 m (0.5%), and for the grain with 6 segments the same change of diameter gives increasing of range for about 180 m (0.45%).

3.5 Delay of grain ignition

The range dependance on delay of ignition is presented on figure 11. For grains with 3 and 6 segments ignition delay of

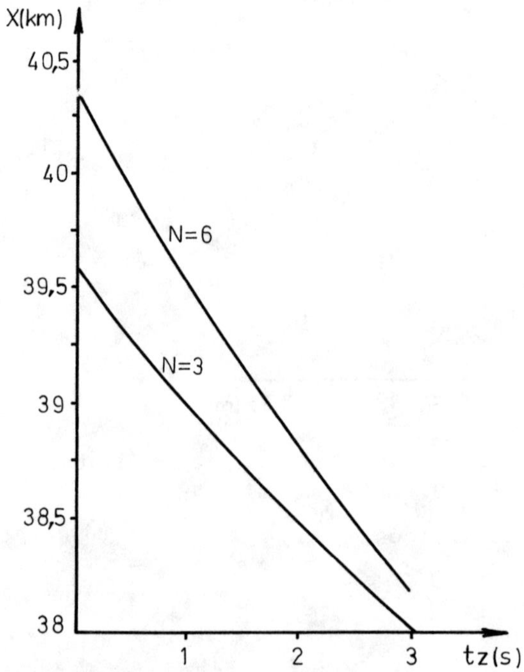

FIGURE 11. Range dependance on grain ignition delay

3 s gives decreasing of range by 1600 m and 2200 m, i.e 4% and 5.5% respectively.
These results show the significance of successful working of base bleed ignitor.

4. CONCLUSIONS

On the basis of the previous discussion it can be concluded:

1. The developed method for range calculation of artillery projectiles with base bleed enables determing the effect of basic grain characteristics on range

2. Using given method it is possible to perform selection of the best grain characteristics and decrease the number of experimental testings for their determing.

3. On the basis of considered effects it is pointed on the range in which the caracteristics should be kept.

Nomenclature

a - burning law constant
A_B - surface of projectile base
A_g - area of burning surface
A_i - surface of base bleed orifice
A_{ref} - reference projectile surface
B - constant given by equation (15)
b - coefficient in burning law
C_D - total drag coeficient
C_{DBB} - projectile base drag coefficient during burning of grain
ΔC_D - decreasing of projectile base drag
C_{D0} - projectile total drag coefficient without working of base bleed
C_{LIM} - constant in equation (10)
C_K - quantity given by equation (10)
D - drag
d_i - diameter of base bleed orifice
d_u - internal diameter of grain
e - burnt web of grain at time t
e_o - total web grain
e_1 - burnt web in the first phase of burning
e_2 - burnt web in the second phase of burning
I - impuls defined by equation (6)
I_{LIM} - limited value of impuls
j - experimental constant
k - specific heat ratio of burning products
l - arc coresponding to the angle
l_z - grain length
$\dot{m}_B$ - mass flux of the air at the base of projectile
$\dot{m}_g$ - mass flux of burning products from burning surface
M - molecular weight of burning products
M_∞ - Mach flight number
n - exponent in burning law
N - number of grain segments
O - burning surface circumference
p_g - pressure in base bleed
p_d - base pressure
R - external grain radius
r - burning rate
r_o - internal grain radius
R_1 - grain radius at the crossing of burning from the first into the second phase
R_g - gas constant of burning products
t - time
t_g - workig time of base bleed
t_z - ignition delay
T_g - burning products temperature
V_∞ - projectile velocity
x_1 - half distance between grain segments
y - geometric parameter defined on figure 3

y_2 - geometric parameter defined on figure 4.
α - angle at time t defined on figure 2.
β - half ange of projectile afterbody
ρ_0 - grain density
ρ_∞ - air density

REFERENCES

1 Murthy, S.N.B. (ed.): Aerodynamics of Base Combustion, Progres in Astronautics and Aeronautics, Vol 40, New York, 1976.

2 Lundhal K., Bartelson N.: Base Bleed, A System to Increase Range and Decrease Balistic Dispersion, Proceedings of the 3rd International Symposium on Balistics, Karlsrue, 1977.

3 Hellgren R.V.: Range Calculation for Base Bleed Propellants, Proceedings of the 6th International Symposium on Ballistics, Orlando, 1981.

4 Fox J.H.: A generalised Base Flow Analysis with Initial Boundary Layer and Chemistry Effects, AEDC-TR-79-46, ADA 072683, 1979.

5 Sahu J.: Numerical Computations of Supersonics Flow over Cylindrical Afterbodies with Base Bleed, Proceedings of the 9th International Symposium on Ballistics, Shrivenham 1986.

6 Schilling H.: A simple Theoretical Approach for Base Bleed Calculations, Proceedings of the 9th International Symposium on Ballistics, Shrivenham 1986.

7 On the Development of Base Bleed Motors for the SRC-ERFB-MK10 MOD 2 Projectile, SRC-BB-R/D-1710, 1983.

8 Hellgren R.V.: Basflodesprojektiler Bonberakninqsprogram och Offektivitetsjamforelser (Base Bleed Projectiles: Range Calculation Program and Eficiencu Comparisons) Research Institute of National Defence, 1983, Stockholm.

9 NATO Surface-to-Surface Artillery Panel the Modified Point Mass Trajectory Model Document AC7225 (Panel IV) D/161.

10 Bownan S.E. and Clayden W.A.: Reduction of Base Drag by Gas Ejection, RARDE report 4/69, 1969.

COMMENT

D.Lyster, SRC,Belgium. Have you done any test to confirm your predictions for the effect of number of segments, grain length and the other parameters you have examined? If so, what are the results.
Author´s Reply. We carred out tests with 152 mm base bleed projectiles to confirm the effect of number of segments and propellant burning rate. Experimental results confirm our predictions.

Discussion for Two Important Parameters of a Base Bleed Projectile

XIFU GUO
Department of Engineering Thermal Physics and Aerodynamics
East China Institute of Technology
Nanjing, PRC

ABSTRACT

The burning rate of a propellant and the ignition time of an igniter are two important parameters of base bleed projectile. These two factors have dominant effects on the extented range and the dispersion of the projectile. So far there is no reference related to determining the effect of these two parameters theoretically.
This paper has given a principle and a method used to determine the burning rate of a propellant and the ignition time of an igniter. It has theoretical and applicational value for us to design base bleed projectiles.

1. DETERMINATION OF BURNING RATE OF PROPELLANT OF BASE BLEED PROJECTILE

1.1 Presentation of Question

In the course of designing a base bleed projectile, we often reference the burning rate obtained to select and design the burning rate of this projectile. This method takes the absolute magnitude of the burning rate as bases, but does not consider the specific structure and launch condition of the base bleed projectile, therefore it is not all–round. Based on the base flow character of the base bleed projectile and the reducing drag mechanism, this paper describes the selection principle and specific method of burning rate of a base bleed projectile.

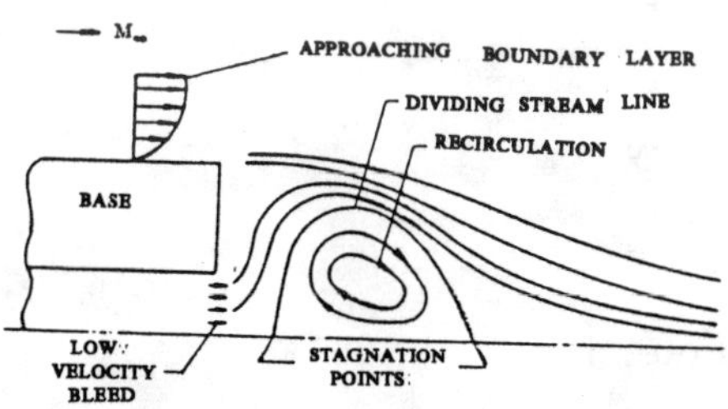

FIG. 1. Schematic illustration of base flow with base bleed (Ref. 2)

1.2 Significance of Selecting Burning Rate

The burning rate equation of the base bleed propellant is

$$r_f = a_1 + a_2 P_{mot}^{a_3} \tag{1}$$

where r_f—burning rate; a_1, a_2,—burning equation coefficients; a_3—exponent of burning equation; P_{mot}—pressure inside the base bleed motor.

The gas of the base bleed propellant exhausts into the base "dead–air" region with very low velocity, so that the base pressure can be raised and the base drag could be decreased. The mass flow of the base bleed gas is

$$\dot{m} = r_f \, \rho_f \, A_f \tag{2}$$

where ρ_f—density of base bleed propellant; A_f—burning area of the base bleed propellant. As shown in Fig. 1. after the gas ejects into dead air region, the flow state in the trailing flow region could be changed, so that the dividing streamline of the dead air region could be extended, flow deflection angle could be diminished, then the base pressure could be raised and the base drag could be decreased. However, the overall effect on raising the base pressure by the base bleed is related to the state of trailing flow region without base bleed. When the structure of the projectile is specified, the state of trailing flow region is related to the state of boundary layer of afterbody and the Mach number of incoming flow. The larger the Mach number of incoming flow, the larger the deflection angle of dividing streamline, the lower the base pressure. Therefore, the base ejecting gas could change the state of flow in trailing flow region, extend the dividing streamline and decrease the deflection angle if the mass flow would be increased. As stated above, the effect of base bleed on increasing base pressure and decreasing base drag depends on not only the mass flow $\dot{m}$ of the bleed but also the state of flow in trailing flow region without bleed, in other words, it depends on M_∞, V_∞, ρ_∞, We measure the overall influence of the two factors with the bleed parameter I. It is defined as.

$$I = \frac{\dot{m}}{\rho_\infty V_\infty A_b} \tag{3}$$

where A_b—base area of the projectile; ρ_∞—density of incoming flow; V_∞—velocity of incoming flow. Relationship between I and P_b is shown in Fig. 2.

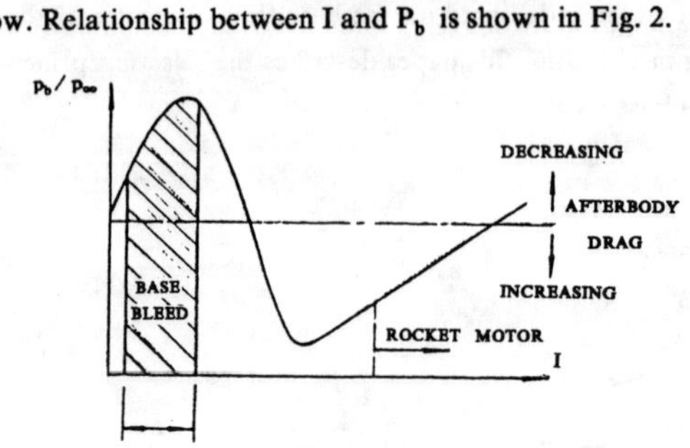

FIG. 2 Base pressure vs base bleed parameter (Ref. 2)

It is known from above analysis that the effect of bleed is expressed by $\dot{m}$. It is related to not only r_f but also ρ_f and A_f. But overall effect of bleed on raising the base pressure depends on the bleed parameter I. Therefore, we could get the following conclusions:

1. Only if the structure and launch parameters of projectile, as well as the character of bleed propellant are the same, it is significant to consider the effect of burning rate on extended range and compare with each other.

2. For projectiles with defferent structure (including base bleed unit), bleed propellant and launch conditions, it is not appropriate to compare burning rate. For example, there are two base bleed units, the burning rate of one unit is 1.2 mm / s, the burning area is 100 mm^2, the burning area of other unit is 200 mm^2. If their other conditions are the same and they both keep the same $\dot{m}$ and I, then tne reduced drag effect of the second projectile would be the same as that of the first one, but its burning rate is only 0.6 mm / s. So, it is possible to make mistakes in design.

3. It is reasonable to compare the bleed effect with the value of I in any case.

1.3 Selection Principle of Burning Rate

The selection principle of burning rate should be: in the condition of certain launch parameters, the overall effect of base bleed along the overall trajectory is to obtain the maximum extended range rate. Launch parameters mainly include the initial velocity, and angle of departure. These two parameters influence the velocity, Mach number, air density and pressure along the trajectory, that is, V_∞, M_∞, ρ_∞ and P_∞. At up leg, with time increasing, V_∞, M_∞, and ρ_∞, would decrease. Since M_∞ is changeable along overall trajectory, so that the drag coefficient is changeable along overall trajectory M_∞ would influence the effect of reducing drag.

At up leg, in the process of the projectile firing towards the apex of trajectory, air pressure decreases. Since the base bleed is subsonic flow, change of ambient pressure influences the change of the motor pressure. When the former decreases, the latter also decreases. So, the burning rate is reduced.

We know that

$$I = \frac{\dot{m}}{\rho_\infty V_\infty A_b}$$

At up leg ρ_∞, V_∞ decrease gradually. If change of area of the bleed propellant is not large that is, the change of $\dot{m}$ is not large, then in the process of projectile flight, the base drag would be increased. In the process of projectile flight, if we change the burning arca, so that and in ρ_∞ V_∞ , A_b would be changed, but we make the value of I always be $(P_b / P_\infty)_{max}$, in other words, $I = I_m$, then we can gain the effect maximum extended range. The burning area formula which satisfies this condition is:

$$\frac{\dot{m}}{\rho_\infty V_\infty A_b} = I_m = const$$

$$A_f = \frac{\rho_\infty}{\rho_f} \cdot \frac{V_\infty}{r_f} I_m A_b \qquad (4)$$

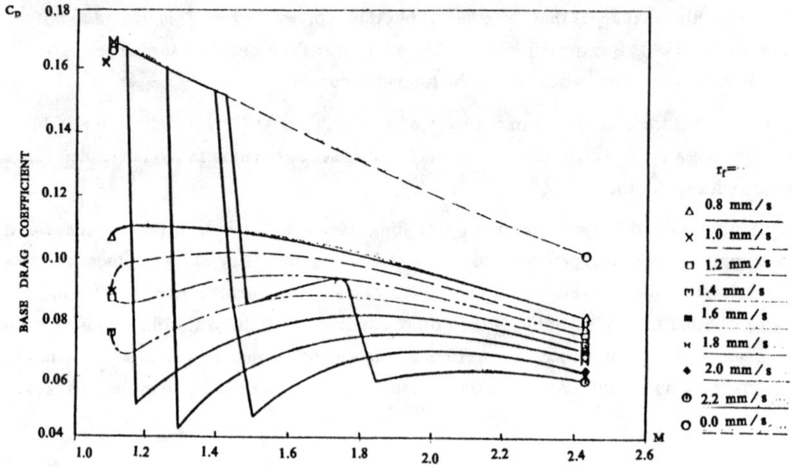

FIG. 3 Drag coefficient vs flight Mach number (Ref. 2)

Since ρ_∞ / ρ_f decrease, I_m A_b is constant, A_f decreases gradually. That is why we design the burning area of the base bleed propellant which reduces along the trajectory. At the trajectory, if the value of I has some change, in order to get larger extended range effect, we must keep I is adjacent to I_m.

For the bleed propellant with certain weight, the burning rate is related to the burning time of base bleed unit. The higher the burning rate, the shorter the burning time. The lower the burning rate, the longer the burning time. The combined influence of base bleed on the extended range rate is concerned with two parameters, they are the reduced drag rate and burning time. As shown in Fig. 3, the burning rate influences the base drag and burning time. Its combined influence on extended range rate is not certain to large. In order to express this influence, we use the concept of impulse of reduced drag rate. It is defined as:

$$S_{RI} = \int_0^{t_b} \frac{R_n - R_B}{R_n} dt \tag{5}$$

where R_n–projectile drag without bleed; R_B–projectile drag with bleed; t_b–burning time. Substituting drag expression to E_q. (5), we get

$$S_{RI} = \int_0^{t_b} I_R dt \tag{6}$$

where

$$I_{RI} = \frac{C_{Dn} - C_{DB}}{C_{Dn}} \tag{7}$$

C_{Dn} –drag coefficient without bleed; C_{DB} –drag coefficient with bleed. In approximate calculation, we set

$$S_{RI} = I_{RCP} t_b \tag{8}$$

$$I_{RCP} = \frac{\sum_{i=1}^{n} I_{Ri}}{n} \tag{9}$$

where i is any point
n is number of round
or we take integrated average value

$$I_{RCP} = \frac{\int_0^{t_b} I_R dt}{t_b} \tag{10}$$

In comparison with other, if S_{RI} reaches maximum, then, this scheme is the optimum one obtaining the maximum extended range effect. In the stage of scheme demonstration rate. The influence of spin on burning rate is related to the structure of base bleed unit and spin velocity. For general spin stabilized projectile, the spin velocity may increase the burning rate about thirty percent. In a word, the principle of selecting burning rate could be expressed as the following: under the condition of specific projectile launch parameters, considering the influence of spin effect, we acquire the burning rate corresponding to the maximum extended range rate. Mathematically, it can be expressed: we get the burning rate which corresponds to the maximum impulse of reduced drag rate S_{RI}

1.4 Method of Determination of Burning Rate

Method of determination of burning rate may be completed by using computer program. In order to understand the calculation procedure clearly, we explain it step by step:

1. Determine the launch parameters of projecitle: (1) Initial velocity; (2) Angle of departure; (3) Structure of projectile; (4) Weight of projectile.
2. Determine the structure of base bleed unit initially: (1) Structure size of base bleed unit; (2) Propellant weight; (3) Propellant structure and burning mode; (4) Propellant density.
3. Calculate ballistic elements and meteorological elements of base bleed projectile by given parameters.
4. Calculate $\dot{m}$ from point to point along trajectory.
5. Calculate I from point to point along trajectory.
6. Calculate $p_b / p_\infty \sim I$ curve.
7. Calculate curve of base drag coefficient vs M or t.
8. Calculate curve of overall drag coefficient vs M or t.
9. Calculate S_{RI}.
10. Calculate the overall range.

11. Compare schemes and determine burning rate.

Example.

Suppose: projectile length 900mm, boattail length 90mm, projectile diameter 155mm, base diameter 145mm, projectile mass 43.5kg, initial velocity 827 m / s. Structure of base bleed propellant: two segments, radial burning, outside diameter of the base bleed unit 120mm, inside diameter 45mm, By using procedure mentioned above, obtain the optimum initial burning rate corresponding the maximum range, which is between 1.8 mm / s and 2.0 mm / s.

2. DETERMINATION OF IGNITION TIME OF IGNITER OF BASE BLEED PROJECTILE

2.1 Presentation of Question

The propellant of base bleed projecitle is ignited by the high temperature and high pressure gas in bore of a gun. When projectile leaving the muzzle, pressure rapidly drops from hundreds of atmospheres to one atmosphere, the burning propellant blown out is possible. In order to ensure the propellant ignition, we install igniter inside the base bleed unit. It can ignite the propellant once again outside the gun tube to make the propellant burn normal. This is so–called secondary ignition. Based on exterior ballistics and dispersion theory of shot, this paper discusses burning time of igniter as well as theory and method which are used to determine the dispersion scope of the secondary ignition time of propellant.

2.2 Principle of Determination of Burning Time of the Igniter

The principle of determination of the burning time of igniter is:

1. assure that the propellant can be ignited reliably; 2. decrease the influence of secondary ignition time dispersion of the propellant on overall range dispersion as small as possible. In view of the ignition burning process of the propellant of base bleed projectile, the burning time of igniter must be not smaller than the secondary ignition time of the propellant, otherwise, the propellant can not be ignited once again. That is not assure that the propellant can be ignited reliably. Suppose the maximum secondary ignition time of the propellant is t_{b2max}, the minimum burning time of the igniter is t_{imin}, to assure that the propellant is ignited reliably, there must be

$$t_{imin} > t_{b2max} \tag{11}$$

TABLE 1. Relationship between E_X and t_i, E_{tb}

E_x(m)	80	100	150
t_i(s)	0.35	0.41	0.57
t_b(s)	0.05	0.063	0.094

Because of the propellant composition, the pressure inside the base bleed unit, performance of igniter, environment conditions and dispersion of initial velocity, the secondary ignition range has certain dispersion. In order to decrease the influence of secondary ignition range dispersion on overall range dispersion the igniter should assure that the secondary ignition range dispersion of the propellant basicaily does not influence the overall range dispersion. Suppose the secondary ignition range is x_b, average velocity in scope of x_b is v_{cp}, the secondary ignition time is t_{b2}, then

$$x_b = v_{cp} t_{b2}$$
$$\Delta x_b = \Delta v_{cp} t_{b2} + \Delta t_{b2} v_{cp}$$

Suppose the probable error of average velocity dispersion is E_{vcp}, the probable error of secondary ignition time dispersion is E_{tb2}, the probable error secondary ignition range dispersion is E_{xb}, then

$$E_{xb} = \sqrt{(t_{b2} E_{vcp})^2 + (V_{cp} E_{tb2})^2}$$

From calculation, we know

$$(V_{cp} E_{tb2})^2 \gg (t_{b_2} E_{vcp})^2$$

Therefore, when considering the secondary ignition range dispersion we only need to consider the secondary ignition time dispersion. That is

$$E_{xb} = V_{cp} E_{tb2} \tag{12}$$

2.3 Method of Determination of Ignition Time of Igniter.

Now, based on two above–mentioned principles, we determine the ignition time of ignitier.

1. Determination of maximum burning time of igniter

Suppose the average value of the secondary ignition time of the propellant is t_{bcp}, probable error of the secondary ignition time dispersion is E_{tb2}. So, the maximum ignitiom time of igniter t_i is

$$t_i = t_{bcp} + 5E_{tb2} \tag{13}$$

where

$$t_{bcp} = \frac{\sum_{i=1}^{n} t_{b2i}}{n} \tag{14}$$

n–number of round; t_{b2i}–secondary ignition time of the i–th round, i = 1, 2, ···, n.

2. Determination of secondary ignition time dispersion of propellant.

The probable error of the overall range dispersion of shot is E_Σ. It can expressed as follows

$$E_\Sigma^2 = E_X^2 + E_{xtb}^2 \tag{15}$$

Where E_X–the range probable error of gun, E_{xtb}–the range probable error of base bleed projectile is related to the drag coefficient C_D, that is

$$X = f_1(C_D) \tag{16}$$

When the secondary ignition time is differert, the drag coefficient is also different. That is

$$C_D = f_2(t_b) \tag{17}$$

The relationship between the probable error E_{tb} of the secondary ignition time of propellant and the range probable error E_{xtb} is

$$E_{xtb} = \frac{\partial X}{\partial C_D}\frac{\partial C_D}{\partial t_b}E_{tb} \tag{18}$$

where $\partial X / \partial C_D$–the sensitivity factor of range on the drag coeffficient; $\partial C_D / \partial t_b$–the sensitivity factor of drag coefficient on the secondary ignition time. Suppose $E_{xtb} = 0.1E_x$, it is ensured that E_{xtb} does not influence E_Σ. Substitute it into the formula, we have

$$E_{tb} = \frac{0.1E_x}{\dfrac{\partial X}{\partial C_D}\dfrac{\partial C}{\partial t_b}} \tag{19}$$

The formula (19) can be used to calculate the secondary ignition time dispersion. After determination of tactical and technical targets, overall scheme of gun system is also determined. E_x, $\partial X / \partial C_D$, $\partial C_D / \partial t_b$ are all known. Thus, we can determine E_{tb} by the formuala (19).

2.4 Calculation of $\partial X / \partial C_D$ and $\partial C_D / \partial t_b$

1. Calculation of $\partial X / \partial C_D$

a. Method based on ballistic equations

Using the motion equations of the projectile, and taking C_D to change one unit, we can calculaate $\partial X / \partial C_D$

b. Method based on correction coefficient

The expression of ballistic coefficient c is

$$c = \frac{id^2}{G}10^3$$

we can get

$$\frac{\partial X}{\partial C_D} = \frac{\partial X}{\partial C}\frac{d^2 10^3}{C_{Dn} G} \tag{20}$$

where $\partial X / \partial C$–the correction coefficient of range on ballistic coefficient; d–projectile diameter; G–projectile weight; C_{Dn}–drag coefficient under the condition of standard projectile.

(a)

(b)

FIG. 4. Photograph with a burning igniter and inert propellant and photograph with a burning igniter and a burning propellant

2. Calculation of $\partial C_D / \partial t_b$
we know

$$C_D = f_3(M),\ M = f_4(t_b)$$

So,

$$\frac{\partial C_D}{\partial t_b} = \frac{\partial C_D}{\partial M}\frac{\partial M}{\partial t_b} \tag{21}$$

By using the above formula, we have two methods to calculate$\partial C_D / \partial t_b$
a. Direct calculation of $\partial C_D / \partial t_b$
Based on the aerodynamics or experiment. we can get $C_D \sim M$ curve, From this curve we obtain $\partial C_D / \partial t_b$
b. Calculation of $C_D / \partial M$, $\partial M / \partial t_b$. recalculation of $C_D / \partial t_b$
By using ballistic equations we calculate $M \sim t_b$ and $\partial C_D / \partial t_b$
recalculate $\partial C_D / \partial t_b$

2.5 Examples

Suppose diameter of a base bleed projectile d = 155mm, its weight G = 45kg, initial velocity $V_0 = 950$ m / s, departure angle $\theta_0 = 45$ ° , flight up leg time $t_s = 46$s, when range probable error $E_x = 80$ m, 100 m, 150 m, find the maximum burning time t_i of igniter and probable error of its dispersion. E_{tb}. Using above–mentional methods we get t_i and E_{tb} as shown in table 1.

2.6 Determination of Secondary Ignition Range of Propellant by Experiment

1. Method using smear camera

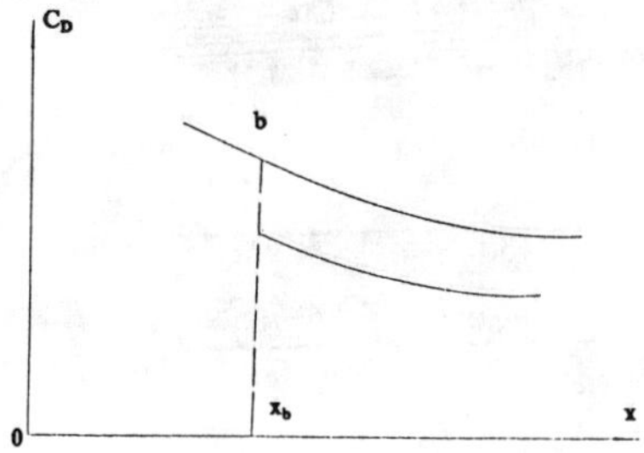

FIG. 5. $C_D \sim x$ curve   Point b is the secondary ignition location of the propellant:

At the side of departure line, among the distance of 50~60m, begin with 10m from the muzzle, we locate smear cameras at 10m interval, By comparison of photographs one with a burning igniter and an inert propellant and the other with a burning igniter and burning propellant, we can determine whether the secondary ignition of the propellant would occur, as shown in Fig. 4. a is the photograph only with burning igniter, b is the photograph with both burning igniter and burning propellant.

2. Method using Doppler radar

Use a Doppler radar to measure velocity of the base bleed projectile among the range about 300 ~500m away from the muzzle, draw the $C_D \sim x$ curve, to the curve, determine the secondary ignition distance of th propellant as shown in Fig. 5.

CONCLUTIONS

1. It is reasonable to compare the bleed effect with the value of I in any case.
2. The burning rate corresponding the maximum impulse reduced drag rate is the optimum burning rate.
3. The ignition time of the igniter is no less than the time of secondarg ignition time of propellant.
4. Dispersion of secondary ignition time does not affect the overall range dispersion of shots.

NOMENCLATURE

A_b	base area of the projectile
A_f	burning area base bleed propellant
a_1	burning equation coefficient
a_2	burning equation coefficient
a_3	exponent of burning equation
C_{DB}	drag coefficient with base bleeding
C_{Dn}	drag coefficient under the condition of standard projectile
d	projectile diameter
E_{tb2}	the probable error of the secondary ignition time of propellant

E_{vcp}	the probable error of average velocity dispersion
E_x	the range probable error of gun
E_{xtb}	the range probable error caused by the secondary ignition range dispersion of propellant
E_{Σ}	the probable error of the overall range dispersion of shot
G	projectile weight
I	bleed parameter
I_m	the maximum value of I
M	Mach number
$\dot{m}$	the mass flow of the base bleed gas
P_b	the base pressure
P_{mot}	pressure inside the base bleed motor
P_{∞}	the pressure of incoming flow
$q(y_i)$	the layer weight of drag coefficient at the point yi
R_B	projectile drag with bleed
R_n	projectile drag without bleed
r_f	burning rate
S_{RI}	impulse of reduced drag
t_b	burning time
t_{b2}	secondary ignition time
t_{bcp}	the average burning rate
t_{b2max}	the maximum secondary ignition time of the propellant
t_{imin}	the minimum burning time of the igniter
V_0	initial velocity
V_{cp}	average velocity in scope of X_b
V_{∞}	velocity of incoming flow
X_b	the secondary ignition range
ρ_f	density of base bleed propellant
ρ_{∞}	density of incoming flow
θ_0	departure angle
ΔX_b	the change of X_b
$\partial C_D / \partial t_b$	the sensitivity factor of drag coeffiicient on the secondary ignition time.
$\partial X_D / \partial C_D$	the sensitivity factor of range on the drag coefficient
$\partial X_D / \partial C_{Di}$	the sensitivity factor of range on the drag coefficient of point yi
$\partial X / \partial C$	the correction coefficient of range on ballistic coefficient.

REFERENCES

1. Xifu Guo etc, The Exterior Ballistics of Base Bleed Projectile, Ballistic Research Lab. of ECIT 1986

2. Hellgren. R. V. Range Calculauon for Base Bleed Projectile, Proc. 6th Intem. Symp. Ball. Orland. Fla. 1981.

3. H. Schilling. A Simple THeoretical Approach for Base Chiculation. Proc. 8th Intem. Symp. Ball. 1985.

COMMENTS

Denis Lyster Chief of Analytical Group Space Research Corp. You say that S_{RI} sho uld be maximum to give maximum extended range effect. This is effectively $\int_0^{tb} C_{DN}-C_{DB} / C_{DN}dt$ (Equation 7, 8, 10). It seems to me that the integration should be weighted to give more influence to the base drag reduction during the early burn phase, where the effect on velocity loss, and hence on range, is greatest.

Author's Reply. The suggestion raised by Mr. Lystter is reasonable. The influence of the base bleed drag on the range is related to the position of the projectile. Therefore, we should consider the influence of the layer weight. If the integration is weighted, then the equation (5) becomes.

$$S_{Ri} = \int_0^{t_b} q(yi)\frac{R_n - R_B}{R_n} dt$$

where

$$q(yi) = \frac{\partial X / \partial C_{Db}(yi)}{\partial X / \partial C_{Db}}$$ the layer weight of drag coefficent at the point yi.

$\partial X / \partial C_{Db}(yi)$— the change of range caused by an unit change of the drag coefficient with base bieeding at the point yi

$\partial X / \partial C_{Db}$ — the change of range caused by an unit change of the drag coefficient along the overall trajectory.

In the same reason, the equation (7) becomes.

$$I_R = q(yi)\frac{C_{Dn} - C_{Db}}{C_{Dn}}$$

A Two-Phase Base Bleed Base Combustion Model and Its Calculation

NAN-YAN WANG
East China Institute of Technology
Nanjing, 210014, PRC

ABSTRACT

A two phase base bleed base combustion model has been presented in which the exit area ratio, jet solid–phase particle mass ratio, fuel enthalpy of combustion, molecular weight of the gas, stagnation temperature in the chamber, flight Mach number parameters have been considered. In the present paper, a practical calculating method has been given completely, and varied estimation examples have been taken. The comparison of these results with many available experimental data shows generally good agreement.

1. INTRODUCTION

The total drag for supersonic projectiles consists of shock wave drag, viscous drag and base drag. The base drag is a major contributor to the total drag and can be as much as 50–70% of the total drag. Thus, it is especially important to minimize this part of the drag. Base bleed is an effective means of reducing the base drag. The structure of the flow in base region is changed by a small quantity of low speed gas injected in to the base area.

The best review about base bleed is given by Murthy et al [1]. Bowman and Clayden [2] have reported their experimental studies on supersonic flow over cylindrical afterbodies with gas injection. Recently Sahu [3] has developed a new numerical method to compute the base region flow using Azimuthal–Invariant thin–layer Navier–Stokes equations, in which an iterative boundary condition procedure has been developed to determine the bleed exit boundary conditions. In Sahu's paper [3], the effects of solid phase fuel particle in the exit jet, fuel enthalpy of combustion and gas molecular weight have not been considered but those factors have shown their importance in base bleed function. Recently, Schilling [4] has presented a simple theoretical approach for base bleed calculations in which boat–tail and burning equation of fuel are considered, but solid phase mass ratio and fuel enthalpy of combustion have not been considered.

The present paper describes a new model of base bleed with base combustion in which exit area ratio, flight Mach number, chamber stagnation temperature, jet gas molecular weight, solid mass ratio of exit two–phase mixture gas, enthalpy of combustion etc. have been considered. The predicted results show general agreement with the available experimental data.

2. BASE BLEED BASE COMBUSTION GASDYNAMIC MODEL

For a common projectile without base bleed function, the flow stream pattern near the base section can be simply sketched out at Fig.1–a. The base area is separated by the separation atreamline from the outside potential flow. One has plenty of reason to think that there is an interior separation streamline separating the base area form the base injection flow in the case of the projectile with a base bleed function, as shown in Fig.1–b. At the end of the base area, there is section area A_2 which is not equal to zero, so that base area becomes a hollow head–cut cone. The schlieren photographs from Brown et al may support the statement above, as shown in Fig.2.

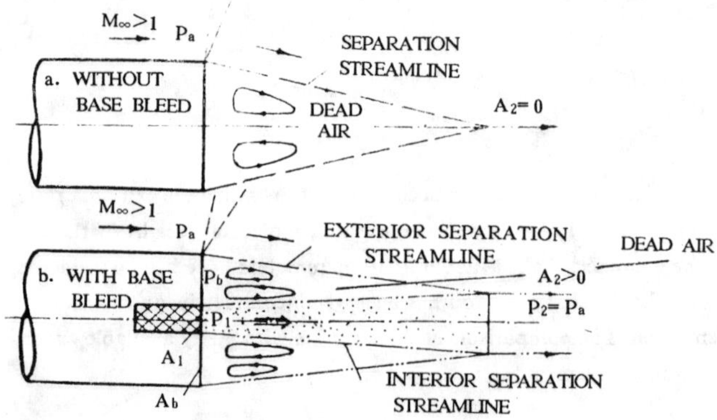

FIGURE 1. Qualitative pattern in the base region

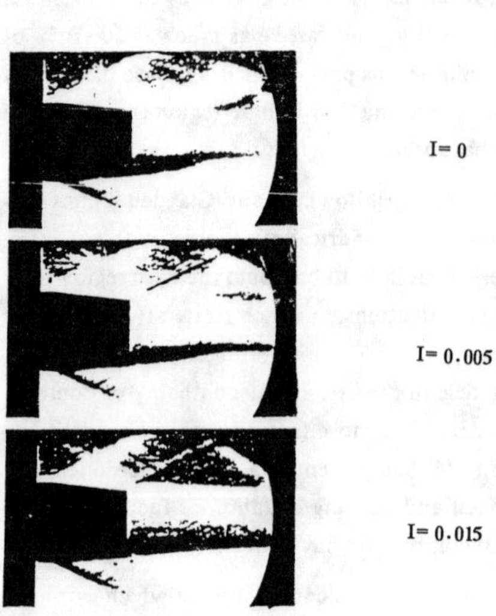

FIGURE 2. Schlieren pictures in the base region

The secondary burning phenomenon are observed in the tests very often. Solid–phase fuel in particle form mixed with gas–phase is burning along the jet path. One can consider that dynamic equilibrium has been reached at section A_2, where $P_2 = P_a$, and the burning of solid–phase fuel has been completed already.

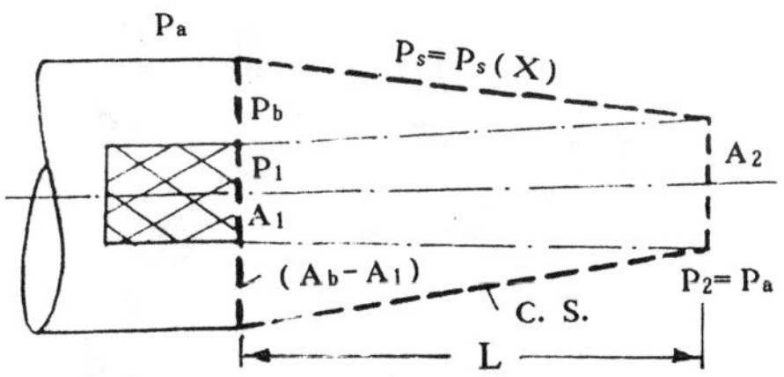

FIGURE 3. Sketch of control volume for base bleed model

Taking a control volume between section A_1 and section A_2 as shown in Fig.3. Governing equations are shown as follows:

Continuity equation

$$\dot{m}_1 = \rho_{m1} u_1 A_1 = \dot{m}_2 = \rho_2 u_2 A_2 \tag{1}$$

Momentum equation

$$P_b(A_b - A_1) + P_1 A_1 - 2\pi\int_{r_2}^{r_b} r \cdot P_s(r)dr + 2\pi\int_0^L r \cdot \tau(x)dx - P_2 A_2 = \dot{m}_2 u_2 - \dot{m}_1 u_1 \tag{2}$$

Energy equation

$$H_{02} - H_{01} = \varepsilon \cdot \Delta H \; - \; \text{Heat lost to surrounding} \tag{3}$$

where solid–phase mass ratio

$$\varepsilon = \rho_p / \rho_m \tag{4}$$

Gas–solod mixture density

$$\rho_m = \rho_g + \rho_p \tag{5}$$

The following equation can be easily transferred from Eq.(1)

$$\frac{P_1 A_1}{P_a A_2} = \sqrt{\frac{R_{m1} T_{01}}{R_2 T_{02}}} \cdot \frac{M_2\left[k_2\left(1 + \frac{k_2 - 1}{2} M_2^2\right)\right]^{\frac{1}{2}}}{M_1\left[k_{m1}\left(1 + \frac{k_{m1} - 1}{2} M_1^2\right)\right]^{\frac{1}{2}}} \tag{6}$$

It is obvious that the gas–solid phase flow at section A_2 should be in an equilibrium state, so that the following expressions can be quoted from Wang [5]:

$$R_{m1} = (1 - \varepsilon)R_g \tag{7}$$

where

$$R_g = R / \mu \tag{8}$$

and

$$k_{m1} = k\left[\frac{1 - \varepsilon\left(\frac{C_p}{(C_p)g}\right)}{1 - \varepsilon\left(1 - \frac{kg \cdot C_p}{(C_p)g}\right)}\right] \tag{9}$$

The vicous term in Eq.(2) is much smaller than the pressure terms so it can be omitted for simplification purpose. Furthermore, the separation streamline can be considered as straight line for the time being. As there exist plenty of reliable methods to estimate pressure distribution in aerodynamics, such as shock expansion method, characteristic method etc, $P_s = P_s(x)$in Eq.(2) can be assured of being solved. Among them, the following equation may be the simpliest one [6].

$$\frac{P_s - P_a}{\frac{k}{2}P_a M_\infty^2} = \pm\left(0.0016 + \frac{0.002}{M_\infty^2}\right)\beta_0^{1.7}$$

It can be rewritten as

$$\frac{P_s}{P_a} = 1 \pm k(0.001 + 0.0008M_\infty^2)\left[\frac{57.3}{2\bar{L}}\left(1 - \left(\frac{\bar{A}_2}{\bar{A}_b}\right)^{\frac{1}{2}}\right)\right]^{1.7} \tag{10}$$

where $\bar{A}_2 = A_2 / A_1$; $\bar{A}_b / A_1$. The length of dead area $\bar{L} = L / 2r_b$. Minus or plus sign chosen in Eq.(10) is dependent on whether A_2 is less than A_b or not. For more general case, define average pressure

$$P_{s \cdot ave} = \frac{2\int_{r_2}^{r_b} r \cdot P_s(r)dr}{r_b^2 - r_2^2} \tag{11}$$

where

$$P_s(r) = f(\bar{L},\ \bar{A}_b,\ \bar{A}_2,\ M_\infty) \tag{12}$$

can be obtained by accurate methods which are available wherever in aerodynamics. Then, Eq.(2) becomes

$$\bar{P}_b = \left(\frac{\bar{A}_2}{\bar{A}_b - 1}\right)\left[1 + \bar{P}_s\left(\frac{\bar{A}_b}{\bar{A}_2} - 1\right) - \frac{\bar{P}_1}{\bar{A}_2} + k_2 M_2^2 - \frac{k_{m1}\bar{P}_1}{\bar{A}_2} M_1^2\right] \tag{13}$$

where $\bar{P}_s = P_{s \cdot ave} / P_a$
Considering the specific heat of gas–solid micture [5]

$$(C_p)_m = (1 - \varepsilon)\ (C_p)_g + \varepsilon C_p \tag{14}$$

and omitting the radiation heat loss to the surrounding from the control volume, Eq.(3) can be rewritten as

$$T_{02} = \left[(1 - \varepsilon) + \varepsilon\frac{C_p}{(C_p)_g}\right] T_{01} + \varepsilon\frac{\Delta H}{(C_p)_g} \tag{15}$$

Most authors in the arer of base bleed like to use injection parameter $I = \dot{m}_j / \rho_a u_\infty A_b$, which the present author adopts. After simple derivation, one can obtain

$$I = \left(\frac{\bar{P}_1}{\bar{A}_b} \cdot \frac{M_1}{M_\infty}\right)\left[\frac{R_a}{k_a} \cdot \frac{k_{m1}}{R_{m1}} \cdot \frac{T_a}{T_{01}}\left(1 + \frac{k_{m1} - 1}{2} M_1^2\right)\right]^{\frac{1}{2}} \tag{16}$$

where R and k are 0.287 kJ / kg • K and 1.404 respectively, for air.

3. SOLUTION OF THE EOUATIONS AND ADDITIONAL ASSUMPTIONS

There exist many semi–empirical expressions for estimating the base pressure without base bleed, one of them is from U.S Army BRL in 1981 [7]

$$\frac{P_b}{P_a} = \left[1 - 0.09M_\infty^2(1 - e^{\bar{L}_{cyl}})\right] \cdot \left[1 + \frac{M_\infty^2}{4}\left(1 - \bar{A}_b^{\frac{1}{2}}\right)\right] \tag{17}$$

The cylinder length $\bar{L}_{cyl}$, boattail shape and flight Mach number are considered in this experssion.

For the case without base bleed, then $\bar{A}_2 = 0$. If the contour of the dead area and Mach number are given, taking iteration to Eqs.(12), (13) and (17), one can obtain the effective length of the dead area $\bar{L}_{eff}$. For the case with base bleed, one can assume the contour of dead area will change only in diameter, but not in the effective length.

According to Eq.(15), one can obtain the stagnation temperature of jet flow at section A_2.
Since most of the base bleed structures so far contain a subsonic or sonic nozzle, the pressure at exit section is equal to the base pressure P_b . As the exit Mach number M_1 is given, the parameters $\bar{A}_2$, $\bar{P}_s$ and $\bar{P}_b$ can be solved by iteration method based on Eqs.(6), (11) and (13).

For the purpose of comparison with experimental data, injection flow rate ratio I is calculated by Eq.(16).

The whole calculating process has been completed by Program BLEED.

4. EXAMPLE

A projectile with 5 calibre cylinder without boattail. $(C_p)_g = 1.0 kJ/kg \cdot K$, $C_p = 0.7 kJ/kgK$, surrounding temperature $T_a = 200K$.

The following parameters have been considered as independent parameters in the calculation examples: flight Mach number M_∞, jet gas molecular weight μ, total temperature in the chamber T_{01}, solid–phase mass ratio ε, fuel enthalpy of combustion ΔH, and base bleed exit area ratio $\overline{A}_b$. The results and the comparisons with experiments are shown in the following sections.

4.1 The Effects of Exit Area Ratio

Taking $\varepsilon = 0$, it means that pure gas or no secondary burning is considered. So that, $T_{01} = T_{02}$. $\overline{A}_b = 16$, 8 and 4 have been accounted for in this example, the length of dead area $\overline{L}_{eff}$ are obtained 10.296, 7.272 and 5.14 respectively. The results are shown on Fig.4. The pure gas is He whose molecular weight is 4, and total temperature T_{01} is taken 300K. The results of the wind tunnel test with a center nozzle by Reid and Hastings [2] are shown on Fig.5. Although they did not give the molecular weight and temperature, one can realize that the theoretical model works well indeed.

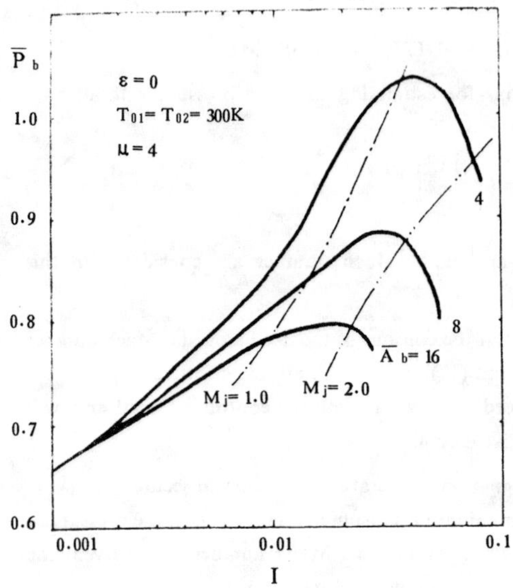

FIGURE 4. The effect of exit area on base pressure

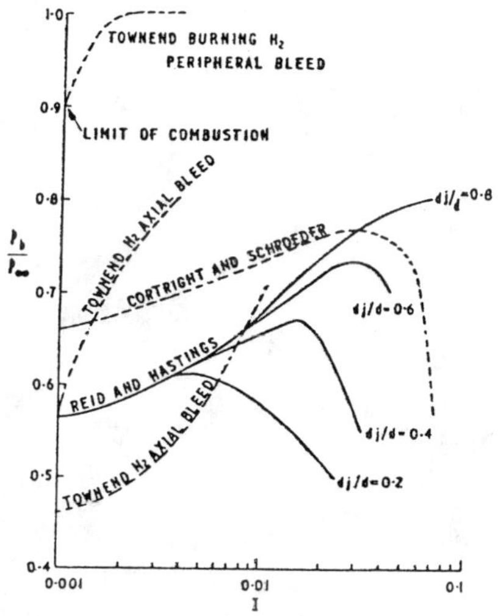

FIGURE 5. Reid and Hastings [2] experimental data

As the value of I is less than 0.003, the effect of the exit area on base pressure is negligible.

The similar curves can be found from Murthy et al. [1] (see Fig.62) and Schilling [4] (see Fig.4).

4.2 The Effects of Jet Gas Molecular Weight

Two gases with molecular weight 4 and 28 respectively have been studied. Taking $\overline{A}_b = 9$, pure gas ($\varepsilon = 0$), $T_{01} = T_{02} = 300K$. The effective length of the dead area, $\overline{L}_{eff} = 7.73$, is obtained. The numerical results are shown in Fig.6, from which one can see that the lighter jet gas molecular weight will bring a better effect of drag reducing under the condition of flow rate I is not too large. Many experiment results have been found to support this conclusion. The experiment data from Brown et al. are shown in Fig.7 and 8.

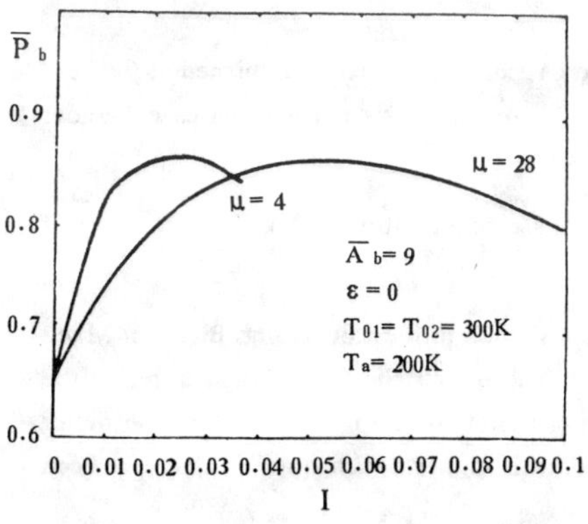

FIGURE 6. The effect of jet molecular weight on base pressure

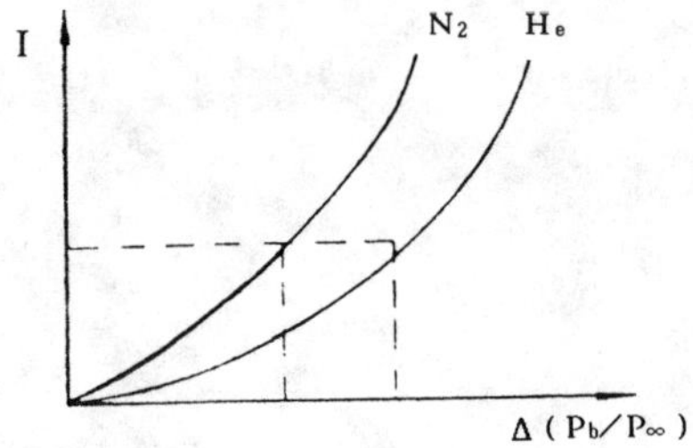

FIGURE 7. Experiment data of base bleed with N_2 and He [7]

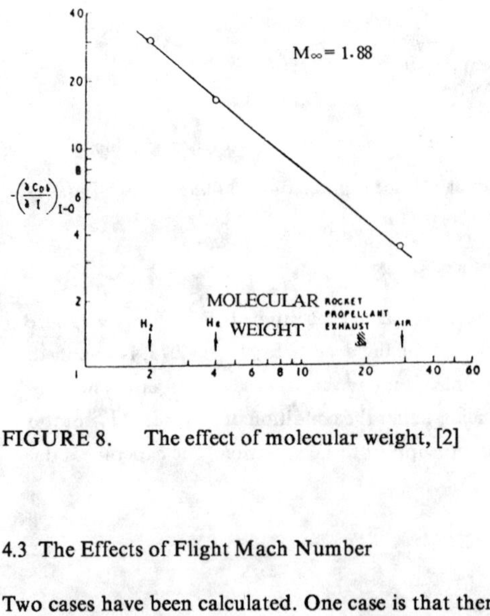

FIGURE 8. The effect of molecular weight, [2]

4.3 The Effects of Flight Mach Number

Two cases have been calculated. One case is that there is solid fuel particle unburned at the base bleed exit section, as well as the secondary burning phenomenon exists. The other case is under the condition that all the fuel is burnt out at the exit section.

Taking $\bar{A}_b = 16$, $T_{01} = 3000K$, $T_a = 300K$, solid–phase mass ratio $\varepsilon = 0.5$, fuel enthalpy of combustion $\Delta H = 2000 \mathrm{kJ/kg}$, jet gas molecular weight $\mu = 30$.

Flight Mach number 1.1, 1.5, 2.0, 2.5, and 3.0 are considered in the calculating. Bleed jet Mach number is changed smoothly between zero and 1.0. The higher the flight Mach number, the larger the absolute value of $(\partial \bar{P}_b / \partial I)_{I=0}$ that one can find from Fig.9–a. So that the better effectiveness of drag reducing can be accomplished with higher Mach number that has been pointed out by wind tunnel tests.

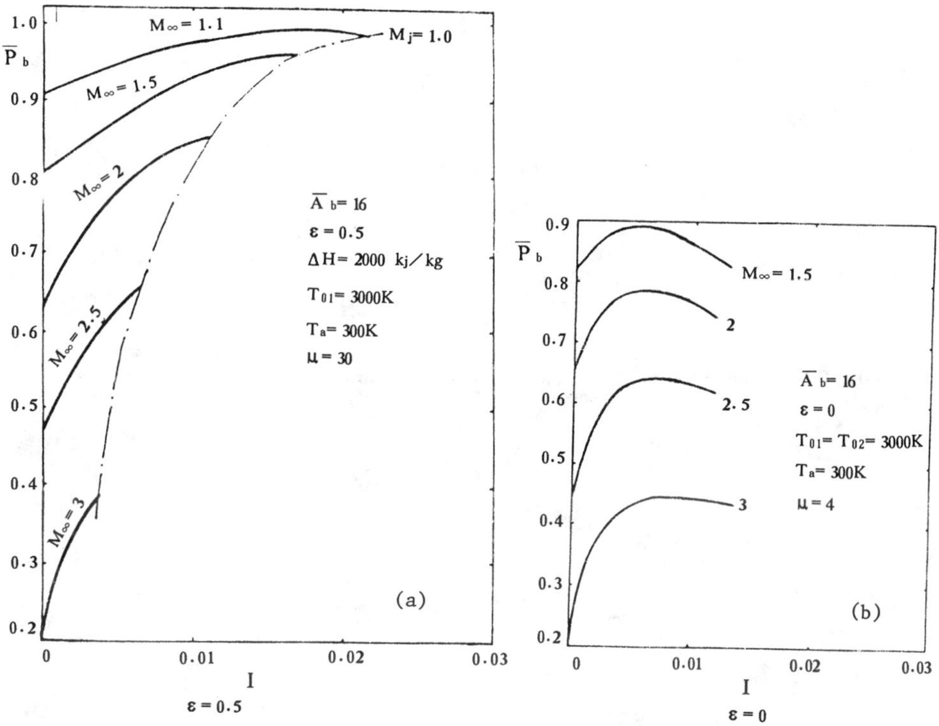

FIGURE 9. The effect of flight Mach number on base bleed

For the other case, $\overline{A}_b = 16$, $T_{01} = T_{02} = 3000K$, jet gas molecular weight $\mu = 4$. Four Mach numbers, 1.5, 2.0, 2.5, and 3.0 are calculated, the effective length of dead area are obtained with 11.3, 10.3, 9.8 and 9.4, respectively. Results are shown in Fig.9–b.

Fig.10 shows the wind tunnel test results, based on Qiu [8] in which Mach number was taken between 0.98 and 2.525.

Alternative test results one can find from Bowman [2] are quoted in Fig.11. Both Fig.9 of the present study and Fig.11 of experiments show the same tendency.

4.4 The Effects of Stagnation temperature in combustion Chamber

Taking pure gas with $\mu = 40$ and $\overline{A}_b = 7$, flight Mach number $M_\infty = 2$, $T_a = 200K$. The stagnation teperatures are 300K, 1000K and 3000K are considered respectively. Fig.12 shows the calculating results in which one can find the higher stagnation temperature in chamber will bring the more effective base bleed function that has been confirmed by plenty of experiments [1] [2].

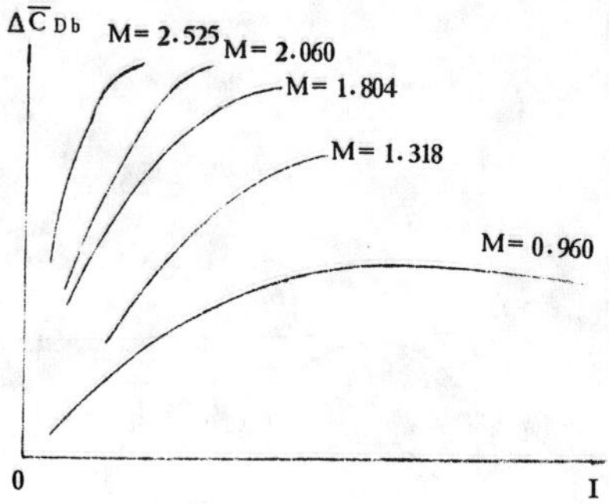

FIGURE 10. The effect of Mach number on drag reduction Qiu [9]

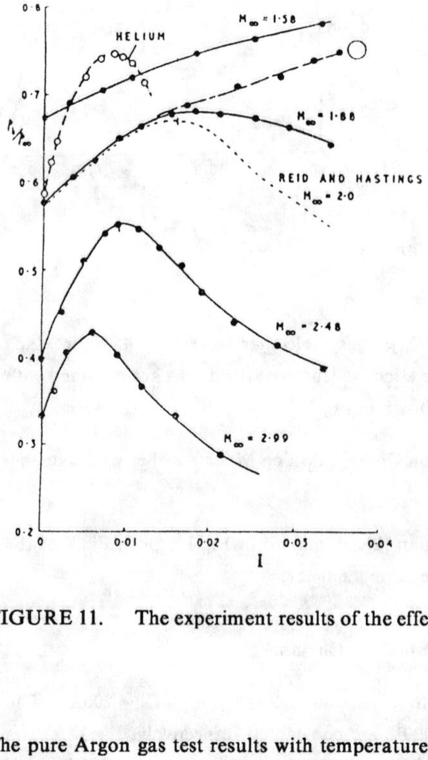

FIGURE 11. The experiment results of the effect of Mach number

The pure Argon gas test results with temperature of 290K and 300K are given in Fig.13–a. As neither M_∞ number nor surroounding tempperature T_a is reported in Bowman [2] the comparison with Fig. 13 is limited qualitatively, yet it is generally satisfactory. Five different temperatures of Argon gas were taken in the test of which the results are shown in Fig.13–b.

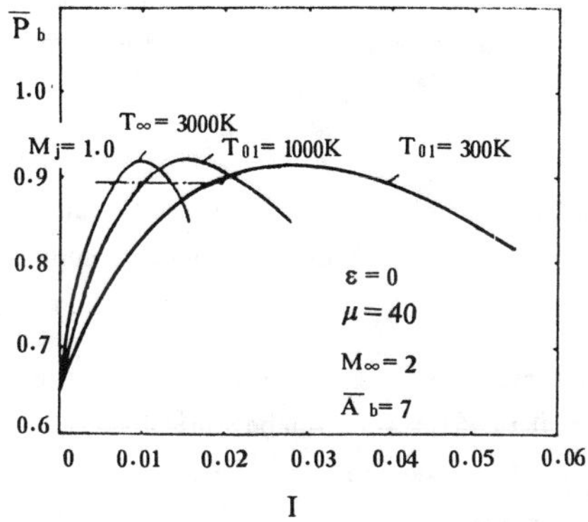

FIGURE 12. The effect of stagnation temperature in chamber on base bleed function

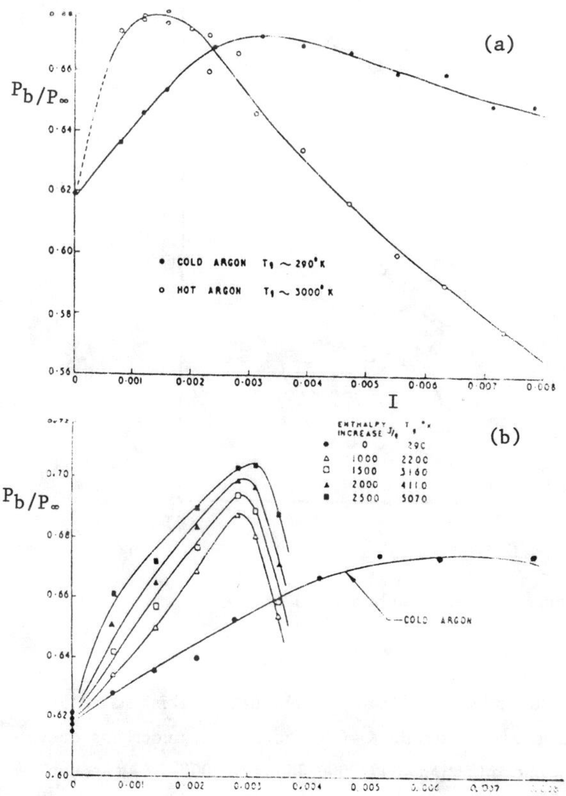

FIGURE 13. The effect of stagnation temperature on base pressure, [2]

4.5 The Effects of Solid–Phase Fuel Particle Mass Ratio and the Fuel Enthalpy of Combustion ΔH

In the calculating, taking $\bar{A}_b = 16$, $M_\infty = 2.0$, chamber stagnation temperature 2000K, jet gas molecular weight $\mu = 30$, abiabatic exponent $k_g = 1.3$, fuel enthalpy of combustion $\Delta H = 3000 kJ / kg$.

Two cases of solid–phase mass ratio $\varepsilon = 0.5$ and 0.8 are considered. The results are given in Fig.14, from which one can find that the increasing of the base pressure will be more obvious for the larger mass ratio case. One can also find that the base pressure may be higher than surrounding atmosphere pressure in the case of certain large value of I.

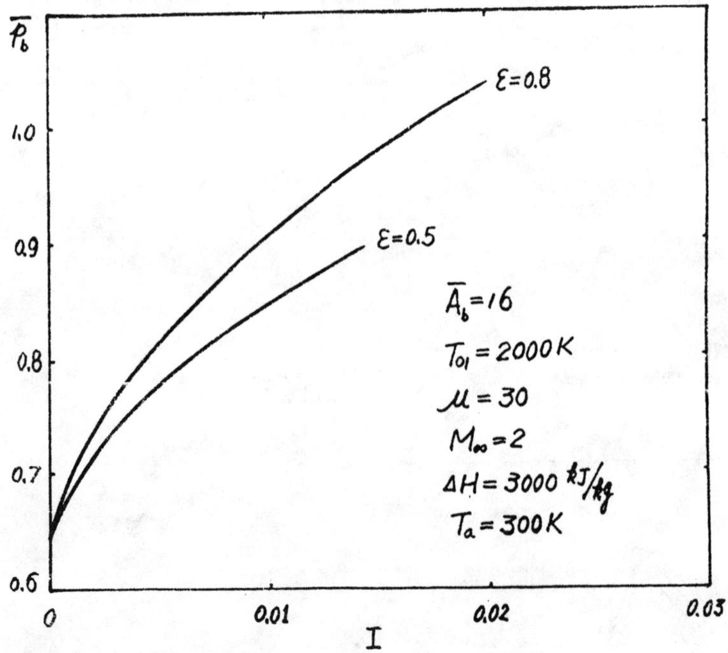

FIGURE 14. The effect of solod–phase mass ratio on base bleed function

The next calculation is for the case of alternative fuel enthalpy of combustion, taking stagnation temperature $T_{01} = 1000K$, and soled–phase mass ratio $\varepsilon = 0.5$, and the rest underlying data same as before. Four cases of enthalpy of combustion 1000, 2000, 3000 and 4000kJ / kg are considered in the calculating. (see Fig.15) The case with higher fuel enthalpy of combustion will be followed by better base bleed effectiveness. Fig.16 is quoted from Din [10].

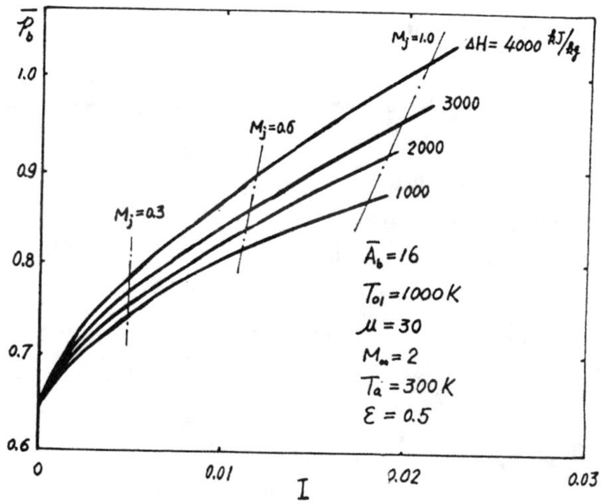

FIGURE 15. The effect of fuel enthalpy of combustion on base bleed function

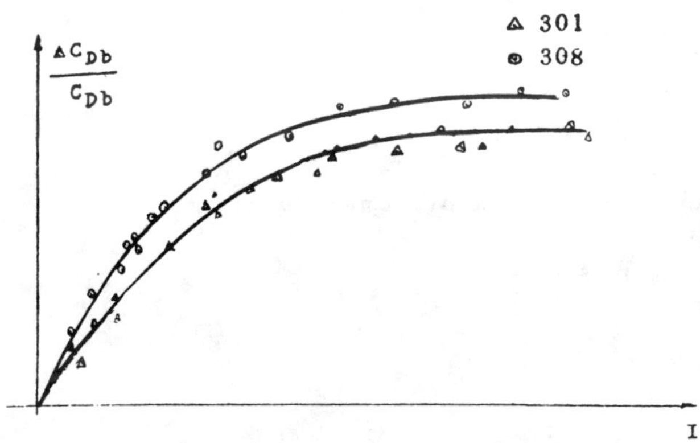

FIGURE 16. Experiment results of two different fuels

5. CONCLUSIONS

A. A practical base bleed base combustion model has been developed. In the present model the following factors have been considered: base bleed exit area, flight Mach number, stagnation temperature in chamber, solid mass ratio in jet, fuel enthalpy of combustion and jet gas molecular weight.

B. In the case of small jet mass flow rate, the effect of exit area on base pressure is negligible. In the case of certain mass flow rate, the larger exit area will be followed by better base bleed effectiveness.

C. The lower jet gas molecular weight of a base bleed structure will be accompanied by a higher base bleed effectiveness.

D. The acceptance of base bleed structure is encouraged at higher flight Mach number, in which condition the larger absolute value of $(\partial \bar{P}_b / \partial I)_{I=0}$ is obtained.

E. When the stagnation temperature in the chamber is rising, higher performance base bleed will be reached.

F. Either high value of solid phase mass ratio ε, or high value of fuel enthalpy of combustion will promote the base bleed function.

LIST OF SYMBOLS

A	Section area
$\bar{A}$	$\bar{A} = \frac{A}{A_1}$, Nondimensional area
C_p	Fuel specific heat
$(C_p)_g$	Gas specific heat
H	Heat enthalpy (kJ / kg)
ΔH	Fuel enthalpy of combustion (kJ / kg)
I	Mass flow rate parameter
k_{m1}	Equivatent adiabatic exponent of gas–solid equilibrium mixture
kg	Gas adiabatic exponent
L	Length of cylinder; Length of dead area
M	Mach number
$\dot{m}$	Mass flow rate
P	Pressure
R	Universal gas constant
R_{m1}	Equivalent gas constant of gas–solid equilibrium mixture
R_2	Gas constant of jet gas at section A_2
r	Radial coordinate
u	Flow velocity
ε	Solid phase mass ratio
ρ_p	Fuel density in gas–solid mixture
ρ_R	Gas density in gas–solid mixture
ρ_m	Gas–solid mixture density
μ	Gas molecular weight
β	Semi–angle of conic dead area

Subscripts

1	Exit section
2	End section of dead area
a	Atmosphere
ave	Average value, see Eq.(11)
b	Base section
eff	Effective
g	Gas phase
m	Mixture
s	Side surface of dead area
o	Stagnation condition
∞	Free stream

REFERENCES

1. Murthy, S.N.B. at al "Aerodynamics of Base Combustion". MIT Press, 1974
2. Bowman, J.E. et al."Reduction of Base Drag by Gas Ejection". R.A.R.D.E. Report 4 / 69, 1969
3. Sahu, J. "Numerical Computions of Supersonic Flow over Cylindrical Afterbodies with Base Bleed", 9th ISB, 1986
4. Schilling, H. " Simple Theoretical Approach for Base Bleed Calculations", 9th ISB, 1986
5. Wang, Nan–yan. "A Analysis on the Impulse Loss of Solid–Propellant Rocket Engines", ACTA ARMAMENTARII (in Chinese) vol.3 1987.
6. Klasnov, N.V. "Aerodynamics of Revolution", Science Press, (translated into Chinese) 1965
7. Gou, X.F "Base Bleed Exterior Ballistics" (in chinese), ECIT, 1986
8. Qiu, Guang–chun et al. " A Study on Base Combustion Drag–reducing Properties and Aerodynamic Estimate Method", Ballistics (in Chinses) vol.1, 1987
9. Din, Ze–sheng et al. " An Experimental Study on Supersonic Base Combustion Drag–reducing", Ballistics, (in Chinese) vol.2 1985.

COMMENTS

K.K. Kuo, Pennsylvania State Univ., USA. You have presented a very interesting paper. The two–phase effect in terms of ε (solid–phase to mixture mass ratio) has been considered at various places in the model, however, in the momentum equation (2), you did not separate the momentum of particle phase from that of the gas–phase. Did you assume that there is no slip velocity between the two different phases? Also, the values of ε in the range of 0.5 and 0.8 in Fig.14 and 15 seem high for non–metalized propellants, would you please comment on this point?

Author's Reply.

$$V_s\left(\frac{dV_s}{dx}\right) = F(V_g - V_s) \tag{1}$$

$$\frac{V^2}{L} \qquad F \cdot \Delta V$$

where F is reciprocal relaxation time. Under the Stokes' law,

$$F = \frac{18\mu}{d_s^2 \rho_s} \tag{2}$$

From Eq.(1), one can see the slip velocity ratio $\frac{\Delta V}{V}$ has the order of $\frac{V}{LF}$. Suppose tipical values: $\mu = 6 \times 10^{-5}$ N.S / m^2, $d_s = 2 \times 10^{-6}$ m, $\rho_s = 1000$kg / m^3, L = 5cm, V = 500m / s, one can get $\frac{\Delta V}{V}$ $\approx 4\%$. So the assumption of no slip velocity between the 2 phases will not make sabstantial error in the result.

The pyrotechnic with equivalent ratio $\Phi = 1.6$ (rich fule) is used in our tests. In this case, around 50% of pyrotechnic fuel has to be burnt owtside the chamber, so that value of $\varepsilon = 0.5$ is reasonable. And $\varepsilon = 0.8$ is also acceptable for comparison purpose.

Base Burn Projectile French Trajectory Model

D. CHARGELEGUE and M. T. COULOUMY
Direction des Armements Terrestres
Etablissement Technique de Bourges
France

ABSTRACT

The basic trajectory model is the standardized Modified Point Mass trajectory model where, in the first phase, the base drag reduction and the gas generator mass flow rate are calculated. The burning rate is expressed as a function of the upstream atmospheric pressure and operating generator temperature. The base drag reduction coefficient is expressed as a function of Mach number and characteristic mass flow rate. Following several firing campaigns, the coherence of the model with firings has been verified.

1. INTRODUCTION

Most of the studies on base-burn conducted in France concern special areas such as aerodynamics, constitution and combustion of propellant or interference between combustion and flow in the vicinity of the projectile base. These studies are very interesting in the aim of such projectiles design and development. Our purpose is different, we have to establish firing tables for the users. So we need a trajectory model which includes base-burn particularities. This model sould be also used by the army in artillery calculators and consequently it must not be too sophisticated.

Hellgren [1] first established such a range calculation program, but it calculates motor pressure and it needs many input data, some of which like base drag coefficient are very difficult to know.

We think that a simpler model can be used. The basic model is a modified point mass trajectory model [2] which is generally used in the preparation of firing tables. This basic model is completed by specific terms which modelise base-burn main aspects : drag reduction and mass flow rate.

Drag reduction and mass flow rate can be written in the following general expressions :

$$R_{BB} = 1/2\ \rho\ SV^2\ Cx_{BB}\ (M, I, P, \theta_p, ...) \quad (1)$$

$$dm/dt = V_C\ (M, I, P, \theta_p, ...)\ \rho_p\ S_C\ (m_{CB}) \quad (2)$$

The time, for which the burnt propellant mass which is calculated by equation (2) is equal to the initial mass of propellant, is an output data. This time can be compared to the measured burning time and is also the separation between the two phases of the trajectory. The terms (1) and (2) take part into motion equation system only in the first phase, during which the drag reduction occurs.

From the general formulation, the problem is to identify parameters which affect drag reduction coefficient Cx_{BB} and burning rate V_C and finally to determine the functions of these parameters. These functions must be accurate enough to provide calculated trajectories and burning time suitable to measured trajectories and burning time whatever the firing conditions (wind, muzzle velocity, elevation, cannon altitude, etc...) are.

To this end, specific tests and corresponding suitable means of measurement have been set, and are described in §2. Chapters 3 and 4 explain the way of interpreting the results and getting the modelisation of the mass flow rate (chapter 3) and the drag (chapter 4) before concluding.

2. TESTS

2.1 Tests Planning

To identify the parameters affecting the drag reduction and burning time, the principle of these tests is the following : for two identical values of a single parameter, the other parameters which can affect the drag coefficient or combustion rate are different.

Two French base-burn 155 mm projectiles have been tested: the first one in June 1986 (155 RTC) and the second one in September 1987 (155 DTC) . The test planning was the same for the two campaigns but the second one took advantage of many means of measurement and especially a trajectory radar.

According to the principle above stated, the test planning is defined as :

25 rounds with propellant block at 21°C
    5 charges x 5 elevations
25 rounds without propellant block
    5 charges x 5 elevations
24 rounds with propellant block at different temperatures
    4 temperatures : -31°C, -5°C, 36°C, 51°C,
    for each one, 3 charges x 2 elevations.

2.2 Means of Measurement

The measurements made during firing tests provide all the data used as inputs and as parameters which allow to develop and adjust the model.

Before each firing, the following measurements are made:
- inertial characteristics of each projectile : mass, position of the centre of gravity, moments of inertia
- for each round : elevation and azimut angles, time of firing, projectile temperature .

It is also very important to know as accurately as possible all the meteorological parameters (air pressure, temperature, humidity, and wind velocity and direction versus altitude). One sounding per hour is required, then wind data are computed by a model of statistic interpolation versus time and distances. So this statistical model can provide sounding as trajectory model input data for each round.

During its flight, the projectile is fully tracked by a trajectory radar, and in addition to that, is tracked by Doppler radar during the beginning of the trajectory (untill roughly 20 s of flight).
Each round is also equiped by a radio-electronic fuze to deliver burning time. Some projectiles carry yaw sondes.The muzzle velocity is measured by two independent means of measurement.

3. MASS FLOW RATE AND BURNING TIME MODELISATION

3.1 Calculation of Mass Flow Rate

The so called "burning time" calculated by the model is the very moment for which the last propellant gramm has been burnt. So it is linked to the mass flow rate which can be expressed by equation (2).

At any given time, the mass of burnt propellant is :

$$m_{CB}(t) = m_0 - m(t) \tag{3}$$

where m_0 is the mass of the round when fired
$m(t)$ is the mass of the round at time t
and we know that at initial time i.e. $t = 0$, $m(0) = m_0 - m_{CB0}$
where m_{CB0} is the burnt propellant mass in the barrel and depends on the charge.
At each step the burning area S_C is calculated with a linear fonction of the burnt propellant mass m_{CB} :

$$S_C = a_i + b_i\, m_{CB} \quad \text{for} \quad m_{CB\,i} \le m_{CB} \le m_{CB\,i+1} \,, \tag{4}$$

generally six linear fonctions (i.e i=0 to 5)are needed to describe in a suitable manner the evolution of the burning area .
Note that ρ_p and the function $S_C(m_{CB})$ is provided by the propellant manufacturer.

3.2 Measurements

First before the campaign, some tests have been conducted on propellant in pyrotechnical laboratory. Particularly we are interested in combustion rate obtained from the strand-burner under several pressures and temperatures. These data may not fully representative of the combustion during round flight, but can allow to determine the combustion rate of reference V_{C0} (at standard pressure and 21 degres celsius) for the lot of the tested propellant blocks . This also give some idea about the behaviour of burning rate with pressure and temperature.
During round flight, the burning time is directly measured by using special fuze.In fact, each projectile with propellant block is equiped with a radio-electronic fuze which emit a signal. As long as the combustion goes on, this signal is perturbated, which permits to obtain the burning time with an accuracy estimated at about 0,1s (see figure 1)

Figure 1 : 155 DTC Measured burning time

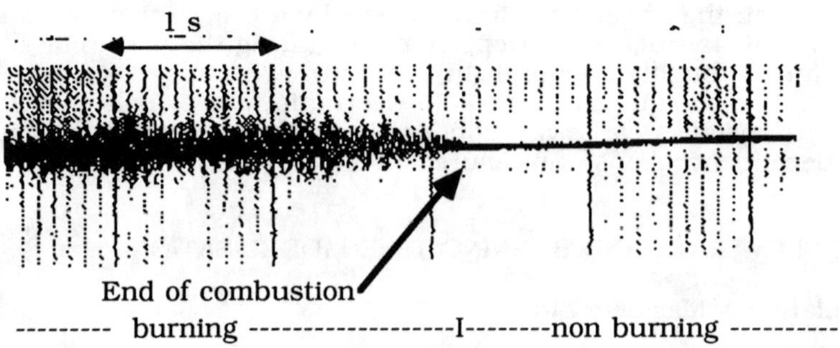

3.3 Identification of Parameters Affecting Burning Time

We have found good correlation of combustion rate with upstream pressure and also propellant temperature, as shown by the following results obtained through the two firing campains.

<u>Correlation with pressure</u>. The rounds with propellant block at 21°C have been exploited with the model where mass flow rate is calculated as :

$$dm/dt = - K(M, I, P, \theta_p, \ldots)V_{C0}\ \rho_p\ S_C\ (m_{CB})\ . \tag{5}$$

For each round, K is adjusted to match the measured burning time. Of course the model prints at every step the following data : Mach number M, characteristic mass flow rate I ($I = (dm/dt)/\rho SV$), air pressure P; it also calculates the average pressure P_{moy} , the average Mach number M_{moy}, and the average characteristic mass flow rate I_{moy} over the whole base-burn phase.
So we can draw K as a function of M_{moy}, P_{moy}, or I_{moy}.

As shown on figure 2-1 and 2-2 using a logarithmic scale , we have only found correlation with upstream pressure with a law on the well known form[3]:

$$K = A.P^{\alpha} \quad (6)$$

Fig. 2-1 155 RTC Combustion rate versus pressure

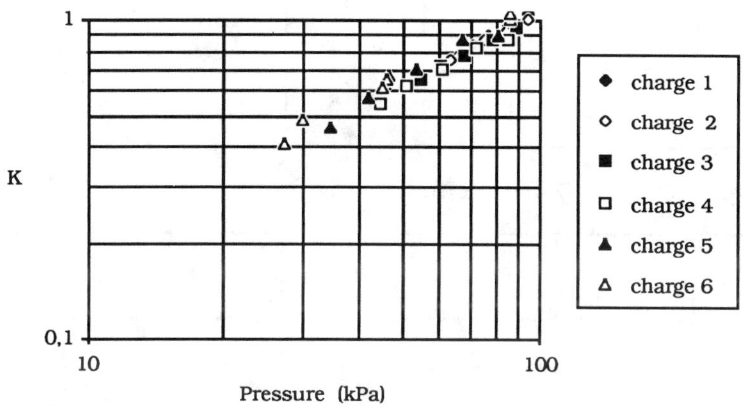

Fig. 2-2 155 DTC Combustion rate versus pressure

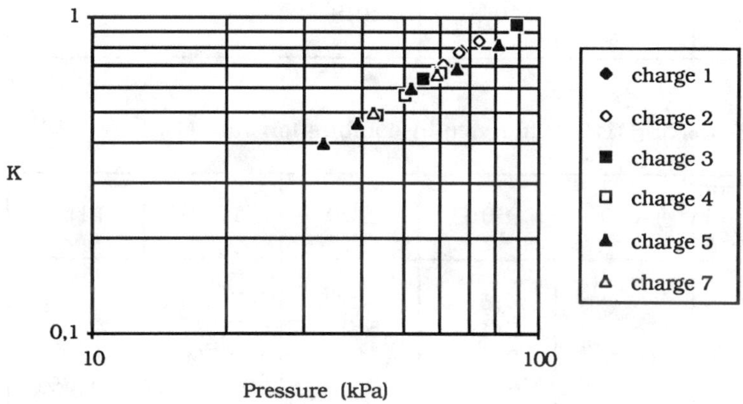

Simulations are done with for mass flow rate expression :

$$dm/dt = - A.P^{\alpha} V_{CO} \; \rho_p \; S_C \; (m_{CB}) \quad (7)$$

Table 3-1 represents the corresponding computed burning time and Table 3-2 represents the difference between computed and measured burning time. The corresponding difference in combustion rate (in %) and the consequences on the range (in meters) are respectively shown in Tables 3-3 and 3-4.

Table 3-1 155 DTC Computed burning time (s)

Elevation mil. / charge	400	600	800	1000	1150
1		30,8	33,3	36,4	38,9
2		31,7	35,3	39,1	41,7
3	29,3	32,6		42,2	45,9
4		34,7	41,4	48,7	55,1
5	31,5	38,2	47,4	60,5	71,0
7		41,8	55,6		

Table 3-2 155 DTC Difference in burning time (s)

Elevation mil. / charge	400	600	800	1000	1150
1		1,3	2,3	2,8	3,1
2		1,6	2,2	3,1	2,2
3	0,2	0,0		0,4	2,5
4		-0,6	0,3	-0,3	-1,1
5	-1,7	-1,1	1,1	1,2	2,0
7		0,5	0,8		

Table 3-3 155 DTC Difference in combustion rate (%)

Elevation mil. / charge	400	600	800	1000	1150
1		4,6%	6,9%	7,4%	7,2%
2		5,4%	5,9%	7,3%	4,5%
3	0,8%	0,0%		0,7%	4,4%
4		-1,5%	0,6%	-0,4%	-1,6%
5	-5,1%	-2,9%	1,9%	1,6%	2,4%
7		1,0%	1,2%		

Table 3-4 155 DTC Difference in downrange (m)

Elevation mil. / charge	400	600	800	1000	1150
1		9	20	24	15
2		16	28	26	12
3	1	0		3	15
4		10	7	0	5
5	23	19	19	14	9
7		22	15		

One may note that for the largest charges the differences are less than 2 s for burning time, 5% for combustion rate, and 22 m for range, which is very satisfactory because it is about the dispersion we can expect with propellant blocks.
For smaller charges the differences are larger but base bleed projectiles will be probably used essentially with larger charges.

<u>Correlation with propellant temperature</u>. In this case, the rounds with propellant blocks at different temperatures were exploited with the model where mass flow rate is :

$$dm/dt = - K\, A.P^{\alpha}\, V_{CO}\ \rho_p\ S_C\ (m_{CB}) \tag{8}$$

K is adjusted to meet the measured burning time and drawn versus θ_p, this graph represented in figure 4, permits to determine the function of propellant temperature which is a third degree polynome :

$$K(\theta_p) = a_0 + a_1\theta_p + a_2\theta_p^2 + a_3\theta_p^3 \tag{9}$$

Fig. 4 155 RTC Combustion rate versus propellant temperature

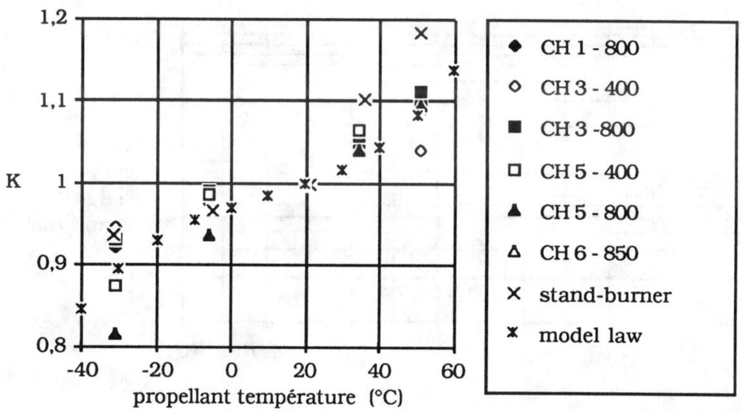

Then as before, we can compare computed and measured burning times (Table 5).

Table. 5 155 RTC Combustion rate versus propellant temperature
Difference between computed and measured burning time

Charge	Elevation mil.	Propellant Temperature -31 °C	-6,4 °C	34 °C	50,7 °C
3	400	0,4	-0,7	-0,6	-2,6
5	400	-3,0	-0,5	-1,0	-1,9
1	800	2,4	1,6	1,6	1,1
3	800	1,0	0,3	0,3	0,0
5	800	-4,0	-0,7	1,4	1,2
6	850	1,0	---	---	2,1

The difference between computed and measured burning times is generally higher at the above temperature than those at 21°C. The dispersion in burning rate seems also to be higher, especially for the higest and lowest temperatures tested. So we think that the modelisation of the combustion rate versus temperature is good enough with a third degree polynome but as this formulation has no physical meaning , as indicated by several experts in chemical propulsion ,we will use in the future a law on the well known form[3]:

$$K(\theta_p) = e^{a(\theta_p - \theta_{p0})}. \qquad (10)$$

As shown on the figure 6 we found a good correlation with such a law .

Fig. 6 155 DTC Combustion rate versus propellant temperature

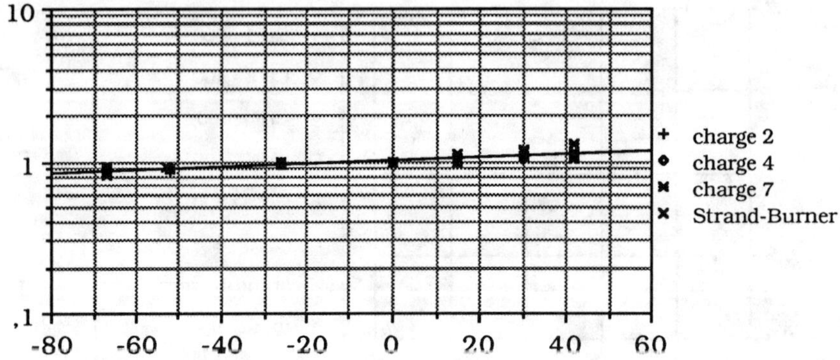

4. DRAG REDUCTION MODELISATION

The additionnal term in motion equation due to drag reduction can be written in the following general expression :

$$R_{BB} = 1/2 \ \rho \ SV^2 \ Cx_{BB} (M, I, P, \theta_p, ...) . \quad (11)$$

The identification of parameters and their functions which affect drag reduction is allowed by the great diversity of firing conditions and the suitable means of measurements described in §2.

4.1 Measurements

At the begining of its flight (about 20 s) each projectile is tracked by a Doppler radar which permits to measure velocity as a function of time. Then these data are exploited by a program developed by ETBS [4] which provides drag coefficient Cx as a function of Mach number. (One value of Cx is calculated for one Mach number per second, so about twenty values of Cx are calculated for twenty values of Mach number for each round)
The decrease of the mass flow rate occurs after the end of the Doppler radar tracking. So in order to observe the effects of the mass flow rate on drag reduction at the end of the base-bleed phase, we must use a trajectory radar tracking.

4.2 Identification of Parameters Affecting Drag Reduction

Correlation with Mach number. Figure 7 presents global drag coefficients calculated from doppler radar data for DTC projectiles with propellant block at 21°C (Cx_{ABB}) .We can see that Cx_{ABB} only depends on Mach number because the dispersion obtained on these coefficients is not greater than it is for a classical projectile.

Fig. 7 Drag coefficient vs Mach number with propellant block at 21°C

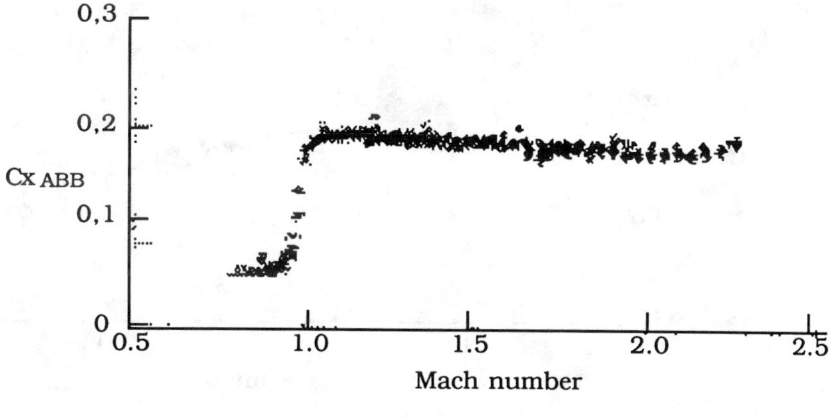

As the propellant temperature affects combustion rate, one may think that drag will be affected. This is not the case as shown on the graph in figure 8 which represents drag coefficients calculated from doppler radar data for DTC projectiles with propellant block at different temperatures (Cx_{ABB}). Temperature effect does not increase the drag dispersion. These two results indicate that an optimal mass flow rate does exist above which the global drag coefficient only depends on Mach number.

Fig. 8 Drag coefficient vs Mach number with propellant block at different temperatures

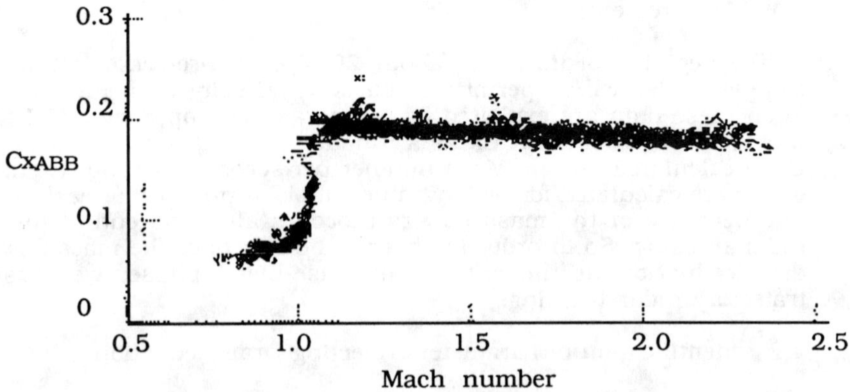

Drag coefficient for DTC rounds without propellant block (Cx_{SBB}) is drawn in figure 9.

Fig. 9 Drag coefficient vs Mach number without propellant block

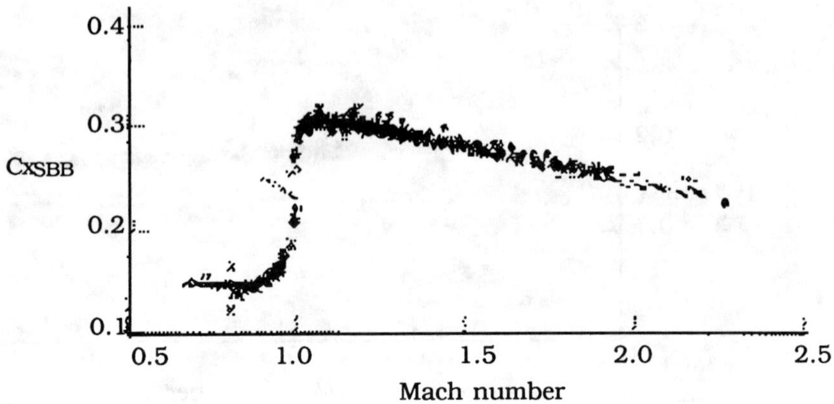

Then the drag reduction coefficient Cx_{BB} used in the formula (11) §4 is calculated as :

$$Cx_{BB} = Cx_{SBB} - Cx_{ABB} \quad (12)$$

The three graphs Cx_{ABB} , Cx_{SBB} and Cx_{BB} as functions of Mach number were plotted on figures 10-1 and 10-2 for each of the two tested projectiles. Cx_{BB} (M) is expressed as some polynomial functions of Mach number (degree < 4).

Fig 10- 1 155 RTC Drag coefficients versus Mach number

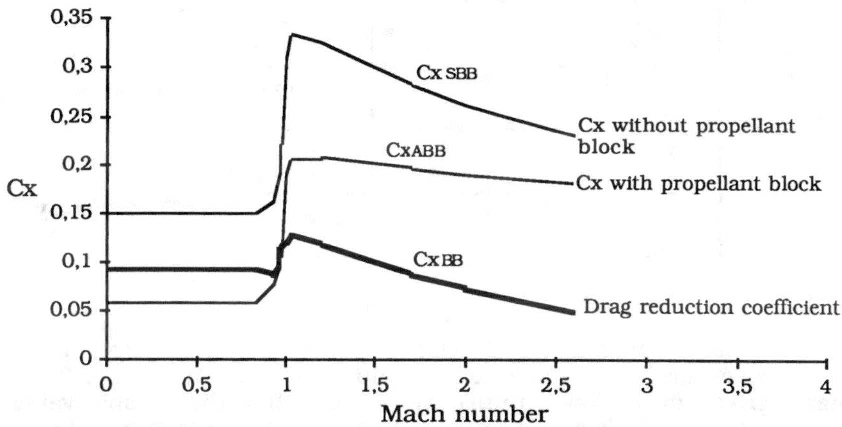

Fig 10-2 155 DTC Drag coefficients versus Mach number

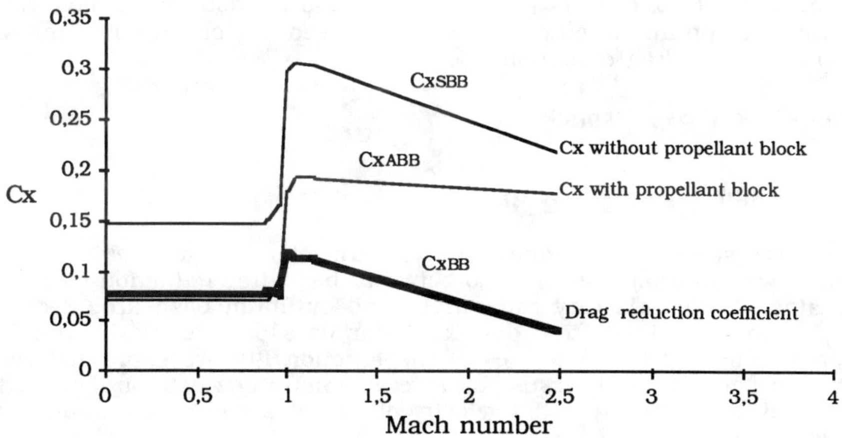

Influence of characteristic mass flow rate. During the "emptying phase" of base-burn motor, it is clear that drag reduction decreases progressively. As a matter of fact, the projectile drag goes from the Cx_{ABB} curve to the Cx_{SBB} curve.

The second firing campaign, with trajectory radar tracking, permit us to show the influence of characteristic mass flow rate as we can see for exemple on the figure 11 .

Fig. 11 Measured base drag reduction coefficient from trajectory radar - characteristic mass flow rate influence .

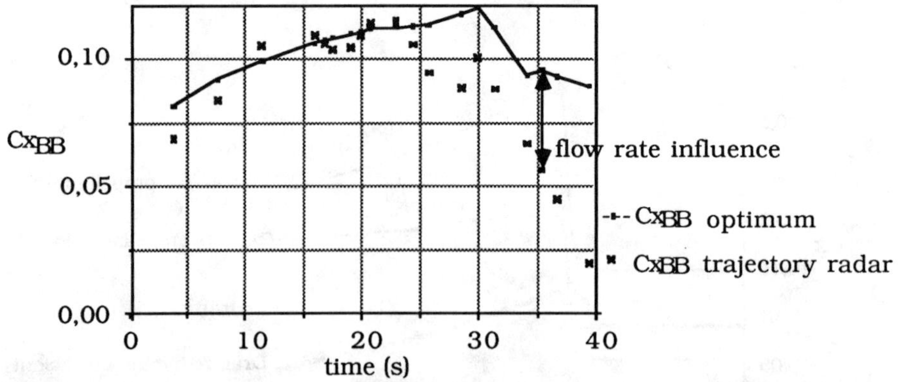

As shown on the figure 10 , the characteristic mass flow rate is greater,or at least equal , than the optimal value until about 25 s . Then the characteristic mass flow rate become smaller than the optimal value and the drag reduction decrease until 0 at the end of the burning phase wich occurs at about 40 s . Obviously the time lenght of these two parts of the base bleed phase trajectory depends on the firing conditions but we verified that the "emptying phase" is always significant .

Comparisons between experimental and calculated impacts from the first firing campain allows us to point out a characteristic mass flow rate I_0 permitting optimum efficiency of base-bleed motor and the following expression for drag reduction :

$$R_{BB} = 1/2 \ \rho S V^2 \ Cx_{BB}(M) \, f(I) \qquad (13)$$

where $f(I) = I/I_0$ if $I \leq I_0$
$f(I) = 1$ if $I \geq I_0$.

From the second firing campaign, with trajectory radar tracking,for each round we can calculate the ratio between base drag reduction coefficients obtained from trajectory radar data and optimum base drag reduction coefficients obtained from doppler radar data at same Mach numbers. These ratio , wich give values of the function f(I), are expressed versus time on figure 12 and versus characteristic mass flow rate on figure 13 for a round fired with an initial velocity of 700 m/s and an elevation of 600 mils.

Fig. 12 f(I) (CxBBr/CxBBopt) versus time.

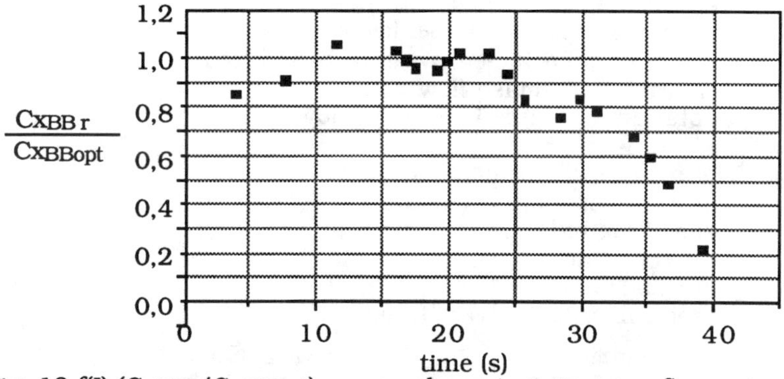

Fig. 13 f(I) (CxBBr/CxBBopt) versus characteristic mass flow rate .

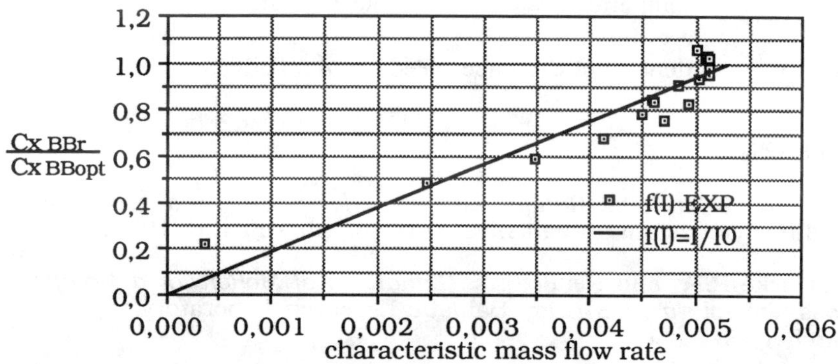

As shown on the figure 13 , the reduction of the results obtained during the second campaign seems to confirm the results from the first campaign and , up to now , do not permit to find a better modelisation of the caracteristic mass flow function f(I).

5. CONCLUSION

These test campaign results have pointed out that one can consider that upstream pressure and propellant temperature are the main parameters which affect the mass flow rate.
By another way, as long as characteristic mass flow rate is greater than an optimal value, drag reduction only depends on Mach number and is independant of propellant temperature. During the emptying phase, characteristic mass flow rate becomes lower than this optimal value. The function of characteristic mass flow rate used for modelling the falling off of the drag reduction, which appears during this phase is still not completely defined. Nevertheless comparisons between computed and experimental trajectories and burning time conducted up to now are very satisfactory.

NOMENCLATURE

Cx_{ABB}	global drag coefficient with base bleed
Cx_{BB}	base drag reduction coefficient
Cx_{SBB}	global drag coefficient without base bleed
I	characteristic mass flow rate
I_0	optimum characteristic mass flow rate
M	Mach number
m_{CB}	burnt propellant mass
P	local atmospheric pressure
R_{BB}	base drag reduction
S	reference area
S_C	burning area
V	projectile velocity w. r. t. air
V_C	local burning rate ($V_C = K(\theta_p).A.P^{\alpha}V_{C0}$)
V_{C0}	standard burning rate in standard conditions (pressure , temperature)
θ_p	temperature of the base bleed grain
ρ	air density
ρ_p	density of the base bleed grain

REFERENCES

1. Hellgren Rune v., *Range calculation for Base Bleed projectiles* , Proc 6th Int. Symposium on ballistics, Orlando, pp 190-196, 1981.

2. Lieske R.F., and Reiter M.L., *Equations of Motion for a Modified Point Mass Trajectory* , US Army Ballistic Research Laboratory report n°1314, March 1966 .

3. Davenas A. , *Technologie des propergols solides* , Editions Masson 1989 .

4. Runfola Y., *Determination du Cx d'un projectile à partir de mesures de radar Doppler* , Colloque franco-allemand, MEBAL 85, Saint-Louis, rapport R 116/85, pp S.II.9, 1985.

COMMENTS

C. J. NIETUBICZ U.S. Army Ballistic Research Laboratory , USA. As was mentionned yesterday, a paper at the APDA meeting in San Diego, CA presented ground test data which shows an effect of spin on burning rate. This data obtained at BRL was for a specific propellant configuration What was the configuration (shape) of your propellant and is the effect of spin dependant on the configuration ?

Author' s Reply . The propellant block configuration was for the two projectiles tested, a classical one : a cylinder passed through by a three projecting pointed star hole. But the base shape was different. So the flow in the vicinity of the base and also the combustion were different, therefore the coefficients in the temperature and pressure functions must be determined for each projectile.
The effect of the spin may depend on the configuration, but we do not note any significant influence of the spin on the combustion rate from our tests.

•

A. BAROUCH Ramat Hasmaron , ISRAEL. You designed your unit so that drag reduction is independant of mass flow rate. But burning time is determined directly by the mass flow rate. Therefore, if the base bleed unit is cooled, range will increase. Has this conclusion been corroborated experimentally ?

Author's Reply . For cold or hot base-bleed unit, when you consider the range corrected from the muzzle velocity you can see few differences which depend on firing conditions.
This can be explain by two effects :
-the first one is the increase of the range for cold base-bleed unit because the base bleed phase is longer,
-the second one is the influence of the characteristic mass flow rate. In the begining of the trajectory,during which the propellant ignition depends on the temperature, the value of the characteristic mass flow rate is smaller than the value,called Io, which permit the optimum efficiency of the base-bleed motor.

•

A. BAROUCH Ramat Hasmaron , ISRAEL. If range depends on the base bleed temperature, how do you introduce the necessary correction in range table ?

Author's Reply . In the firing tables, to take into account base-bleed temperature, the range corrections are provided in a table F(III).

•

A. BAROUCH Ramat Hasmaron , ISRAEL. Since burning time depends on the ambiant pressure, which depends on the meteorogical conditions, it follows that in base bleed projectiles we have an additional effect of the meteorogical conditions on the range. How do you allow for it in the range table corrections ?

Author's Reply . There is no doubt for us that pressure should be introduced in firing tables for range corrections.
Up to now, we do not work on this specific topic.

•

E.M.G. COOKE Royal Ordonnance , U.K. Does the composition tested contain large particles ? Our experiments have shown 12% combustion rate enhencement in static spin tests. Our composition contains particles of average size 30 μm, with little or no particles greater than about 100 μm.

Author's Reply . The particles size is controlled by industry. In our tests, we never noticed significant variations in combustion rate.

•

Jim FLEMING , FCI , U.K. You mention that the rate of burning does not vary much between 60 and 250 rev/s. But what is the change between static conditions and 60 rev/s ?
E.M.G. COOKE Royal Ordonnance , U.K. Is there an enhencement in combustion rate between 0 and 60 rev/s ?

Author's Reply . We have tested many firing conditions :

- muzzle velocity from 370 m/s to 810 m/s
- elevation from 400 mils to 1150 mils
- propellant temperature from -31°C to +51°C

Therefore, the spin varied between 60 rev/s to 260 rev/s. As shown by the graphs (fig 2-1, 2-2, 3-1, 3-2 of our paper) represented the combustion rate versus pressure, we do not note any significant influence of the spin on the combustion rate.
Some tests have also been done on propellant in Stand-burner at different pressures. The corresponding points have been added on the graph below.
Up to now, our tests do not allow us to point out an influence of spin on combustion rate, nevertheless if it exists, it is much less significant than the upstream pressure or the propellant temperature influence.

Combustion rate as a function of pressure

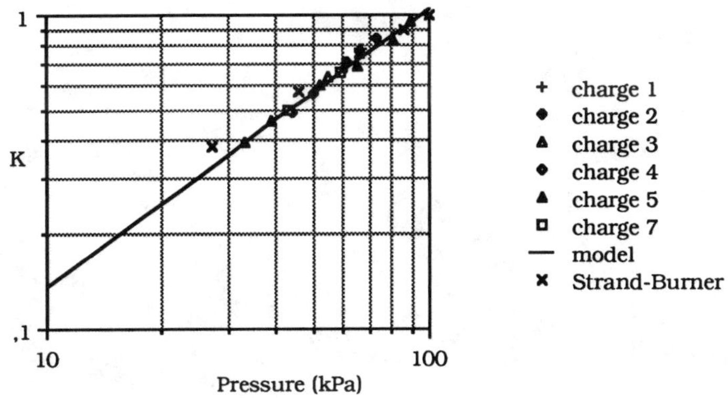

Trajectory Modeling for Base Bleed Projectiles

GERHARD LAUBE
Defense Technology Agency
Ballistic Division
Thun, Switzerland

ABSTRACT

This paper presents some ballistic results of a 155 mm Basebleed projectile which was evaluated in the years 87-88.
It explains a method of data reduction leading to a simple ballistic range calculation.

INTRODUCTION

Basebleed projectiles have become widely used as an efficient mean of increasing the range of artillery projectiles. Therefore there is a strong need for a trajectory model that fits into existing fire control computers. Such a model should avoid additional integration variables which would lead to longer calculation times and should fit into existing calculation techniques in order to keep software maintainance costs down. Our artillery fire control computer uses trajectory integration within a mass point model [1].

THEORETICAL MODEL

The well known "swedish model" for basebleed calculation, see eg. [2] uses a representation of the effective drag in the form

$$Cd_{eff} = f*Cd_{base} + Cd_{frict} + Cd_{front} \quad (1)$$

where the drag reduction factor depends on the mass flow I, the Mach number M, and projectile related geometrical data A. It is factorized into

$$f(I,M,A) = f1(I)*f2(M)*f3(I,M,A) \quad (2)$$

The mass flow depends on the burning law of the basebleed propellant and therefore on the chamber pressure which may be related to the ambient base pressure. Furthermore, the burning characteristics may be affected by the spin rate. Hence, mass flow calculation is rather complicated.

But, if a suitable initial guess of the burning time is available - depending on muzzle velocity, ambient atmospheric conditions and propellant temperature, f1 an f3 in formula (2) may be expressed as funtions of time instead as functions of mass flow. Moreover, with

$$g = (f*Cd_{base} + Cd_{frict} + Cd_{front})/Cd_{tot} \qquad (3)$$

the total drag Cd may be used as a reference to the drag reduction factor:

$$Cd_{eff} = g(t,M,A)*Cd_{tot} \qquad (4)$$

DATA COLLECTION

In order to get data samples covering a wide range of parameters, a test plan was set up covering firings from low zones (380 m/s) to high zones (915 m/s), elevations from 200 mils to 1100 mils. The purpose of using high elevations in the upper register was to expose the active projectile to low ambient pressure conditions which would influence the burning rate and burning time.

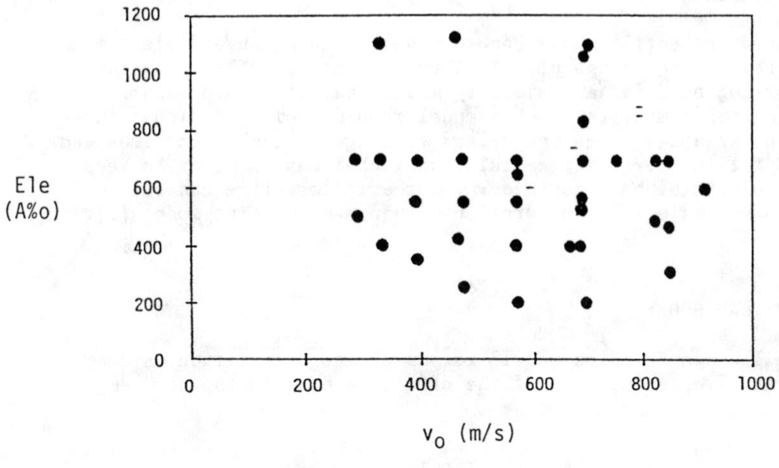

Figure 1: Summary of the main firing conditions

Unfortunatly, low atmospheric pressure and high Mach numbers tend to be correlated, because high Mach numbers are achieved by high muzzle velocities. High muzzle velocities generally give trajectories at high altitudes, i.e. at low air pressure. We tried to overcome this situation by firing at high elevation combined with low muzzle velocities, but it turned out, that the separation of Mach number and atmospheric effects will remain difficult. Variation of the mass flow parameter was also achieved by using rounds conditioned at low or high temperatures.

Inert shells, i.e. shells with inactive basebleed units, were used for reference purpose, so that the effectivness of the drag reduction could be observed in a direct comparison under the same conditions. Due to test site range restrictions, this direct comparison was not always possible, since not always two different target areas were available at the same time which would have been required for firings with both projectiles. A non-tracking Doppler radar was used to get the velocity data. Meteorological data were collected by standard met sonde technique.

DATA REDUCTION

The drag coefficient was caluculated from the doppler radar data using a point mass model. It is not necessary to calculate with decreasing mass, the aeroballistic effect of the mass decrease can be incorporated in the efficiency factor g. This figure shows raw drag data. Drag curves for different muzzle velocities. Of cause, a basebleed fired at 800 m/s will be burnt out long before the projectile reaches the subsonic region, whereas a basebleed fired at 400 m/s will give a reduced drag in the subsonic region too.

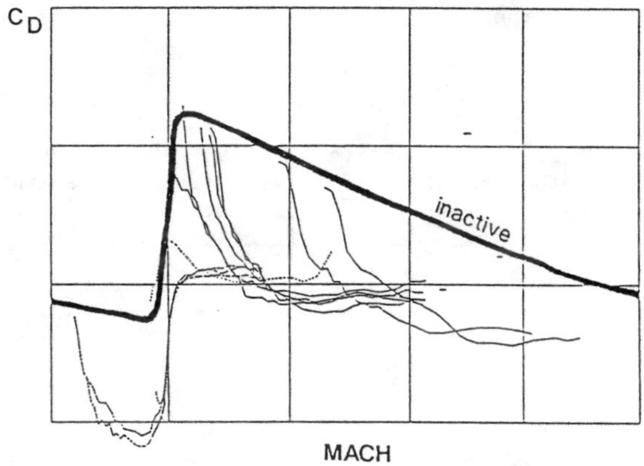

Figure 2: Measured drag data from basebleed projectiles.

In the next step, the drag reduction factor is derived as ratio of $g = Cd_{active}/Cd_{inactive}$, see Figure 3.

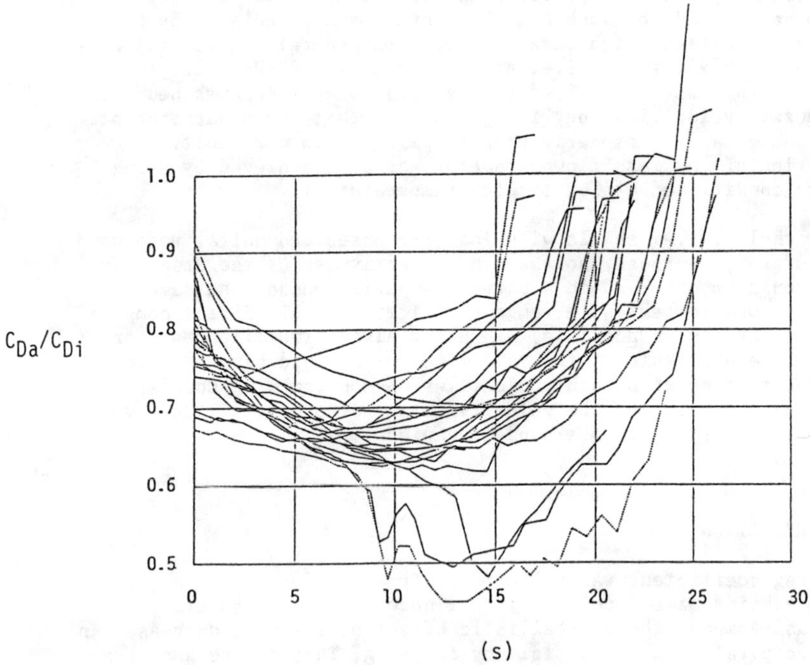

Figure 3: Drag reduction factor as as function of time, for various Mach numbers.

For the interpretation of the data, burnout time is very important. Optical observation of burnout time turned out to be rather erratic. A still active unit may have no optical signature and, on the other hand, ineffective burnout of remaining propellant may be clearly visible. Therefore, we decided to define the burnout time as the end of efficiency, measured by the doppler radar.

Now, having the burnout time t_B or a suitable guess of it, we are able to rearrange the drag reduction data as a function of the relative burning time t/t_B. This procedure eliminates the variation of burnout time due to the various influencing factors, such as high propellant temperature and high spin rate which result in an increase of the burning rate, and therefore in a decrease of the burning time. Figure 4 shows the result of this reduction step.

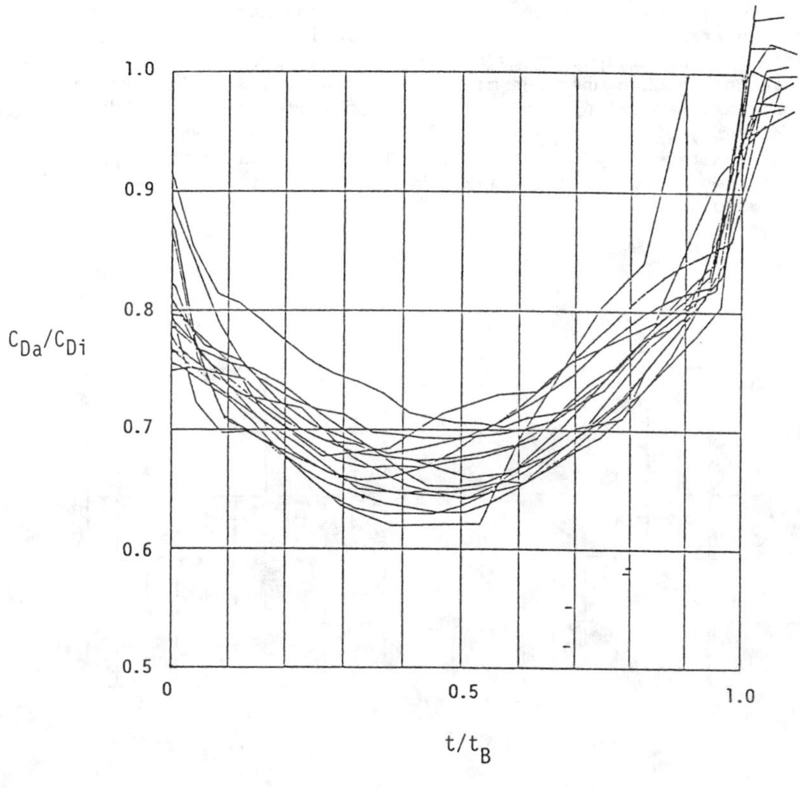

Figure 4: Drag reduction factor vs relative burning time t/t_B

Obviously, the main effect - the effect originating from the mass flow - has been eliminated. The effectiveness, i.e. the drag reduction factor, of a well designed basebleed unit is not very sensitive to variations in the mass flow. It has to be outlined, that Figure 4 contains also measurements on high and low temperature conditioned projectiles, for which the burning time varies more than ten percent. The mass flow will change by approximately the same amount, but the effectiveness varies by far less than ten percent.

Remember that, because of the usage of the total drag as a reference (4), combined effects of mass flow and Mach number have to be expected. It becomes now difficult to identify these remaining effects. Higher spin rates result in higher burning rates. Therefore both Mach number and mass flow correlate with high muzzle velocities. Separating the drag reduction factor curves for different zones gives a separation of Mach number and mass flow parameter simultaneously. The resulting drag reduction curves separated for zones 8 and 10 (700 m/s and 920 m/s respectively) are shown in Figure 5-6.

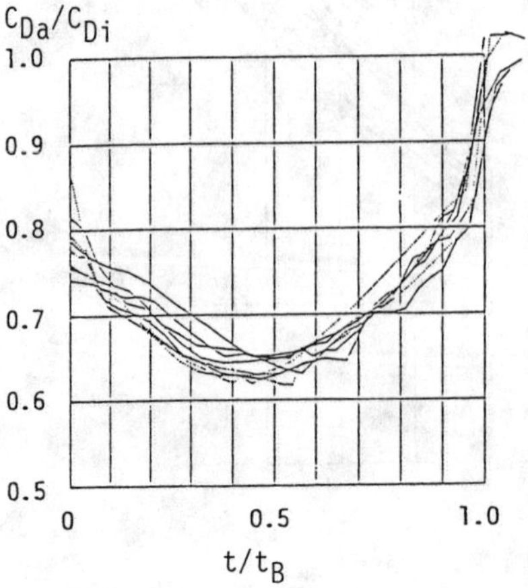

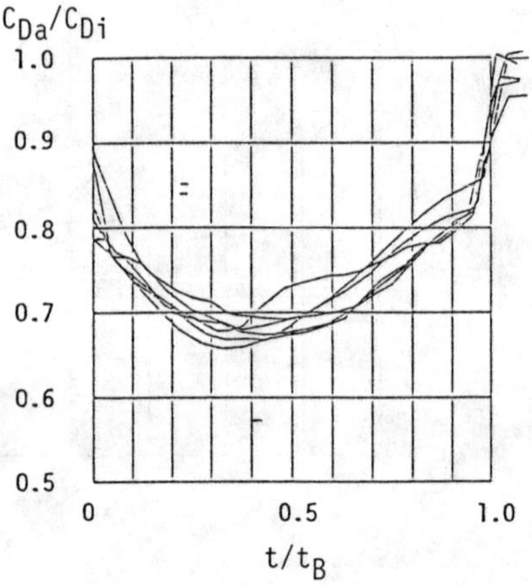

Figure 5:
Drag reduction for zone 8

Figure 6:
Drag reduction for zone 10

BURNING TIME

Up to now, an empirical formula based on test firing data is used, a function of muzzle velocity and elevtion angle. Simple approaches of correlating burning time with free air trajectory data were not successful.

Firing tests, where base pressure and basebleed unit chamber pressure are to be recorded in flight, are in preparation.

ACCURACY OF THE MODEL

Errors in the burning time (as a result of a bad guess or an inappropriate formula) will result in a range error and a time-of-flight error. It has been calculated, that a ten percent error in burning time, using unaltered drag reduction data results at zone 8 only in a 200 meters range error. Hence, fortunately, the problem of burnout time is not that delicate.

The drag reduction curves of Figure 5-6 show some bandwidth. To get a feeling for the yet remaining error, we performed calculations with the lower and upper envelopes of these figures instead of the mean curve, of Figure 7. This is a worst case assumption, since no actual curve is lowest or highest over the whole domain.

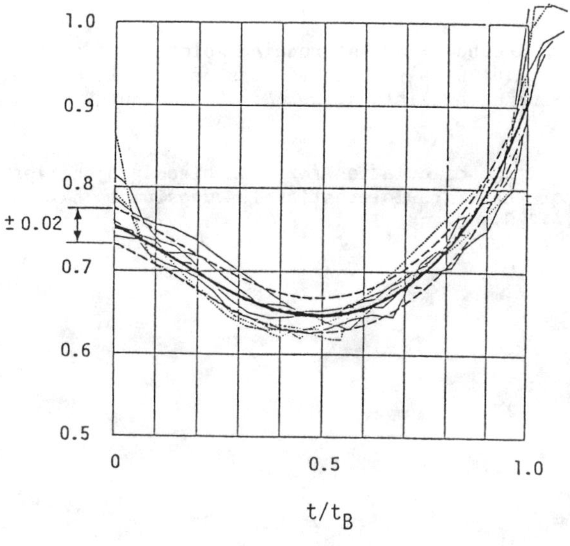

Figure 7:
Drag reduction curves used for range sensitivity calculation

The following table shows the results of this calculation for an example with zone 8. The actual firing distance was 21.9 km, the basebleed unit burning time was 21.7 secs.

Basis: actual firing	range error	time-of-flight
mean drag reduction curve	- 66 m	- .42 s
lower drag reduction curve	+218 m	+ .07 s

Basebleed projectiles are tactically used only over a limited spectrum of the possible parameters. They are used preferrably with high zones at long ranges. Therefore usual fitting techniques with heavy weights on high zones and long ranges are applicable and will further improve the accuracy of the calculation.

CONCLUSIONS

It was proven that with sufficient accuracy a simple extension of the standard integration formula does the job of trajectory calculation for the purpose of fire control. Further work has to be done, especially in the field of predicting the burnout time.

REFERENCES

1 G. Laube and B Kneubühl: A fast running solution of the exterior boundary value problem. Proceedings of the sixth international symposium on ballistics, pp 486-491, Orlando, 1981,

2 R.V. Helgren: Range calculation for base bleed projectiles. Proceedings of the sixth international symposium on ballistics, pp 214-220, Orlando 1981.

COMMENTS

M.A. Hasselaar, MOD, Netherlands. The model uses the measured burning time as an input variable of the model. What are the different conditions for the determination of the burning time?
Author's Reply. Actually we are using an approximation of the burning time as a function of muzzle velocity and elevation, together with a small correction for air pressure at muzzle altitude. This formula was derived from experimental data fired at conditions shown in Figure 1 of the presentation.

T. Wijker, Muiden Chemie, Netherlands. In your model, do you assume that the base bleed is temperature-insensitive?
Authors reply. The evaluated base bleed unit in fact is not temperature sensitive with respect to range. In test firings at practical maximum range of zone 8, a grain temperature variation from -20 to +50 degrees Celsius resulted in mean range variations of only -3 to +20 meters. This is far below range dispersion. Therefore we did not incorporate grain temperature in trajectory calculation. We neglect the fact, that burning time and therefore mass flow rate are temperature dependent, because for this particular base bleed desing, it has no practical effect on range and time of flight.

We think insensitivity to grain temperature is important. How should the artillerist measure the correct grain temperature? It is impracticable. In this context, I would like to draw attention to another practical problem. Muzzle velocity is usually measured by radars. These radars take velocity samples on the first part of the trajectory and then extrapolate back to muzzle. This means, that the radar reading is a theoretical value, based on the true muzzle velocity and the performence of the base bleed unit at the very beginning of the trajectory. When calculating velocity error correction tables, we should keep this fact in mind, otherwise incorrect correction tables would result.

SESSION III:
EXPERIMENTAL AND BALLISTIC PERFORMANCE OF BASE BLEED PROJECTILES AND GAS GENERATORS

Base Burning Performance at Mach 3

WARREN C. STRAHLE
School of Aerospace Engineering
Georgia Institute of Technology
Atlanta, Georgia 30332, USA

ABSTRACT

Results are presented for base bleed effects on base drag reduction using hydrogen fuel and several diluents as the injectants. Also, preburned hydrogen with oxygen is investigated as a fuel-rich, hot gas injectant. It is found that the hot gas injectant is most effective in producing base drag reduction for a fixed injectant flow rate and that impressive specific impulse may be obtained even at high base drag reduction.

INTRODUCTION

Over the past four decades there have appeared spurts of interest in the use of combustion or mass addition either from the side or from the base of a projectile in supersonic flight. The purpose has been for generation of control forces or thrust (drag reduction) for these projectiles or missiles. A brief list of references (not exhaustive) illustrating the chronology is given as Refs. 1-7. This symposium represents a new phase of interest in the problem, but is specialized to the use of base bleed for drag reduction.

Prior work at this laboratory [7,8] has focused on experiments in base bleed for base drag reduction (or elimination) using various injectants, with hydrogen as the combustible. The experiments were limited to a facility of fixed Mach number of 3, but sufficient variables were changed to provide much physical insight into the problem. In particular, this paper will focus on the advantages of using a hot fuel rich bleed gas as the injectant, as opposed to either a pure cold fuel or inert hot gas.

FACILITY

The facility used was a blowdown windtunnel designed to simulate the base flow for a bluff-base, axisymmetric projectile with a fineness ratio of about 6. The Mach number was 3 at the projectile tailplane. Test section details are shown in Fig. 1.

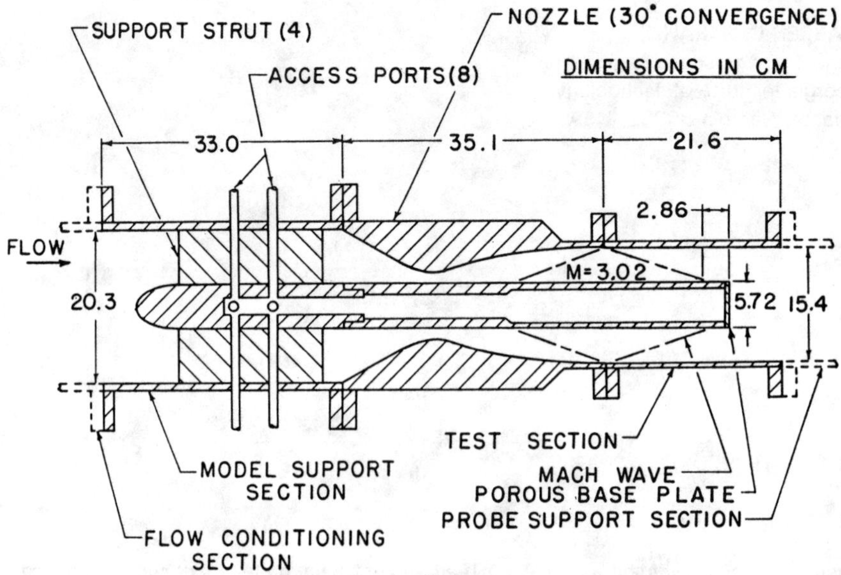

FIGURE 1. Base burning test section schematic

The windtunnel characteristics are described in Table 1.

TABLE 1. Wintunnel Characteristics

Mach number at base tailplane	3.02
Stagnation pressure	9.3×10^5 Pa
Stagnation temperature (drifts during run)	250-280 K
Reynolds number based on base diameter	3×10^6
Boundary layer momentum thickness at tailplane	0.67% D
Centerbody base diameter (D)	5.72 cm
Tunnel area at base plane	161 cm^2
Maximum run time	5 min
Test section static pressure	24600 Pa
Test section static temperature	100 K
Mass flow rate	0.82 kg/s

The hollow cylindrical model extends through the windtunnel nozzle and is supported by four streamlined struts in the upstream portion of the windtunnel where the Mach number is 0.07. Gases for base injection and instrumentation leads were ducted through the support struts.

Two base configurations were used. The plane base shown in Fig. 1 is a plane sintered metal base plate. This plate was used for base bleed with hydrogen injection with diluents of He, CO_2 and H_2. The second base configuration is shown in Fig. 2. This was used for the combined preburning of pure oxygen with diluted hydrogen before injection into the wake.

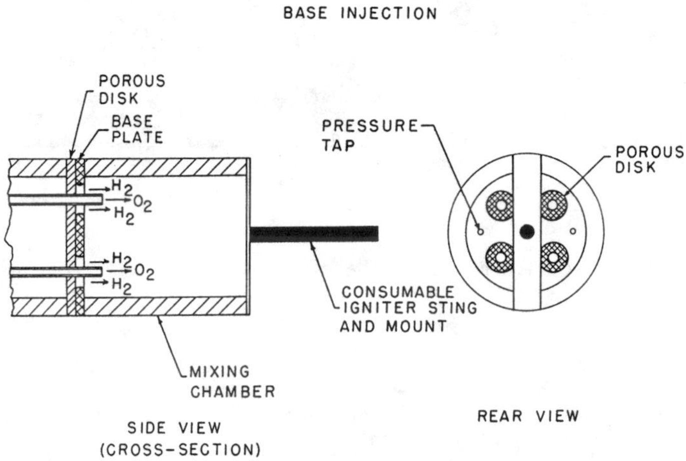

FIGURE 2. Base configuration for preburning

Ignition for all cases was obtained by igniting a solid propellant consumable igniter sting as shown in Fig. 2. Base pressure was measured as the average of four or five base pressure static taps drilled through the plate nearest the base. In all uses the injectant momentum flux was less than 5% of the base force rise achieved, assuring a base bleed situation as opposed to a thrusting system with the addition of base bleed.

PHYSICAL INTERPRETATION

From a variety of sources [7], both theoretical and experimental, the mean flow field may be expected to look as depicted in Fig. 3. The shaded area downstream of the body contains the complex fluid mechanics and combustion phenomena. There are many ways to physically reason what may be expected with injection and base burning. The author prefers, for simplicity, arguments based upon injection effects on the streamline passing through the rear stagnation point. This streamline must, by definition, stagnate, so the static pressure plus the dynamic head govern this process. If the Mach number is reduced on this streamline, the stagnation pressure falls so the static pressure must rise. This results in a base pressure rise. The Mach number in combusting flows is primarily determined by the speed of sound because flow speeds are mainly set by the freestream speed. Since the dominant species in the flow is N_2, mass diluent effects of various molecular weights will only have a weak effect on the stagnation pressure. Consequently, combustion heating is the dominant cause of base pressure changes.

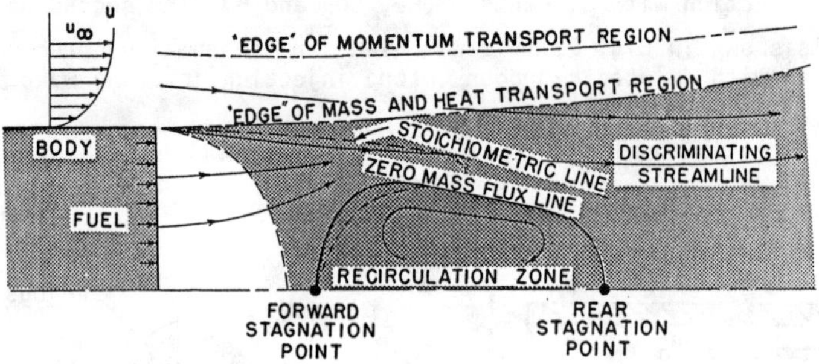

FIGURE 3. Presumed flow model for base burning

Another source of heating of the subject streamline would be the injection of hot gas, and this is what is accomplished in the preburning case. Control of the injectant temperature, however, alters the entire shape of the flow field and it is not clear what the optimum (if any) mix would be when considering an H_2/O_2 ratio of injectants.

In consideration of the results below, there are two performance parameters of paramount interest. These are the base pressure, P_b, rise with injection. The no-injection base pressure will be denoted P_{bo}, and the freestream static pressure at the tailplane will be denoted P_1. The relevant force change on the base is then $(P_b-P_{bo})A_b$, where A_b is the base area. The overall mass injectant rate is that of $H_2 + O_2$ + diluent, denoted $\dot{m}$. The effective specific impulse is then

$$I_{sp} = (P_b - P_{bo})\, A_b/\dot{m} \qquad (1)$$

In order to facilitate comparison of various injectants it is also useful to quote the effective heating value of the fuel plus diluent combination. Pure hydrogen has a value of 1.21×10^5 J/g and a typical hydrocarbon has values in the range $2 - 2.5 \times 10^4$ J/g.

One final parameter used to characterize the injectant rate is the injection parameter I, defined as the ratio of mass injection rate to the "capture" mass flow rate of the free stream, $\rho_1 V_1 A_b$, where ρ_1 is the freestream density and V_1 the freestream velocity. For base bleed situations I is always a very low number compared with unity.

RESULTS

Shown first are the results of pure and diluted hydrogen, using the base plate of Fig. 1. In Fig. 4 the Isp is based upon the flow rate of hydrogen alone, regardless of the diluent. Where diluent was used, the amount was chosen to bring the effective heating value of the mixture to values which would be practical for a fuel rich gas generator. The reason for presenting the results on the basis of the H_2 flow rate alone are to emphasize the main results of Fig. 4 that the diluent makes little difference. The basic performance depends only upon the amount of fuel poured into the base region (the heat dump into the wake). This result has been qualitatively argued above.

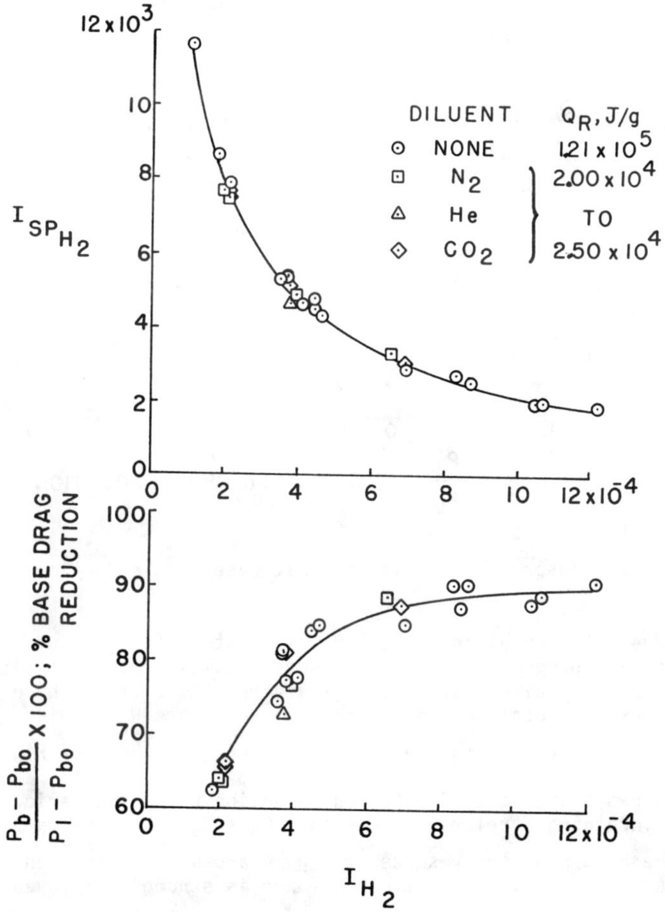

FIGURE 4. Base burning results with diluted and undiluted hydrogen.

An interesting presentation of results occurs when the specific impulse is plotted as a function of the relative base drag reduction, as shown in Fig. 5. Here the true I_{sp} is shown on the ordinate, but scaled by the effective fuel heating value. For pure H_2 injection the solid straight line results. With diluent roughly the same line is found; however, there is a mild effect of the diluent. Two major points here are (1) impressive I_{sp} results may be attained at base pressure reduction values as high as 90%, but (2) with this injection method 100% base drag elimination cannot be achieved.

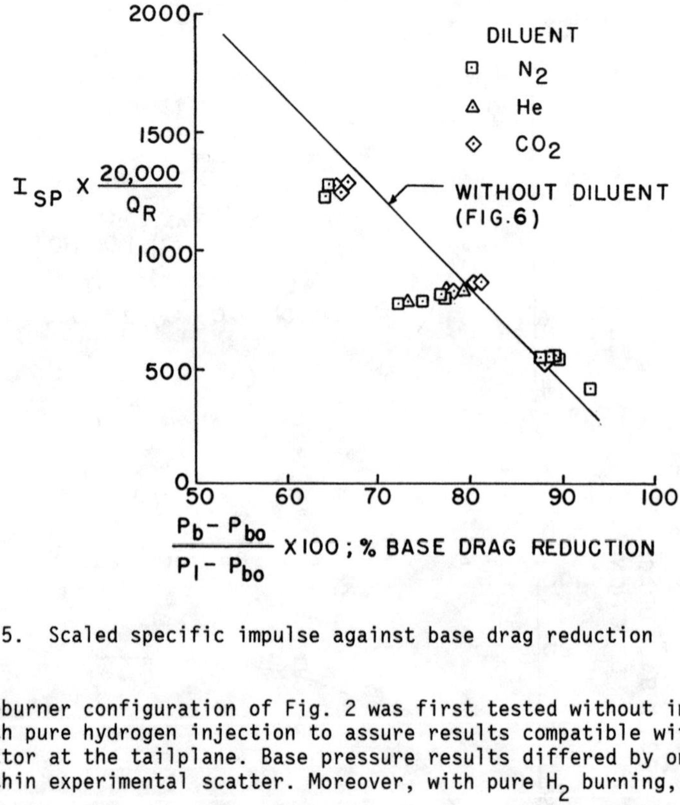

FIGURE 5. Scaled specific impulse against base drag reduction

The preburner configuration of Fig. 2 was first tested without injection and with pure hydrogen injection to assure results compatible with the Fig. 1 injector at the tailplane. Base pressure results differed by only 1%, which was within experimental scatter. Moreover, with pure H_2 burning, the results for the preburner configuration did not differ from those of Fig. 4.

Figure 6 shows the effect of hydrogen injection rate with various values of $O_2 + N_2$ maintaining a relatively constant $O_2 + N_2$ mass flow rate. For comparison the base pressure results of Fig. 4 are also plotted on Fig. 6. The rather startling result is that performance is strongly improved with preburning.

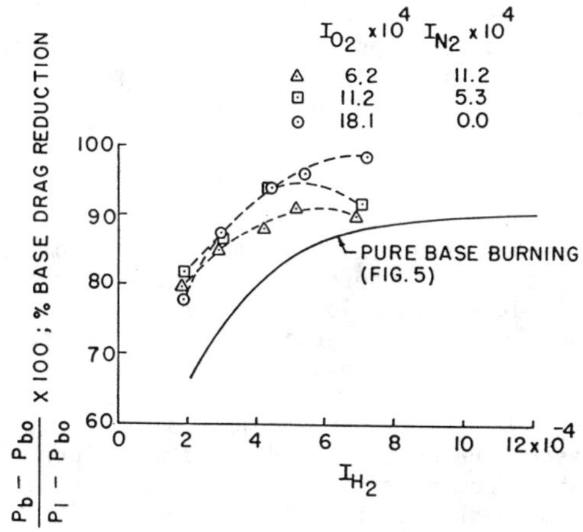

FIGURE 6. The effect of preburning on base pressure reduction

The net performance gains attainable by preburning are shown in scaled I_{sp} vs base drag reduction in Fig. 7. The dashed line represents an envelope of upper limit performance attainable. Impressive values of I_{sp} are attainable with nearly 100% base drag elimination. It was also checked that no effect of inert diluent appeared when the H_2/O_2 ratio was held fixed.

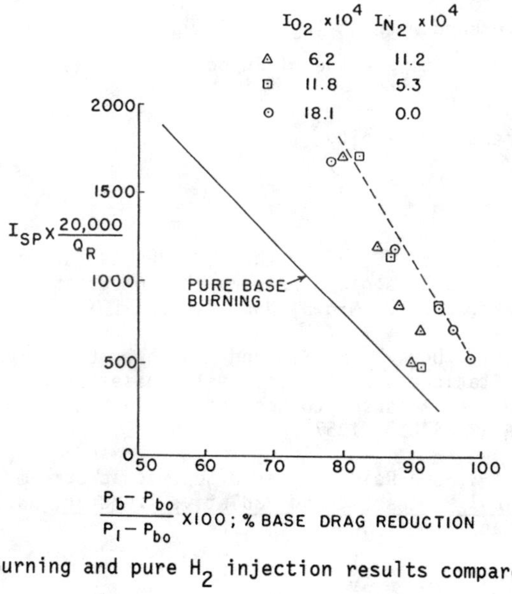

FIGURE 7. Preburning and pure H_2 injection results compared

It is unclear from a physical viewpoint and without a theoretical model why preburning causes such a large performance increase, as compared with pure or diluted hydrogen injection. Nevertheless, the performance gain is impressive and makes base bleed an extremely attractive propulsive device.

CONCLUSION

Pure base burning performance with hydrogen fuel is dominantly determined by the hydrogen flow rate and is therefore dependent upon mainly the heat dump rate into the wake, but with a mild effect of the diluent used. This has the approximate consequence that the specific impulse for pure base burning (no preburning) is directly proportional to the heating value of the injectant.

Combined preburning and base burning yields considerably higher performance than with pure base burning of a cold injectant. Impressive performance values are attainable. For example, 90% of the base drag can be eliminated with a specific impulse greater than 1000 using an effective fuel heating value of 20,000 J/g.

NOMENCLATURE

A_b	base area
I	injection parameter
I_{sp}	specific impulse
$\dot{m}$	injectant mass flow rate
P_b	base pressure
P_{bo}	base pressure without injection
Q_R	effective heating value of injectant
V_1	freestream velocity
ρ_1	freestream air density

REFERENCES

1. Baker, W. T., Davis, T., and Mathews, S. E., Reduction of Drag on a Projectile is a Supersonic Stream by the Combustion of Hydrogen in the Turbulent Wake, JHUAPL CM-673, June 4, 1951.

2. Sarafini, J. I., Dorsch, H. G., and Fletcher, E. A., Exploratory Investigation of Static and Base-Pressure Increases from Combustion of Aluminum Borohydride. Adjacent to Body of Revolution in Supersonic Wind Tunnel, NACA RM ES7E15, 1957.

3. Townsend, L. H., and Reid, J., in a Supersonic Stream, in Supersonic Flows, Chemical Processes, and Radiative Transfer, ed. D. B. Olfe and V. Zakkay, Mac Millen, New York, 1964.

4. Billig, F. S., External Burning in Supersonic Streams, JHUAPL TG-912, May, 1967.

5. Strahle, W. C., Theoretical Considerations of Combustion Effects on Base pressure in Supersonic Flight, Twelfth Symposium (International) on Combustion, pp. 1163-1173, The Combustion Institute, Pittsburgh, 1969.

6. Schadow, K. C., and Chieze, D. J., Base Drag Reduction by Combined External Burning/Base Burning, CPIA Pub. 315, March, 1980.

7. Strahle, W. C., Hubbartt, J. E., and Walterick R. E., Base Burning Performance at Mach 3, AIAA Journal, Vol. 20, No. 7, pp. 986-991, 1981.

8. Hubbartt, J. E., Strahle, W. C., and Neale, D. H., Mach 3 Hydrogen External/Base Burning, AIAA Journal, Vol. 19, pp. 745-749, 1981.

Structure and Ballistic Properties of the Rheinmetall RH 49 Base Bleed Projectile

LUTZ BÖRNGEN
Department of Ballistics
Rheinmetall GmbH
Ulmenstr. 127
4000 Düsseldorf 30, FRG

ULF HAHN
Development Artillery Ammunition
Rheinmetall GmbH
Ulmenstr. 127
4000 Düsseldorf 30, FRG

ABSTRACT

In this paper we present our 155 mm Base Bleed Projectile RH 49 with bomblet submunition. This projectile is effective against semi hard targets at ranges up to 31 km. The shell carries 49 grenades of the type RH II with self-destruction device. The Base-Bleed unit which consists of two parts, gives an increase of range of 19 - 26 % when compared with RH 49 Base-Bleed inert at burning times of 26 - 36 sec, corresponding to average burning rates of 1,4 mm/sec at normal conditions and of 1 mm/sec at trajectories for maximum range. From radar measurements we calculate C_D-values as a function of Mach-numbers. A typical curve for a baseburnprojectile is obtained this way.

During the burning of the Base-Bleed the drag coefficient of the projectile is almost independent on the velocity. After burn-out the C_D-values approach the curve of the projectile with inert Base-Bleed unit.

To obtain a precise C_D-curve during baseburning, we used a semi-empirical model developed at Rheinmetall for simulation of the Base-Bleed effect. This simulation is based on the theory of the outlet of a jet of gas from a reservoir. The simulation of the Base-Bleed effect is linked with a point mass trajectory model.

INTRODUCTION

The total drag of an artillery shell consists of wave drag, friction drag, drag due to lift and base drag. Base drag can contribute up to 75 % of the total drag of the projectile. Figure 1 compares the base drag to the total drag as a function of Mach-number for the RH 49-BB round without base burning. This shows that one possibility to achieve extended ranges for fielded 155 mm gun systems is the reduction of base drag (ref. 1). At Rheinmetall the theoretical basis of the Base Bleed Effect (ref. 2 + 3) was elaborated and the projectile RH 49-BB was developed in recent years. As a result of the theoretical work a simple engineering-type model for base-bleed calculations was developed. Another similar model is known (ref. 4) which is more specialized and bases much more on a great number of wind tunnel measurements.

Next the base-bleed simulation is described. Afterwards the construction and firing results of the RH 49-BB projectile are presented. Finally the methods of trajectory calculation are mentioned.

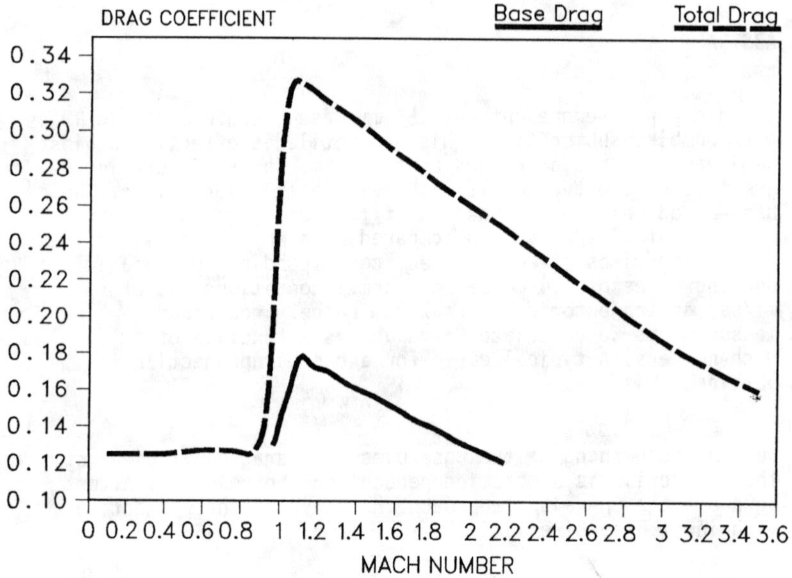

FIGURE 1 Comparison of total drag and base drag of RH 49-BB without Base-burning.

BASE BLEED SIMULATION

The supersonic flow around the base of an artillery projectile is rather complicated. The reason for the great base drag is the low base pressure. The base drag is greatly reduced, if a small amount of gas is blown into the afterbody wake. The gas is produced by a pyrotechnical unit placed in the bottom of the projectile (Base-Bleed-Effect). With the definition of a dimensionless gas flow rate,

$$I = \frac{\dot{m}_N}{\rho_\infty V_\infty A_{ref}} = \frac{\Gamma}{\gamma} \frac{p_B M_N a_\infty}{p_\infty M_\infty a_N} \left(\frac{d_N}{d_{ref}}\right)^2 \qquad (1)$$

it is possible to express the base pressure as a function of mass flow through a nozzle in the bottom of the round (Fig 2). Only the first part of this curve up to the maximum of base pressure (i. e. minimum of base drag) is important for the Base-Bleed-Effect. In this region where the mass flow rates are in the extent of a few percent, the base burn units work.

The Rheinmetall simulation of the Base-Bleed-Effect is based on the theory of the outlet of a jet of gas from a reservoir and the stream tube theory. Figure 3 shows schematicly the flow around the base of a boat tail of an artillery shell with base-bleed.

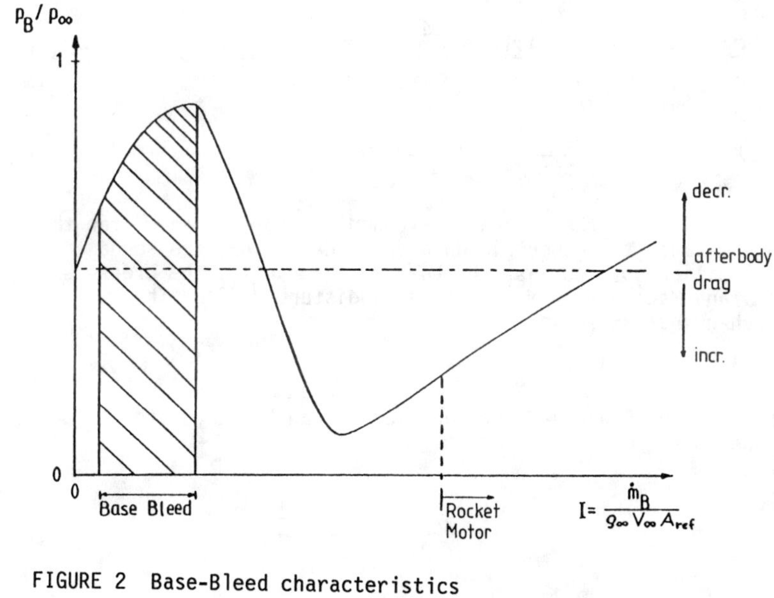

FIGURE 2 Base-Bleed characteristics

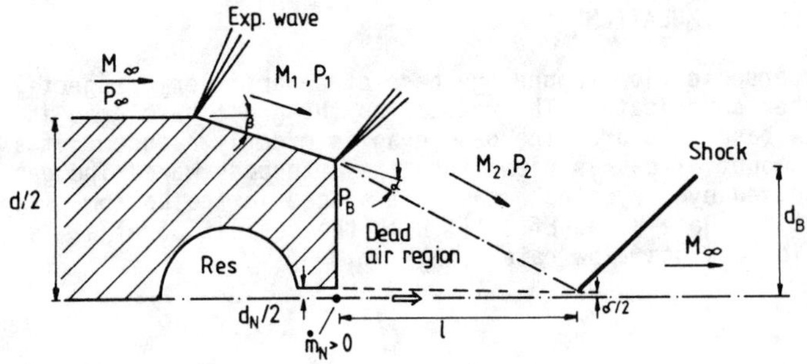

FIGURE 3 Schematic of Supersonic base flow pattern

Using these theories it is possible to deduce the following system of equations, describing mass flow I, dead air angle α, base pressure ratio p_B / p_∞, length of dead air region l :

$$I = \frac{p_B\, M_N\, \Gamma}{p_\infty\, M_\infty\, \gamma} \left(\frac{d_B}{d_{ref}}\right)^2 \frac{a_\infty}{a_{RES}} \sqrt{1 + \frac{\Gamma - 1}{2} M_N^2} \qquad (2)$$

$$\alpha = \nu(M_2) - \nu(M_1) \qquad (3)$$

$$\frac{p_B}{p_\infty} = \left(\frac{2 + (\gamma - 1) M_\infty^2}{2 + (\gamma - 1) M_2^2}\right)^{\frac{\gamma}{\gamma - 1}} \qquad (4)$$

$$l = \frac{d_B - \delta}{2} \cot(\alpha + \beta) \qquad (5)$$

Input data are the adiabatic exponents γ for air and Γ for the propellant, the geometric data d_{ref}, d_N and d_B, the sound velocities a_∞ and a_{Res} the boat tail angle β , the pressure p_∞ and Mach-number M_∞ of the undisturbed flow. The Mach-number M_1 follows from

$$\beta = \nu(M_1) - \nu(M_\infty) \qquad (6)$$

ν(M) is the Prandtl-Meyer-Function. δ can be calculated by the equations:

$$\delta = d_N \sqrt{M_N} \left(\frac{\Gamma + 1}{2 + (\Gamma - 1) M_N^2}\right)^{\frac{\Gamma + 1}{4(\Gamma - 1)}} \qquad (7)$$

To solve this system of equations, the assumption that the length of the dead air region l does not change with base bleeding was made.

The gas flow rate is given by the burning law of the Base-Bleed grain. We assume the following burning law for the simulation.

$$R = C\, P_{RES}^{\ \mu}$$

$$= C\left(1 + \frac{\Gamma - 1}{2} M_N^{\ 2}\right)^{\frac{\mu\Gamma}{\Gamma - 1}} p_B^{\ \mu} \qquad (8)$$

and now

$$I = \frac{a_\infty\, \rho_{PR}\, F_{PR}}{M_\infty\, \gamma\, A_{ref}\, p_\infty}\, R \qquad (9)$$

ρ_{PR} is the constant density of the base-bleed grain. The burning surface F_{PR} depends on the geometry of the base-bleed unit and the burning rate.
The solution of this system of equations gives the change in base pressure and so also change of base drag by base burning. The connection of this simulation to a ballistic flight programme makes it possible to calculate the trajectories of base-bleed projectiles.

PROJECTILE RH 49-BB

The 155 mm RH 49-BB is a long range projectile with bomblet submunition. The round is effective against semi hard targets at ranges up to 31 km fired with the field howitzer FH 70 or 38 km fired with 52 caliber extended range ordnance. The shell carries 49 grenades of the type RH II with self-destruction device. Figure 4 shows the construction of our projectile. The

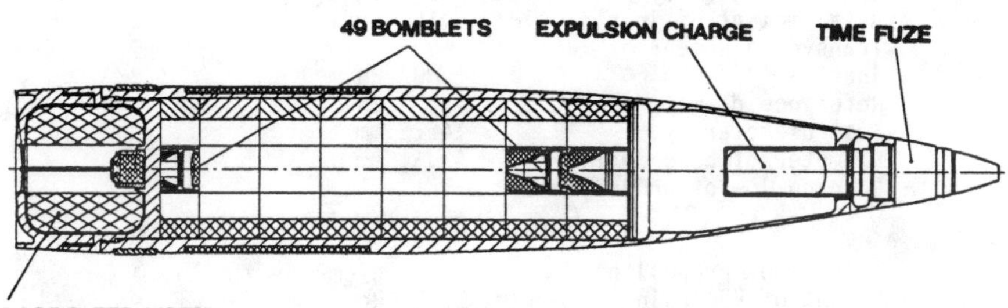

FIGURE 4 Long range projectile 155 mm RH 49-BB

main parts are the one-piece shell, the bottom with the base-bleed motor, the 49 grenades, the expulsion charge and the mechanical time fuze. The base-bleed grain is shown in Figure 5. A summary of the technical characteristics of the RH 49 base-bleed projectile is given in table 1.

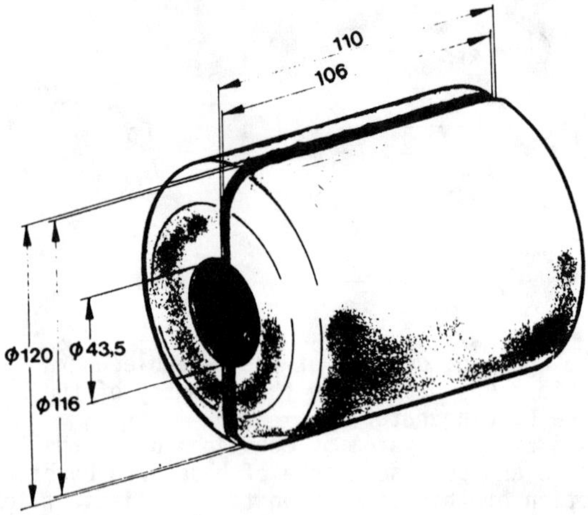

FIGURE 5 RH 49-BB - Base-Bleed-Grain

TABLE 1. Technical characteristics of RH 49 Base-Bleed

Weight	44.6	kg
Muzzle velocities		
Charge 8: L10A1	823	m/s
Charge 7: L 8A1	678	m/s
Center of gravity	319	mm
Axial moment of inertia	0.15	kg m^2
Transverse moment of inertia	1.80	kg m^2
Reference diameter	154.7	mm
Base diameter	146.2	mm
Diameter of base orifice	43.5	mm
Burning low of propellant		
C =	1.4	mm/sec
μ =	0.31	
Density of propellant	1.37	g/cm^3
Weight of BB grain	1.45	kg

FIRING RESULTS

The firing results obtained in recent years have demonstrated that the maximum range of the RH 49-BB projectile is greater than 30 km with a muzzle velocity of 823 m/s. The probable errors are $PE_R < 0.3$ % and $PE_D < 1$ mil. Compared with the inert projectile the base burning produces increasing ranges of 19 - 26 % depending on muzzle velocity and elevation.
Thus the base-bleed motor causes a decrease of base drag of about 60 - 80 %.
The burning time was measured staticly and in flight by optical and radar methods. The static burning time is 26 s at a temperature of 21° C. Depending on the trajectory the burning time increases up to 36 s. The static burning rate is 1,4 mm/sec. At grain temperature of - 32 °C the burning rate decreases 4 %, at 52° C the burning rate increases 8 %. From these experimental data, it is possible to determine the coefficients of the burning law: C = 1,4 mm/sec, μ = 0,31.

From radar measurements the drag coefficient C_D was calculated as a function of Mach-number for the inert projectile and for the base-burning phase. These curves are shown in Figure 6. The

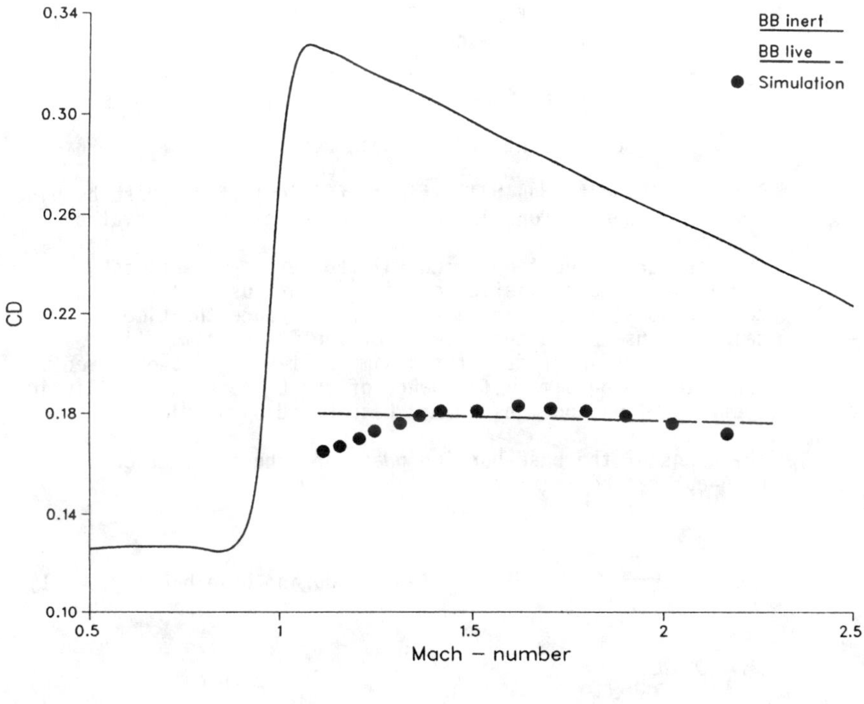

FIGURE 6 RH 49-BB - Total drag coefficient vs. flight Mach-number

value of C_D during base-burning is nearly constant (C_D = 0,18). The curve deduced from radar measurement is compared with results of the base-bleed simulation. The C_D curve of the inert projectile has the typical shape of a modern artillery shell. After burn-out the C_D- value jumps from the base-burning-curve to the inert curve. The Mach-number related to this jump depends on the firing conditions.

SIMULATION OF TRAJECTORIES

The simulation programme of the base-bleed effect was linked to a two-degrees-of-freedom trajectory model. The given base drag of the inert projectile, which comes from empirical equations or from tables like DATCOM (Ref. 5), is corrected in the described way. So the drag can be written.

$$D = \frac{\pi \rho_\infty v_\infty d_{ref}^2}{8m} \left(c_B c_D - c_B' f_B c_D \right) \tag{10}$$

with

C_D - drag coefficient of the inert projectile, function of Mach-number

C_B - Ballistic coefficient of the inert projectile, function of muzzle velocity and elevation

f_B - Coefficient of base drag reduction

C_B' - Additional ballistic coefficient of the base burning phase, function of muzzle velocity and elevation

The base drag reduction is calculated by the base-bleed simulation. The ballistic coefficients are used to fit the calculations to the firing results. To reduce the time of calculations we use the results of the flight simulation with base-bleed as input data for a simpler two-degrees-of-freedom trajectory programme with change of the C_D-curve and ballistic coefficient without gasdynamic base-bleed simulation.

The drags of the base-burning phase and the phase after burn-out can be written

$$D_1 = \frac{\pi \rho_\infty v_\infty d_{ref}^2}{8m} c_{BB} c_{DB} \quad \text{during base-burning} \tag{11}$$

$$D_2 = \frac{\pi \rho_\infty v_\infty d_{ref}^2}{8m} c_B c_D \quad \text{after burn-out} \tag{12}$$

C_{BB} - ballistic coefficient during base-burning, function of muzzle velocity and elevation

C_{DB} - drag coefficient of the base bleeding projectile, constant value of 0,18

This programme allows very fast calculations of trajectories of the RH 49-BB round. The results of these calculations are presented in Figures 7 - 9. In Figure 9 the velocities of the inert and base-burning (burning time 36 s) projectile as a function of time of flight for maximum range are shown.

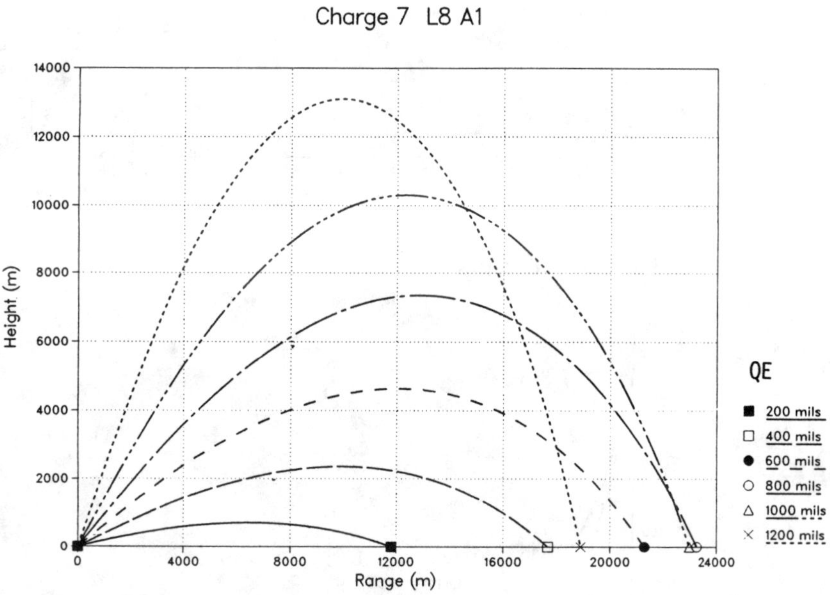

FIGURE 7 RH 49-BB - Calculated trajectories for charge 7 (L8A1), V_o = 678 m/s

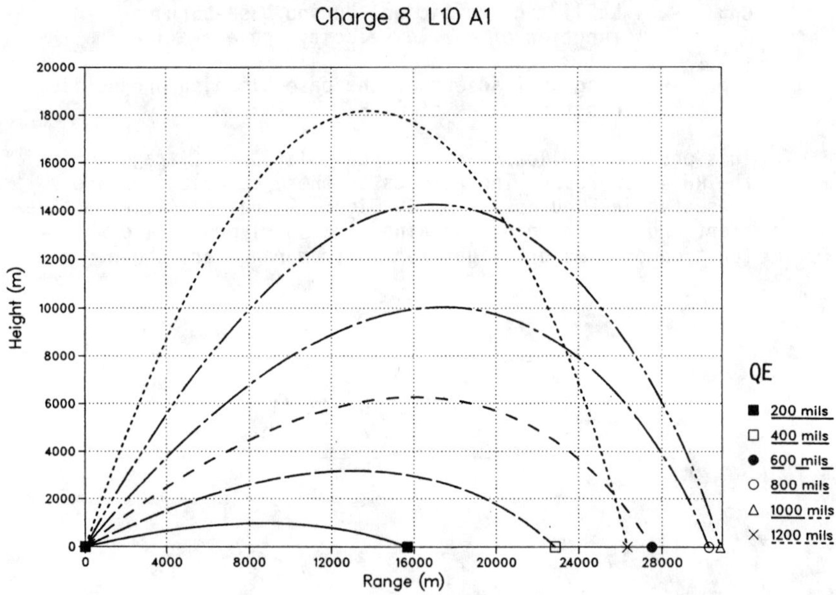

FIGURE 8 RH 49-BB - Calculated trajectories for charge 8 (L10A1), v_0 = 823 m/s

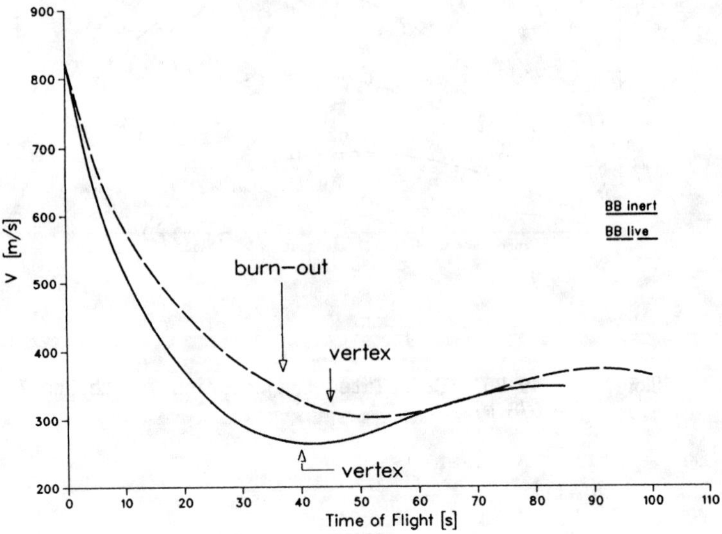

FIGURE 9 RH 49-BB - Velocity curves for maximum range trajectory

FINAL REMARKS

The long range 155 mm projectile RH 49-Base Bleed with bomblet submunition is presented. With the help of a simple theoretical model for calculating base bleed characteristics methods for range calculation during development are explained. In the future, calculations with the modified point mass trajectory (ref. 6) model are planned.

LIST OF SYMBOLS

A	:	Area
a	:	Speed of sound
C	:	Burning equation coefficient
c	:	Coefficient
D	:	Drag
d	:	Diameter
F	:	Propellant surface
I	:	Dimensionless mass flow
l	:	Axial extension of dead air region
M	:	Mach-number
m	:	Projectile mass
$\dot{m}$	:	Mass flow
P	:	Pressure
R	:	Burning velocity
v	:	Velocity
α	:	Dead air angle
β	:	Boat tail angle
δ	:	Diameter of inner flow thoat
Γ	:	Ratio of specific heats of combustion products
γ	:	Ratio of specific heats of air
ρ	:	Density
μ	:	Exponent of burning equation

SUBSCRIPTS

B	:	Base
D	:	Drag
N	:	Nozzle
PR	:	Propellant
ref	:	Reference value
RES	:	Reservoir
∞	:	Undisturbed flow values

REFERENCES

1. Romer, R., Reichweitensteigerung bei Artilleriegeschossen Wehrtechnik. pp. 54-58, Obtober 1976.

2. Schilling, H., Experimental Investigation on the Base-Bleed Effect for Body-Tail Combinations, Proc. 8th Intern. Symp. Ball., Orlando, Fla., 1984.

3. Schilling, H., A Simple Theoretical Approach for Base-Bleed Calculations, Proc. 9th Intern. Symp. Ball., Shrivenham, 1986.

4. Hellgren, R.V., Range Calculation for Base-Bleed Propellants, Proc. 6th Intern. Symp. Ball, Orlando, Fla., 1981.

5. Finck, R.D., et al., USA Stability and Control DATCOM, McDonnell Douglas Co., 1972.

6. Lieske, R.F., Modified Point Mass Trajectory Modeling for the 155 mm, DPICM, M 864 Base-Burn Projectile, BRL, Aberdeen MD 21005-5066.

A. Barouch, IMI, Israel. In formula (11) you define a ballistic coefficient C_{BB} during base burning as function of v_0 and ϑ. Do you have an analytical expression for it and what is its range of variation?

Author's Reply. We have an analytical expression base on the theoretical calculations with the Base-Bleed simulation program. The range of variation is 1.04 - 0.89. The theoretical function has to be confirmed in further firings.

***

G. Laube, GRD, Switzerland. You use ballistic coefficients for range fitting in your ballistic model. Do you also use a time-of-flight correction factor in your model? Remark: When using cargo shells with time fuzes, correct time-of-flight might be as important as correct range.

Author's Reply. No, we don't think that this correction is very important.

***

M. Hasselaar, MOD/MBAI, Netherlands. In your presentation you have stated that you find burning times up to 36 s. How do you handle with that in your model?

Author's Reply. The burning times can be measured or calculated by the Base-Bleed simulation program. It is an input parameter in the simple two-degrees-of-freedom trajectory program.

***

D. Cappelen, Norwegian Defence Research Establishment, Norway. What is the difference between the calculated range and the measured range from firings?

Author's Reply. To answer this question we give three examples:

v_0	QE	(measured - calculated range)
763	674	+ 133
824	720	+ 122
824	820	- 323

***

D. Chargelegue, Etablissement Technique de Bourges, France. What's the value of the boat-tail angle?

Author's Reply. The value of the boat-tail angle is 3 deg.

***

Dr. B. Zeller, SNPE, France. Could you tell the unit for pressure in the burning rate law (equation 8)

$$R = C P_{RES}^{\mu} \quad ? \qquad C = 1{,}4 \text{ mm/s}, \quad \mu = 0{,}31$$

Is it bar?
Is it atmosphere?

Author's Reply.

P_{RES} is a dimensionless quantity:

$$P_{RES} = \frac{P'_{RES}}{P_N}$$

$$P_N = 101{,}325 \text{ kPA} = 760 \text{ Torr} = 1 \text{ atm} = 1{,}01325 \text{ bar}$$

A Study of Drag Reduction by Base Bleed at Subsonic Speeds

ZESHONG DING, YAFEI LIU, and SHAOSONG CHEN
East China Institute of Technology
Nanjing, PRC

ABSTRACT

This paper presents the experimental research results of the drag reduction rate of the base bleed of solid fuel–rich propellent column at subsonic speeds. The experimental range of Mach number was from 0.6 to 0.98. Experiments show that the base drag can be removed by the base bleed of propellent column under the range of M = 0.6–0.8. The Bowman–Clayden's empirical formula, which was widely used, is not applicable in this speed range.

1. INTRODUCTION

The base bleed method is considered as one of the most effective current techniques for drag reduction. The range of large–caliber, long–range and low–drag projectile can be extended to 30% by using the technique of base bleed, and such gain is quite high. However the application of base bleed is now generally limited to supersonic speed only. The research information relevant to subsonic and transonic speeds is very rare. At present, Ref. [1] is the only base bleed information found for three–dimensional body under subsonic and transonic speeds. The experimental results of base bleeding cold air predict that the drag reduction rate of the base bleed of propellent burning products under such speed range is very low.

Now it is neccesary to make further researches on base bleed effectiveness at subsonic and transonic speeds. It is of great importance for developing base bleed technique and applying it to projectile flying at subsonic or transonic speed to reduce its base drag. For this purpose, the wind tunnel tests have been conducted by using the solid fuel–rich propellent columns for base bleed. The test results of drag reduction efficiency at subsonic speeds have been obtained and compared with the data given in literature[1].

2. EQUIPMENT, MODEL AND TEST

2.1 Equipment

The tests was carried out in the base bleed wind tunnel HG–03 of East China Institute of

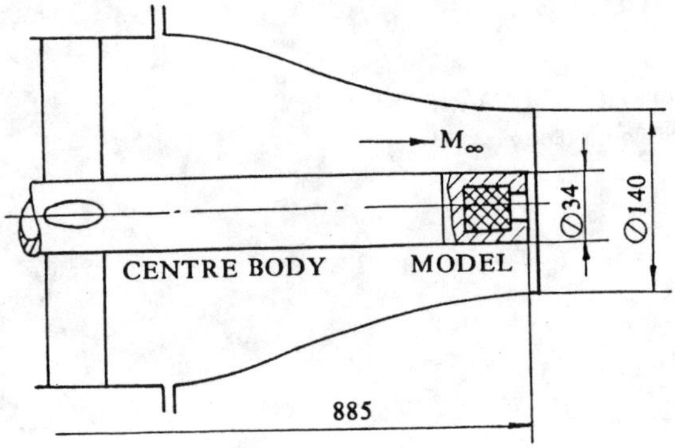

FIGURE 1. The sketch of centre body nozzle (All dimensions are in millimetres)

Technology. This wind tunnel is an open–jet one with a centre body. Mach number varies from 0.6–3. Its static pressure of free stream can reach atmospheric pressure.
The length of this centre body was 0.885m, its diameter was 0.034m. It was supported in a stilling chamber by four radial struts. Measuring pressure tubes came out through struts from the wind tunnel to eliminate the disturbance of the model support on the base flow field.

The flow field of the wind tunnel is of good quality. When $M = 0.6 \sim 0.8$, $\frac{\Delta M_{max}}{M_{cp}} < 0.6\%$. When $M = 0.98$, $\frac{\Delta M_{max}}{M_{cp}} < 1\%$. The centre body nozzle is sketched in Fig.1.

2.2 Model

In tests, a cylindrical afterbody model of 0.034m in diameter was connected with the end of the centre body. The ratio of the diameter D_j of a bleed nozzle to the diameter D of the model base was 0.5.

2.3 Pressure Measurement

The total pressure of the wind tunnel, the static pressure of free stream and the base pressure of the model were measured in tests. The transducer of base pressure is a π–type beam strain one. Its measuring range is $\pm 9{,}800 N/m^2$. All data were acquired and reduced by a computer.

2.4 Base Bleed Propellent Column

Three kinds of different composition propellent column were used in tests. Number 1 propellent column belongs to pyrotechnic type, its burning temperature over 3000℃. Number 2 and 3 propellent columns belong to propellent type, their burning temperature about 1300℃.

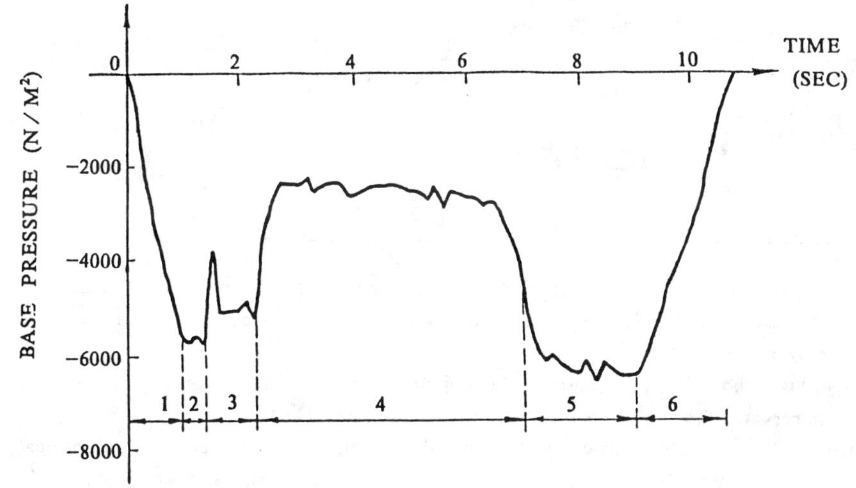

FIGURE 2. Curve of base pressure variation with time (the base pressures are gauge pressures)

2.5 Test Sketch

The tests of four Mach numbers were carried out: M = 0.59, 0.71, 0.82 and 0.98. The variation of free stream Mach numbers depended on the methods of adjusting the total pressure of the wind tunnel.

The propellent column was ignited by using a small high energy igniter. It was mounted on a sting which is located far downstream of the model base. After the flow of the wind tunnel was stabilized, the igniting sting was moved forward into the model base and ignited the propellent column, then moved back to far downstream quickly. In the whole process of the propellent column burning, there was no disturbance of the igniting sting on the flow field of the model base. After extinguishment of the propellent column for 3–4 seconds, the operation of the wind tunnel was stopped.

In the whole process of wind tunnel operation, the model base pressure was being continuously measured. A curve of the base pressure variation with time is obtained and showed in Fig.2. The base pressure in the Fig.2 is the difference pressure for atmospheric pressure. The burning time of the propellent column can be determined from Fig.2. Then the mass burning rate $\dot{m}$ and injection parameter I can be determined from the mass and burning time of the propellent column. The base drag reduction rate of base bleed $\frac{\Delta C_{Db}}{C_{Db}}$ can be calculated from the base pressure variation of the model in the circumstances with or without the burning of the propellent column. A $\frac{\Delta C_{Db}}{C_{Db}} \sim I$ curve at some Mach number can be acquired through several tests.

There is small influence of the boundary layer profile ahead of the base corner on the base

Table 1. R_e Numbers at Different Mach Numbers

M	R_e ($\times 10^7$)/m
0.59	1.64
0.71	2.00
0.82	2.43
0.98	3.26

pressure. However under the condition of turbulent boundary layer, there is no apparent influence of the variation of R_e number and boundary layer thickness on base pressure increase caused by base bleed[2].

The turbulent boundary layer exists on the projectile afterbody under the flight condition. The R_e numbers at different Mach numbers of these tests are shown in Table 1.

These values are far above the critical transitional R_e number (6.5×10^6) of three–dimensional boundary layer, therefore the boundary layer ahead of the model base is in turbulent state. In the case of practical purpose, the influence of R_e number on base pressure and the drag reduction rate of base bleed may be neglected.

3. RESULTS AND DISCUSSION

The curves of the base drag reduction rate of base bleed varied with injection parameter I were obtained from the experimental data at different Mach numbers, and are shown in Fig.3–Fig.6. As seen from these figures, the total variation tendency of the $\frac{\Delta C_{Db}}{C_{Db}} \sim I$ curves of three types of propellent column is similar under subsonic speeds. It is characterized as follows:

(1). Under small value of I ($I < 0.002 \sim 0.003$), $\frac{\Delta C_{Db}}{C_{Db}}$ increases quickly with increasing of I.

At small I, the velocity and momentum of base bleed are small. When I increases, the mass rate of burning gas entering the wake region and the energy released by burning increase rapidly. Then the wake region is obviously enlarged, the base pressure and $\frac{\Delta C_{Db}}{C_{Db}}$ raise greatly.

(2). After $I > 0.003$, as value I increases further, the rising slope of curves gradually decreases. With a greater value I, the mass rate entering the wake region is larger, and its velocity and momentum are greater. The large bleed momentum may destroy the structure of the base recirculation region, reduce the base pressure and partly cancel the profit gained from the increase of bleed mass and energy, then make the curve changed slowly. As value I increases further, the increase of $\frac{\Delta C_{Db}}{C_{Db}}$ is smaller.

(3). At different Mach numbers, the effect of number 1 propellent column on drag reduction is

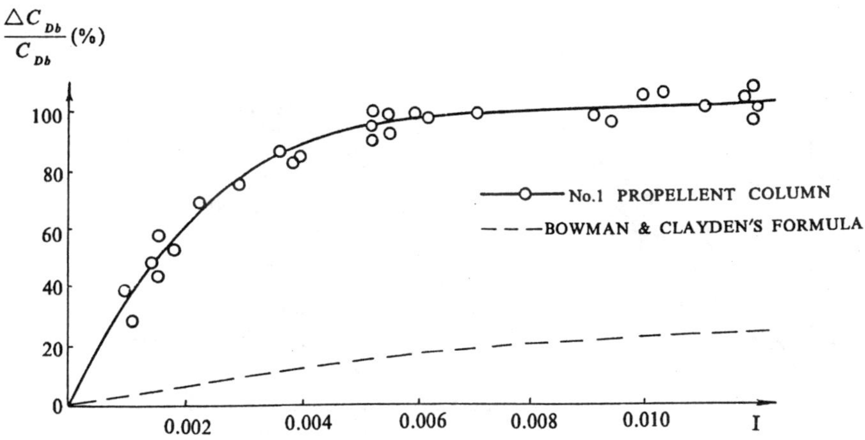

FIGURE 3. Curve of base drag reduction rate by base bleed at M = 0.59

the best.

It is well known that the drag reduction rate of base bleed depends on many parameters, such as free stream Mach number, geometry of the model, injection parameter, bleed gas velocity, molecular weight of bleed medium (or specific volume), bleed gas temperature, etc.

The comparison of the drag reduction rate of three types of propellent column in tests was made under the same conditions of Mach numbers, geometry of the model and injection parameters, so that the difference in the drag reduction rate may mainly result from the molecular weight and temperature of bleed medium. Except for additives, the compositions of number 2 and 3 propellent columns are basically the same. There are small differences in the temperature and molecular weight of burning products so that the difference in the drag reduction rate is also small. For No.1 propellent column, much residue exists in combustion products and the specific volume is less than that of the other two types of propellent, which is unfavourable to drag

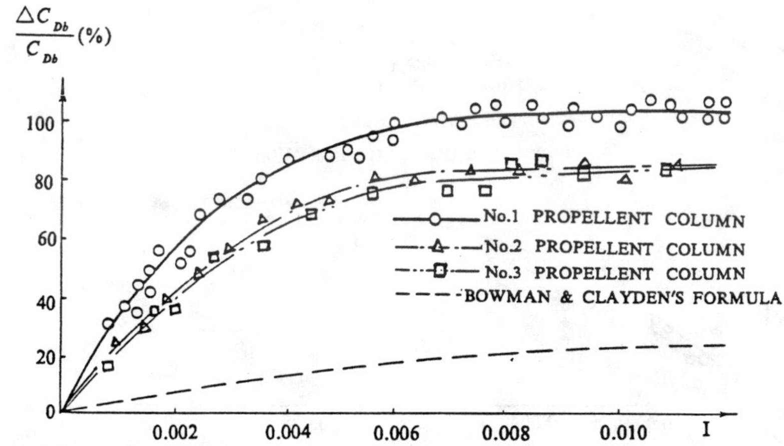

FIGURE 4. Curve of base drag reduction rate by base bleed at M = 0.71

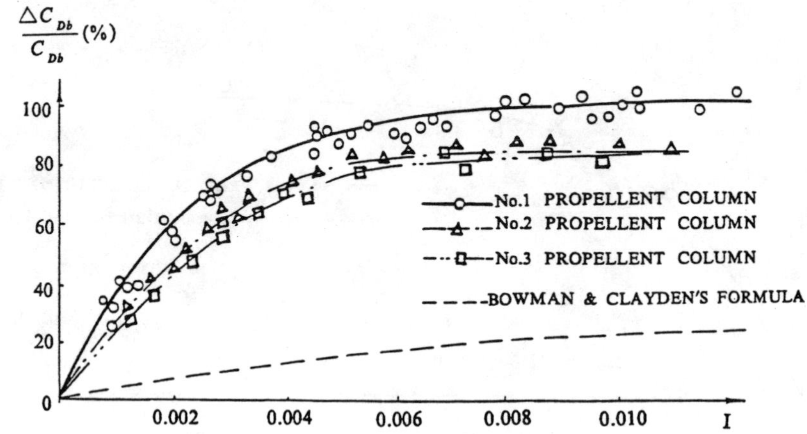

FIGURE 5. Curve of base drag reduction rate by base bleed at M = 0.82

reduction[1]. But its burning temperature exceeds 3000℃, which is far higher that of number 2 and 3 propellent columns (about 1300℃). Obviously high burning temperature is the radical reason for the best effect of the drag reduction of number 1 propellent column.

(4). In the range of M = 0.59–0.82, the drag reduction rate of base bleed of number 2 and 3 propellent columns is about 80%, while that of number 1 propellent column reaches as high as 100%, i.e. the base drag is completely eliminated.

From the results of the experiments with base bleeding cold air, Bowman and Clayden[1] correctly derived the empirical formula with propellent as fuel:

$$C_{Dbb} = C_{Db} e^{-JI}$$

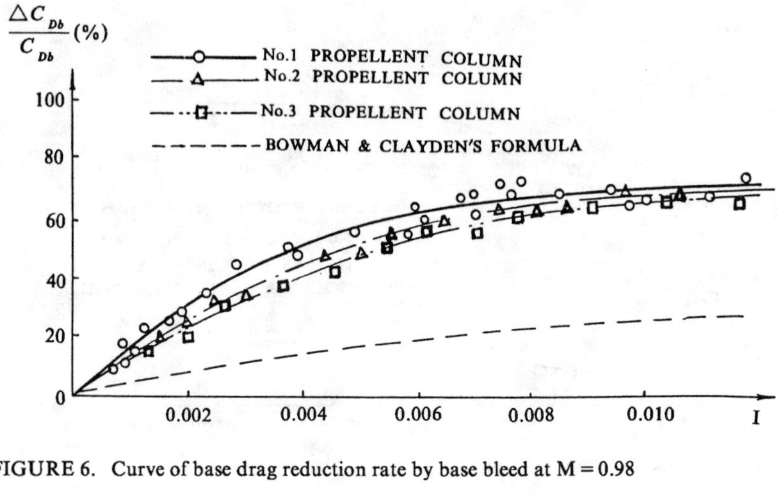

FIGURE 6. Curve of base drag reduction rate by base bleed at M = 0.98

It was widely used. However this research shows that the agreement between its estimated value and experimental value for the drag reduction rate of the base bleed of solid fuel–rich propellent column is poor. The main differences are as follows:

(a) At subsonic and transonic speeds, estimated drag reduction rate by this formula is much lower than the experimental value of solid fuel–rich propellent column (shown in Fig.3 – Fig.6). In the range of $M = 0.59–0.82$, the estimated value is about $\frac{1}{5}$ of the experimental value of number 1 propellent column, and $\frac{1}{4}$ of the experimental value of number 2 or 3 propellent column.

(b) The variation relationship between the drag reduction rate of the base bleed of solid fuel–rich propellent column and injection parameter I is not the exponent relationship specified in Bowman and Clayden's empirical formula. According to this formula, when I increases from 0.007 to 0.012, the drag reduction rate increases as high as 50%. However this research shows that the variation of drag reduction rate is very small.

(c) At subsonic and transonic speeds, the variation of the drag reduction efficiency of the base bleed of solid fuel–rich propellent column with Mach number is not the same as that given by Ref. [1].

Bowman and Clayden used $-\left(\frac{\partial C_{Dbb}}{\partial I}\right)_{I=0}$ to denote the base drag reduction efficiency of base bleed. According to the experimental data on the base bleeding cold air at $M = 0.11$, 0.64 and supersonic speed, they obtained $-\left(\frac{\partial C_{Dbb}}{\partial I}\right)_{I=0}$ at transonic speeds using interpolation method. In Ref. [1], they also introduced the experimental results of the base bleed at transonic speeds carried out by Sykes (shown in Fig.7). Based on Bowman and Clayden's curve, $-\left(\frac{\partial C_{Dbb}}{\partial I}\right)_{I=0}$ is constant at subsonic speeds, its value is very low. It rises rapidly in the range of transonic and low supersonic speeds, and reaches the maximum at approximately $M = 2.5$. The $-\left(\frac{\partial C_{Dbb}}{\partial I}\right)_{I=0} \sim M$ curve of solid fuel–rich propellent column obtained by this research shows that the base drag reduction efficiency of the base bleed at subsonic speeds is very high. Its value in the range of $M = 0.59–0.82$ is equivalent to the value of $M = 2–2.5$. When Mach number increases from subsonic to supersonic speed, the variation tendency of the base drag reduction of base bleed is high–low–high (shown in Fig.7).

It is seen from initial analysis that the defect in the experimental method of Ref. [1] is the reason for the large difference between the estimated value of the empirical formula and the experimental value of drag reduction rate of the base bleed of solid fuel–rich propellent column. The experimental method adopted by Bowman and Clayden to establish the empirical formula was as follows: First they carried out the experiments of base bleeding cold air at different Mach numbers, and obtained the variation law of base drag with Mach number and injection parameter I in such condition. Then they carried out the test with the cold and hot argon of base bleed under the condition of $M = 2$ (using electrical arc heater to heat). They employed the results of the argon test to modify the temperature and molecular weight of bleed medium for the experimental data of base bleeding cold air, and obtain the calculation formula of base drag adopted to the condition of the base bleed of the burning products of propellent column. This

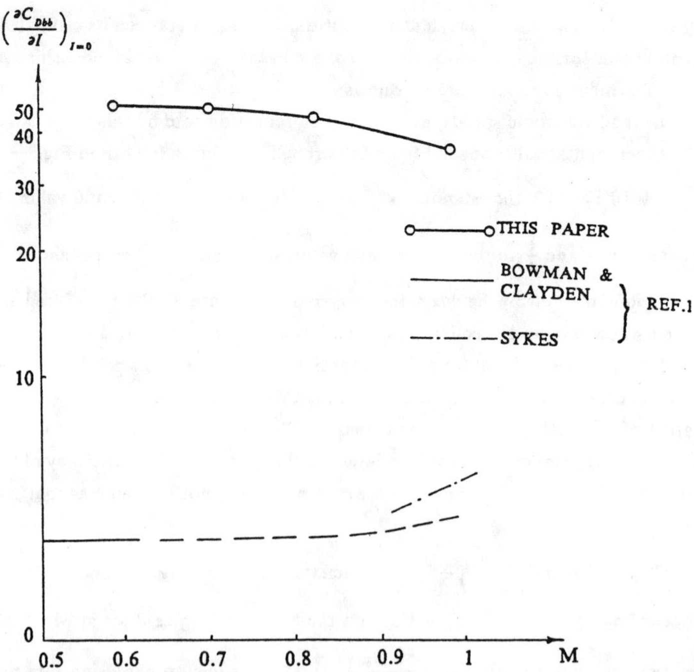

FIGURE 7. Variation of base drag reduction efficiency $-\left(\frac{\partial C_{Dbb}}{\partial I}\right)_{I=0}$ with Mach number

method neglected the secondary combustion effect in the near wake region, influence of solid particle and function of the different release location of enthalpy under the condition of the base bleed of solid fuel–rich propellent column.

It is concluded from this research that using gas (cold or hot) to simulate the base bleed of solid fuel–rich propellent column is not satisfactory for engineering requirement.

4. SUMMARY

This research carried out the tests of the drag reduction of base bleed at subsonic speeds in the wind tunnel using solid fuel–rich propellent column. The main conclusions obtained are as follows:

(1). In the range of M = 0.6–0.8, the drag reduction rate by the base bleed with solid fuel–rich propellent column is very high. The number 2 and 3 propellent columns may decrease base drag by 70–80%, while number 1 propellent column may completely eliminate base drag. It is worthwhile developing the technique of the drag reduction of base bleed in the range of subsonic and transonic speeds.

(2). In the range of subsonic, transonic and low supersonic speeds, the variation tendency of the drag reduction efficiency of base bleed of the propellent column with Mach number is high–low–high. Its change takes place rapidly in the range of transonic speed.

(3). As the estimated value by Bowman and Clayden's formula at subsonic or transonic speeds is about $\frac{1}{5} - \frac{1}{4}$ of experimental value of drag reduction rate of solid fuel–rich propellent column, it is not applicable.

5. NOMENCLATURE

C_{Db}	base drag coefficient without base bleed
C_{Dbb}	base drag coefficient with base bleed
$\frac{\Delta C_{Db}}{C_{Db}} = \frac{C_{Db} - C_{Dbb}}{C_{Db}}$	base drag reduction rate by base bleed
D	model diameter
D_j	bleed nozzle diameter
$I = \frac{\dot{m}}{\rho V S}$	injection parameter
$\dot{m}$	mass burning rate
M	Mach number
R_e	Reynolds number
S	base area
V	velocity of free stream
ρ	density of free stream
$-\left(\frac{\partial C_{Dbb}}{\partial I}\right)_{I=0}$	base drag reduction efficiency by base bleed

6. REFERENCE

1. Bowman, J.E. and Clayden, W.A., Reduction of Base Drag by Gas Ejection, **RARDE** 4 / 69.
2. Ward, J.R. and Baltakis, F.P., Wind Tunnel Experiments of the Effect of Near–Wake Combustion on the Base Drag of Supersonic Projectiles, **BRL IMR**, 379.

COMMENTS

Charles Nietubics, BRL, USA. You have shown only the integrated drag data. How many pressure taps were on the model base?

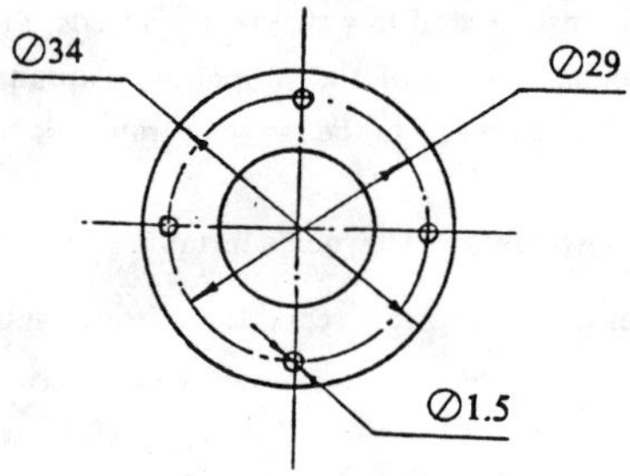

FIGURE 8. The distribution of measuring pressure orifices (All dimensions are in millimetres)

Author's Reply. Four measuring pressure orifices were adopted for base pressure measurement in tests. The orifice diameter is 1.5mm. They distribute uniformly along a circumference in diameter of 29mm, as shown in Fig.8. The radial distribution of model base pressure is unequal, but the difference is small (refer to Ref.[1] and Lyle, D.Kayser "Effects of Base Bleed and Supersonic Nozzle Injection on Base Pressure", **BRL.MR.** 2456). Our research also shows that the base pressure distribution further approaches to uniformity under base bleed condition. Therefore it is possible that the base pressure variation of base bleed measured at four abovementioned measuring orifices may represent the base pressure variation of the whole model base. Ref. [2] and J. Richard Ward et. (**BRL.R.** 1745) all adopted the distribution similar to this research to measure the variation of base pressure under the condition of base bleed.

Dewis Lyster, SRC. What materials were used for the grains 1, 2 and 3?

Author's Reply. Number 2 and 3 propellent column are fuel–rich solid fuel of propellent type. Their main compositions are 20% of fuel, 75% of oxide, but their additives are different. Their burning temperature measured is about 1300℃. Number 1 propellent column is a new type of solid fuel–rich material. Its burning temperature exceeds 3000℃ (Tungsten–rhenium pyrometer couple was eroded in the process of temperature measurement).

Testing of Parts and Complete Units of the Swedish Base Bleed System

NILS-ERIK GUNNERS, KURT ANDERSSON, and YNGVE NILSSON
National Defence Research Establishment
P.O. Box 551
S-147 25 Tumba, Sweden

ABSTRACT

The Swedish Base-Bleed system was invented about twenty years ago. During the research leading to the invention and during the continued development a number of different methods and experimental prodecures for testing parts as well as the complete system have been developed.

The paper describes several of these tests in relation to the functioning and the design of the system. Examples of test results and advices on their application are presented. The first part describes laboratory methods and experimental facilities for testing parts such as propellant grains, igniters and nozzle systems in combustion bombs, spin stand, soft recovery acceleration facility and supersonic wind tunnel. The second part treats parts and complete base-bleed units functioning and performance testing by means of gun firing in aeroballistic range and in actual free flight trajectories.

1. INTRODUCTION

Longer range and greater accuracy are constant objectives when new ammunition is developed or existing ammunition modified. Several methods of achieving these goals may be applied. Most methods can be combined.

Afterbody drag accounts for a substantial part of the total drag, more than 50 % for a slender-nose projectile. Thus, the importance of decreasing the afterbody drag is apparent. By means of a base-pressure-increasing system, such as base-bleed, the base pressure can be increased considerably during an important part of the flight time. This also affects the optimum design of the projectile afterbody.

The Swedish Base-Bleed System was invented in the late 1960-ies, /1/. Characterized by great simplicity and very good performance when correctly designed it has become an internationally adopted prototype for most later designs. It is especially advantageous to modern low drag projectiles. During the research leading to the invention and during the continued development a number of different methods and experimental procedures for testing parts as well as the complete system have been developed. Since these methods and procedures are strongly related to the design and functioning of the system a short resumé of the fundamentals might be of value. Phenomena to which the system is subjected and which give rise to testing will be pointed out.

Environmental tests, such as drop tests, vibration tests, temperature cycling and performance over specified temperature and humidity ranges, are carried out in accordance to actual regulations for conventional ammunition.

Several aging tests are performed of the base-bleed propellant grain and igniter as well as of mechanical parts of the base-bleed unit such as sealings. Aging tests of complete units are also performed. Ultimate tests are gun firings.

In this report neither environmental nor aging tests will be further discussed.

It should be noted that the tests reported here are primarly used for research and development although most of them, with minor alterations, could be used as production tests.

2. FUNDAMENTALS OF THE SWEDISH BASE-BLEED SYSTEM

The purpose of the base-bleed system is to eject, or bleed, a suitably low mass flow of under stoichiometric solid propellant combustion products from the body of a flying object, such as a gun projectile, into the near wake where the combustion is completed with ambient air. This affects the wake flow in such a way that the base pressure is increased and the base drag thereby reduced. The base mass flow should be optimal and should occur during a substantial part of the flight time.

There are many problems associated with the design of base-bleed units for projectiles. The system has to withstand the pressures, temperatures and accelerations in the gun, at the muzzle and during flight. The ignition has to be reliable and consistent and the unit should be designed to work efficiently. Morover the mass and volume should not adversely affect the lethality or cargo capability of the projectile.

The general design of the Swedish base-bleed unit is shown in fig 1. It is a compromise of efficiency, simplicity and low cost and is based on the fact that a controlled low mass flow of mainly gases can be generated by the combustion of a propellant at low pressures in a combustion chamber within the projectile body.

The main parts of the base-bleed unit are the combustion chamber with the propellant grain, the igniter and the base with a suitable nozzle system. In a sense the entire projectile afterbody is part of the unit since it in most cases constitutes the combustion chamber. The optimum design of the afterbody is very much governed by the performance of the base-bleed unit.

The base-bleed unit is in the course of the interior ballistics subjected to high pressures and pressure changes and to strong axial and angular accelerations. These cause mechanical stresses, strains and deformations of all unit parts.

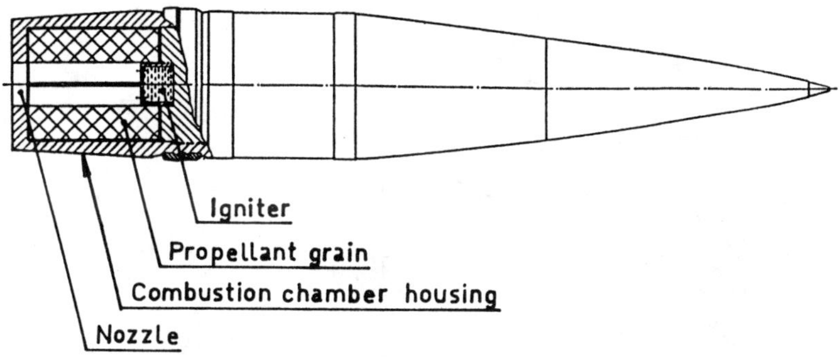

FIGURE 1. Low drag experimental projectile with base-bleed unit.

The propellant compositions should be chosen with great care. A suitable grain design can greatly reduce strains and stresses and the risk of cracks. Spin stand tests and axial acceleration tests are useful for studying mechanical performance.

When the gun is fired, the base-bleed propellant grain, as well as the igniter composition are ignited by the combustion gases in the gun. Due to the very steep pressure drop when the projectile leaves the muzzle, the base-bleed propellant combustion tends to be extinguished. The igniter however has a pyrotechnic composition with very low sensitivity to pressure variations. Thus it is capable of sustaining the combustion of the propellant grain.

It is very important for good range and precision to achieve an instant, reproducible, full effect onset of the base-bleed immediately at the muzzle. Several theoretical and experimental ignition studies and tests have been carried out. These and numerous complete base-bleed projectile firings emphasize the importance of perfect ignition.

The conditions are demonstrated in fig 2. Showing calculated range decrease versus effective ignition delay for a medium caliber, low drag base-bleed artillery projectile. An ignition delay of 0.1 sec results in a range decrease of approximatively 0.4 %. This is of the same order as the total dispersion at maximum range of good projectiles.

For a normally working unit, full onset of base-bleed occurs, with only small dispersion, within a couple of meters from the muzzle. Consequently ignition should normally not involve any problems but, due to the importance, has to be thoroughly tested and controlled.

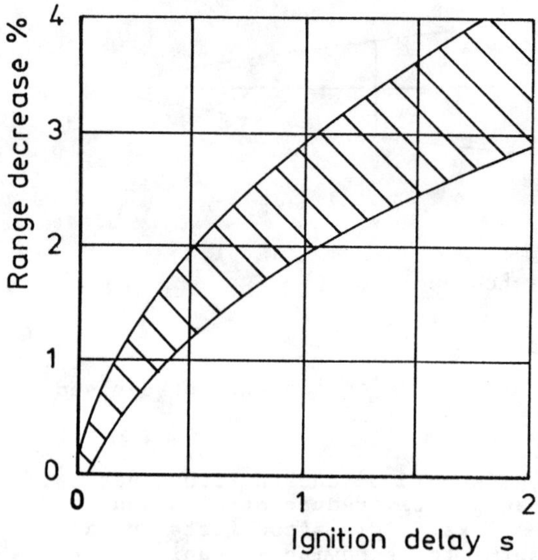

FIGURE 2. Calculated range decrease versus ignition delay for a medium caliber low drag base-bleed projectile. Shaded area indicates possible variations due to the size of the base-bleed unit and due to flight conditions.

Due to high gun pressure the regression of the igniter composition is conciderably higher in the course of the interior ballistics than outside the barrel. Several mm of igniter composition length is usually burned in the barrel. This has to be taken into consideration when designing ingiters.

The first part of the propellant grain combustion takes place at high pressure while the projectile travels along the barrel. The regression rate is high but the duration very short. The propellant layer burned in the gun is generally relatively thin and usually not determinative for the grain dimensions.

During free flight is the mass flow through the nozzle low and subsonic. The pressure in the combustion chamber exceeds the base pressure only slightly. Due to the effect of the base-bleed the chamber pressure differs very little from ambient pressure. The ambient pressure is degressive along the trajectory up to apex. This results in a degressive regression rate of the propellant. It should be noted that regression rates and thus mass flow and duration of burning are functions of elevation.

The projectile might eventually reach an altitude at which the pressure is so low that propellant combustion is no longer possible. It is vital to determine regression rates and to study propellant combustion from sea level atmospheric pressure down to pressures at which extinction occures.

3. DISPOSITION OF TEST METHODS

The base-bleed unit appears simple in its design. It is however evident that several severe requirements, many with small margins, are to be met. Theoretical calculations may be performed and can elucidate some problems but due to the complexity of the over all process only experimental tests can finally ensure reliable and good performance.

During the research leading to the invention of the Swedish base-bleed system and during the continued development a number of different methods and experimental procedures for testing parts as well as the complete system have been developed and applied.

A summary of some methods is presented in fig 3. In this figure is indicated which tests might be considered as essential or of value. As appears, in some cases standard methods already available have been applied, sometimes however with alterations. In some cases new, comprehensive facilities have been designed and built especially for the base-bleed research, development and testing.

METHOD \ OBJECT	Propellant	Propellant grain	Igniter	Mechanical parts	Compl base-bleed unit
Combustion bomb tests	●		●		
High/low pressure bomb tests		●	●		
Tensile tests	●				
Spin tests	○	●	●		●
Linear acceleration tests	○	●		●	
Supersonic wind tunnel tests	○	●			●
Aeroballistic range tests		●	●	●	●
Gun firing tests	○	●	●	●	●
Environmental tests		●	●	○	●
Aging tests	●	●	●		●

● essential

○ of value

FIGURE 3. Summary of test methods.

4. LABORATORY METHODS AND EXPERIMENTAL FACILITIES

4.1 Combustion Bomb Tests

Subatmospheric to high pressure propellant combustion. As indicated earlier the pressure in the base-bleed unit combustion chamber is normally close to ambient pressure. Starting from gun site atmospheric conditions ambient pressure is decreasing up to apex. An advanced projectile trajectory apex might be as high as 15 km. It is however probable that the optimum end of combustion should occur at a lower altitude. The range of interest for combustion studies is mainly at subatmospheric pressures. Regression rates should be determined down to the pressure at which extinction occurs. The combustion behaviour such as occurance of oscillatory combustion should be observed and measured. Although oscillations can be observed in combustion bomb tests, the combustion will tend to be stable in a real base-bleed combustion chamber where the thermodynamic situation is different and usually more favourable. Strand burning regression rates appear to be quite accordant with base-bleed combustion regression provided correction is made for the influence of spin. Combustion tests at subatmospheric pressures are mostly performed in a strand burner window bomb of 500 cm^3 volume connected to a surge tank of 1 m^3 volume as shown in fig 4. The surge tank is connected to a vacuumpump allowing the bomb and tank system to be evacuated to the desired pressure. The system is fitted with valves necessary for operation. Pressures are measured at several locations.

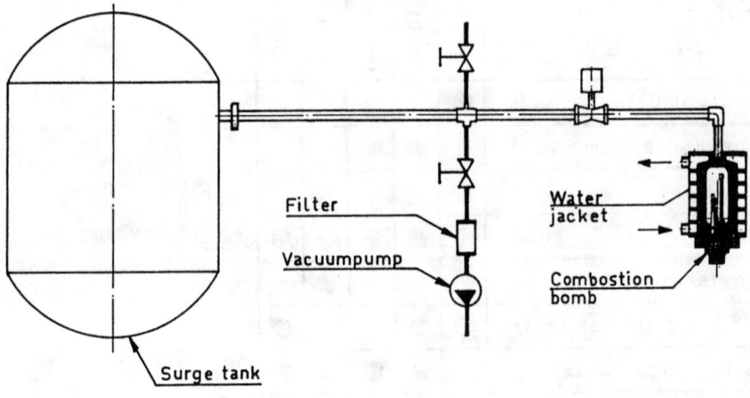

FIGURE 4. Subatmospheric propellant combustion test facility.

A majority of combustion bomb tests have been performed with propellant strands of 5 by 5 mm square cross section and 120 mm length. The strands are inhibited on the sides.

Ignition is accomplished with an electrically heated iron wire. Fine, melting silver wires are used for determination of the average regression rate.

Examples of test results are shown in fig 5 for two different HTPB propellants. Regression rates at pressures higher than atmospheric determined in the same combustion bomb are shown for one of the propellants.

Igniter combustion. During research and development of igniters high pressure combustion bomb tests are valuable. In this case a cylindrical bomb with 100 cm^3 internal volume and maximum 200 MPa pressure is used. The bomb is fitted with transducers for measurements of pressure and light emission.

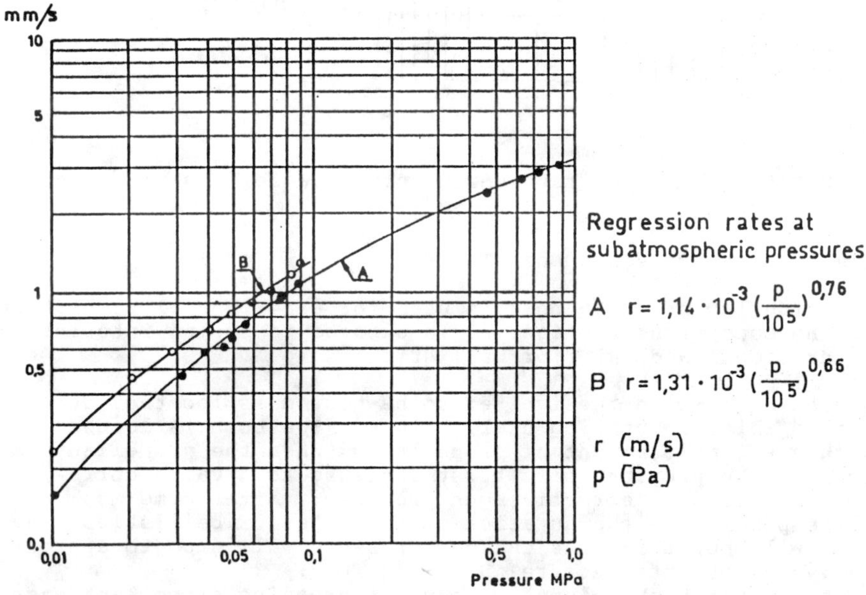

FIGURE 5. Regression rate versus pressure for two different HTPB propellants.

It has also connections for the ignition of the igniters to be tested. Test igniters with net diameter 21 mm and length 15 to 20 mm are generally used.

Influences of different variables such as composition, components particle sizes and compacting pressure have been studied in order to find suitable igniters for different designs. As examples regression rates for two igniters are shown in fig 6.

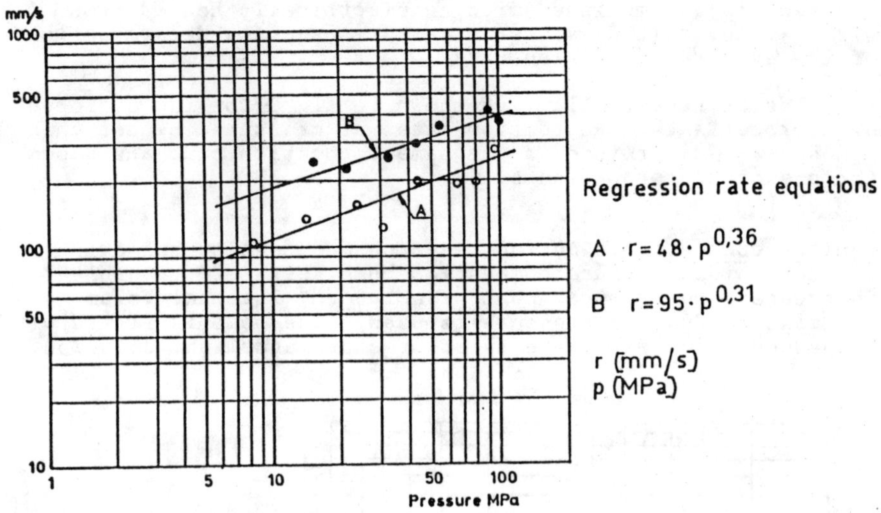

FIGURE 6. Regression rate versus pressure for two different igniter compositions.

4.2 High/Low Pressure Bomb Tests

The high/low pressure bomb is used for experimental simulation of the conditions the igniter is exposed to during interior, transitional and exterior ballistics.

The bomb in which the igniter to be tested is located, is fitted which a bursting disk. Fig 7. The bomb is loaded with a suitable amount of propellant. When the propellant is ignited the pressure will first increase to a value corresponding to some effective gun pressure. After some ms, corresponding to the duration of the interior ballistics, the disk will burst and the chamber pressure will drop to approximatively ambient pressure.
By balancing bomb volume, pressure increasing propellant mass, nozzle area and disk burst conditions an acceptable simulation of the influence on the igniter combustion can be attained.

4.3 Tensile Tests

In the gun the propellant grain is subjected to very strong axial and radial accelerations. In addition the base-bleed system should be working over a wide temperature range, such as −40 to +60°C. Determination of the mechanical properties of the propellant over the actual temperature range is most important. These properties will in turn determine the maximum grain size. It is convenient to have a series of propellant compositions developed and tested in order to make the optimum choice for each design.

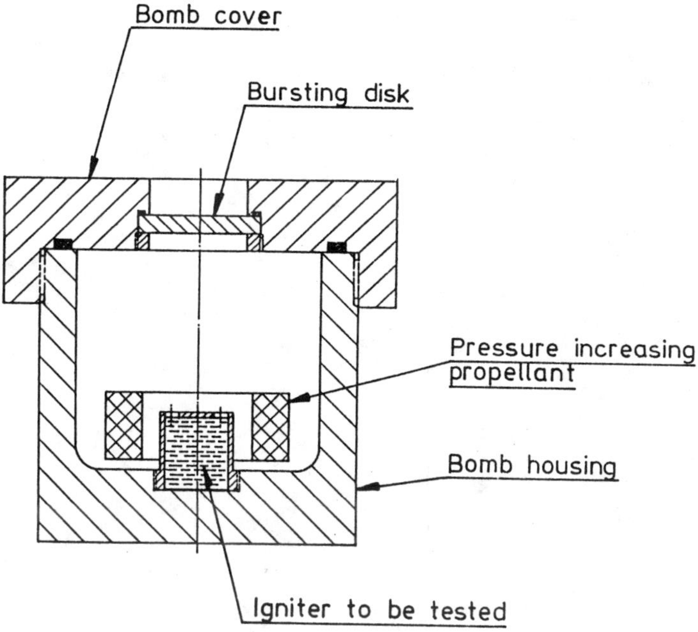

FIGURE 7. High/low pressure igniter combustion test bomb.

Tensile tests are made with JANAF specimens in a tensile test machine with cross head speeds between 1 and 500 mm/s. Tests are made at different strain rates and temperatures. The results are presented in master curves. Examples of results are shown in fig 8 - 10.

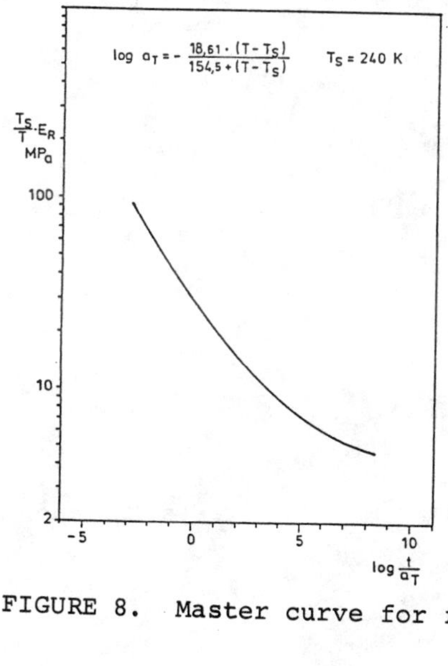

FIGURE 8. Master curve for relaxation modulus

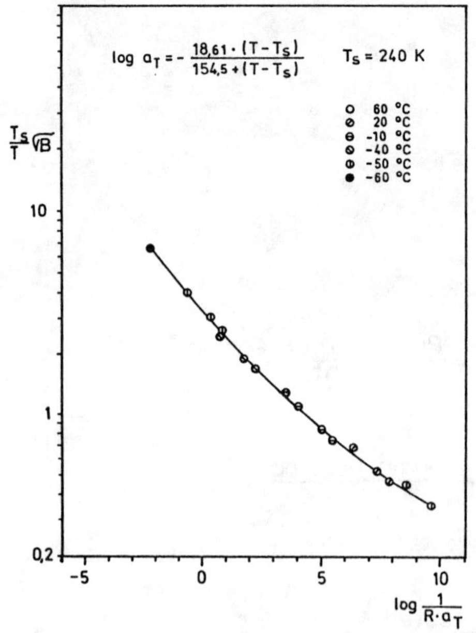

FIGURE 9. Master curve for maximum stress

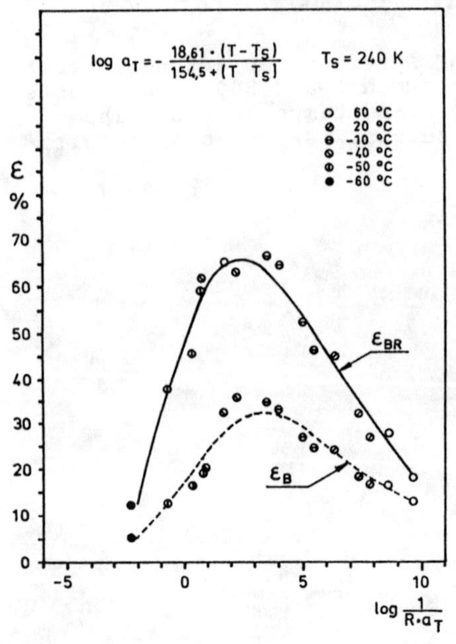

FIGURE 10. Master curve for strain at maximum stress (ε_B) and strain at break (ε_{BR})

4.4 Spin Tests

Spinning solid propellant rocket motors perform differently from nonspinning motors and the same applies to spinning base-bleed units with solid propellant grains. The most important consequences of the spin are increased burning rate due to radial acceleration, altered combustion chamber flow and nozzle flow. Spin also influences the ignition process. Consequently spin tests are most valuable.

A spin stand designed and built at FOA (Swedish Defence Research Establishment) is shown in fig 11. An electric motor rotates the shaft by way of a hydraulic drive at variable spin rates corresponding to actual projectile rotation, up to about 20 000 rpm. The shaft in which transducers are situated, is journalled by two ball bearings. The main end of the shaft carries the test unit to be spun, the other end is connected to a slip ring assembly. The bearing housing is fixed to a steel plate which rests on two spring blades. Pressures can be measured at different locations through pressure taps. The signals from the pressure transducers are carried via the slip ring assembly to the signal processing equipment. It is also possible to use radio transmission for the signal transfer. Ignition is accomplished via the slip ring assembly or via the combustion chamber nozzle.

FIGURE 11. Spin stand facility

Spin tests are usually performed in full scale to study the effect of spin on propellant grain combustion and on chamber and nozzle flow and blocking. Fig 12 shows a photograph from a spin test. Other spin tests may also be performed such as igniter spray tests and grain deformation tests. In the first case the actual igniter is used in combination with a dummy propellant grain. The spray pattern of the igniter jet will be clearly marked on the dummy grain surfaces and can be studied after the test.

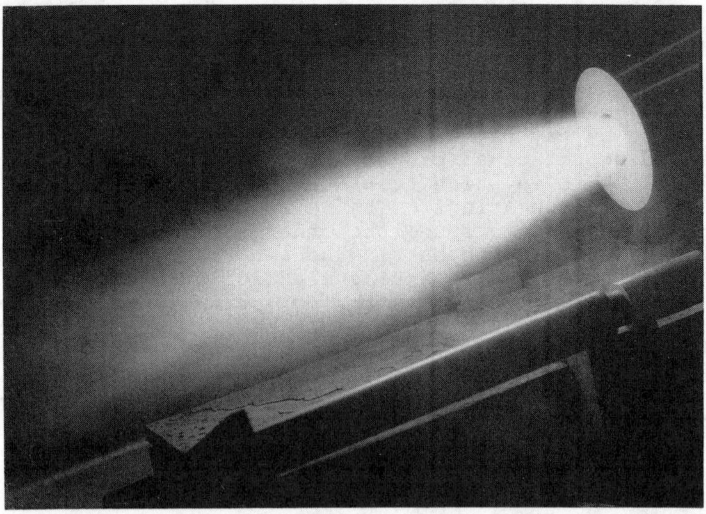

FIGURE 12. Spin stand test of base-bleed unit combustion at 15 000 rpm.

FIGURE 13. Deformation of propellant grain in open ended combustion chamber spinning at 12 000 rpm.

Grain deformation due to spin can conveniently be studied by stroboscopic photographing of a grain in a spinning open ended combustion chamber. The deformation of a segmented grain without central channel is shown in fig 13. External axial flow is generally not provided for in a spin stand of this design. Hence only internal processes of base-bleed units can be studied. On the other hand this can readily be made in full scale.

The interaction of base-bleed and external ambient flow is ordinarily studied in supersonic wind tunnels. At FOA a spinning experimental base-bleed unit has been designed and built for this purpose.

4.5 Linear Acceleration Tests

Soft recovery of gun fired projectiles has always been an interesting and important task. In 1966 was at the proving ground of FOA designed and built a very special linear acceleration and soft recovery facility. It has performed excellently over the years and is still in current use.

The basic element is a recoilless 84 mm caliber smooth-bored gun in a heavy experimental version. This gun is fixed to a long support and the barrel is elongated to 275 m by means of pipes. Fig 14. When the gun is fired an increasing pressure is built up in front of the projectile as it travels along the barrel. The pressure behind the projectile is continously decreasing. The projectile is softly retarding and will come to a stop after about 100 m. Compressed air is used to move the projectile out of the barrel.

FIGURE 14. Linear acceleration test facility

By changing projectile mass and gun propellant mass, linear accelerations up to 400 000 m/s^2 (40 000 g) can be obtained. Numerous tests with propellant samples and grains have been carried out in order to obtain data on mechanical properties of the propellant as well as to study deformation, occurrence of cracks etc of the grains.

Mechanical behaviour of igniters subjected to strong accelerations has also been studied. The facility is well suited to acceleration tests of mechanical parts and electronics.

4.6 Supersonic Wind Tunnel Tests

The supersonic wind tunnel at FOA, used for base-bleed studies, is a blow down type of tunnel with fixed nozzles for M = 1.81 and M = 2.65. Fig 15. The nozzle test section diameters are 125 mm and 170 mm respectively. The afterbody model is supported by a central cylindrical tube extending from upstream the tunnel nozzle. Thus no problems with support interference will occur. Model external diameters are mostly 50 mm.

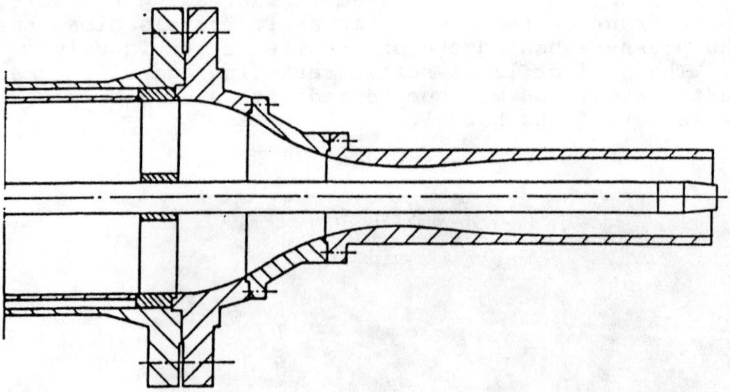

FIGURE 15. Supersonic wind tunnel test section. Nozzle with central support tube and after body model with base-bleed unit. Test section can be open or closed for tests at subatmospheric pressures corresponding to flight at high altitudes.

Several non spinning models of different base-bleed unit designs have been built and tested. The unit is located within the afterbody model and consists of the combustion chamber with the propellant grain and the base with the nozzle system to be tested.

Ignition of the propellant grain is accomplished through the base nozzles by means of the jet from a pyrotechnic igniter mounted on a retractable support.

Pressures in the combustion chamber and on the after-body surface are measured by means of a large number of taps connected to scanning transducers. Lines from the transducers are fed through the support tube.

In a typical test, the tunnel is run some seconds after the base-bleed has ceased. Thus data with as well as without base flow are obtained during the same test run.

A support device for wind tunnel testing of spinning base-bleed units has also been designed and built. The device is mounted to the central axial support tube. Air bearings are used. Spin rates up to about 40 000 rpm are brought about by means of an air turbin and spin rates are measured wiht a photo-cell system. By measuring the pressure in the chamber stabilising the axial position, total axial force and base drag can be calculated.

Several hundreds of base-bleed tests have been carried out with variables such as propellant composition, grain configuration, nozzle configuration, nozzle area and afterbody configuration. Also further advanced systems related to base-bleed, such as external burning, have been investigated.

The large number of test results have made possible the formulation of very reliable performance data which in turn have been introduced into our exteriour ballistics code. Very good agreement between calculated and complete gun firing results are demonstrated over a large range of calibers, sizes and applications.

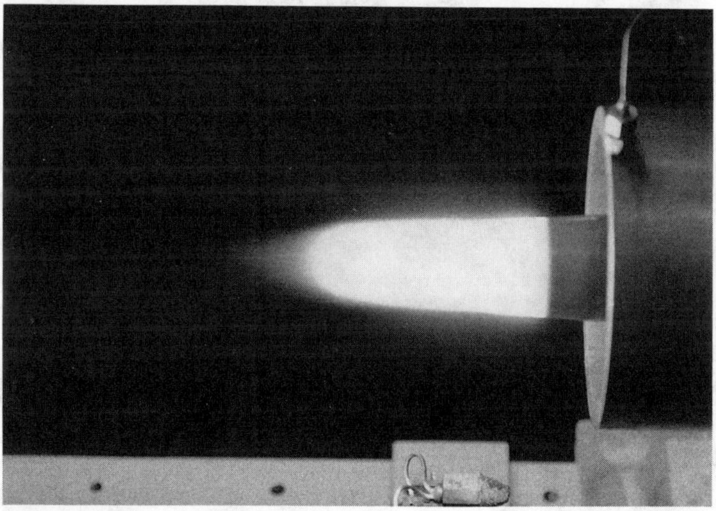

FIGURE 16. Supersonic wind tunnel test of base-bleed unit combustion at M=1.81.

5. GUN FIRING TESTS AND FACILITIES

5.1 Aeroballistic Range Tests

The purpose of aeroballistic range tests is mainly to determine aerodynamic coefficients and stability factors of projectiles in free flight. The flow around free flying projectiles with and without base-bleed can be studied. Since range tests concern the very first part of the trajectory the base-bleed ignition process may be observed simultaneously.

In Sweden is a 240 m long, 11.5 m wide and 6 m high covered aeroballistic range named Nordbanan at the proving ground of FFK in Karlsborg. In this range it is possible to fire rounds of up to 155 mm calibre.

The range of measurements extends over 230 m and includes 93 yaw-cards and 10 photographic stations. Fig 17. Each station is equipped with two Hasselblad 500 EL cameras for horisontal and vertical photographs perpendicular to the trajectory, and flashes of very short duration. By focusing the cameras on white back ground screens the projectile attitude, and the shadowgraph of the flow field can be registered. The camera shutters open shortly before the projectile passes and closes shortly afterwards. The actual exposures are brought about by the very short flashes. When applying the method with open cameras and flashes the strong light from the base-bleed combustion might interfere with the shadowgraph. However there seems to be no difficulties in using yaw-cards for base-bleed projectile firings.

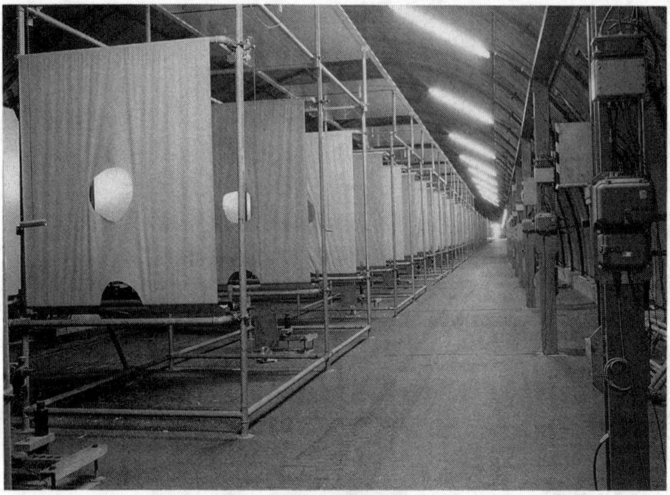

FIGURE 17. Indoor picture of the aeroballistic range at Karlsborg proving ground.

5.2 Ballistic Firing Tests

Ballistic firings offer the ultimate tests of the performance of the complete base-bleed projectiles. The fundamental tests concern range and dispersion at different angles of elevation and muzzle velocities. In Sweden this kind of firings are often made over water. The impact is usually measured by radar and/or cine-theodolites.

Firings in darkness with good visibility offer excellent conditions for studying the performance of base-bleed units during flight. The ignition process as well as the combustion process - duration, stability etc - can be observed and recorded. Registrations are made with open shutter cameras.

During most ballistic test firings Doppler radar measurements are performed. The results are generally presented as drag coefficients, C_D, versus Mach-number and versus flight time and provide important data for use e.g. in aeroballistic calculations.

In general all malfunctions and disturbances of importance of the base-bleed process will cause observable devergences of the drag coefficient from normal values.

In fig 18 is shown measured C_D versus Mach number for an experimental 155 mm base-bleed projectile. As can be seen the base-bleed unit of round 3 is not working correctly. The ignition is retarded, causing high C_D at start. The base mass flow is displaced towards the later part of the process.

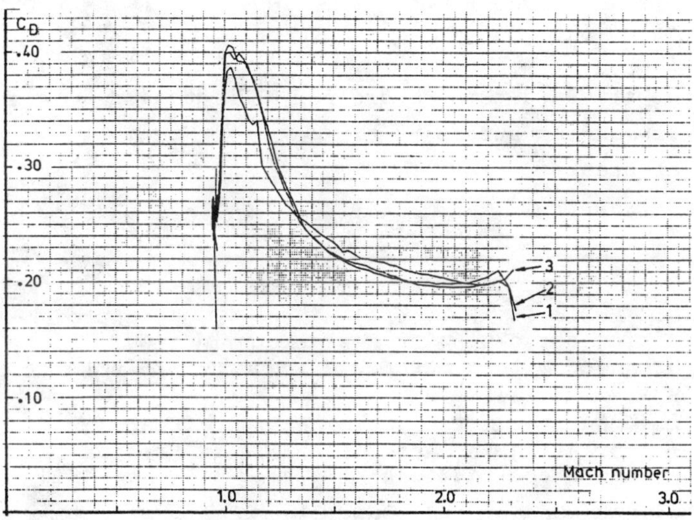

FIGURE 18. Measured drag coefficients versus Mach number for an experimental 155 mm base-bleed projectile.

6. CONCLUSIONS

The base-bleed unit is an apparantly simple system with no moving parts. It is however evident that the over all process is very complex and that several severe requirements are to be met. Analytic design methods and data are still insufficient. Thus experimental tests for research and development of parts and subsystems as well as complete base-bleed units are of utmost importance. Experiments are also necessary for obtaining basic data such as regression rates, mechanical properties and aeroballistic coefficients. However good theories might be developed testing will still be needed for proving reliable and efficient performance.

In this report some of the more important test methods and experimental procedures now in use have been presented. New methods are however currently beeing developed. In later years several laser based methods have been introduced. Fiber optics and micro electronics render possible new transducers, data transfer and data prcessing technique. Larger computers and computer codes will be developed .

7. NOMENCLATURE

a_T	temperature shift factor in the WLF-equation	
C_D	drag coefficient	
E_R	stress relaxation modulus	MPa
p	pressure	MPa
r	regression rate	mm/s
R	strain rate	s^{-1}
T	temperature	K
T_s	reference standard temperature for WLF-equation	K
ε_B	strain at maximum stress	%
ε_{BR}	strain at break	%
σ_B	maximum stress	MPa

8. REFERENCES

1. Gunners, N-E., Hellgren, R.V., Liljegren, T., Kanonprojektil med anordning för ökning av dess skottvidd (Gun projectile arranged with a base drag reducing system) Swedish Patent 340076, 1969

2. Andersson, K., Gunners, N-E., Hellgren, R.V., Swedish Base-Bleed - Increasing in Range of Artillery Projectiles through Base Flow *Propellants and Explosives* 1, Oct 1976

3. Bartelson, N. and Lundahl, K., Base-Bleed - A System to Increase Range and Decrease Ballistic Dispersion *Proc. 3rd International Symposium on Ballistics, March 23-25, 1977, Karlsruhe*, Deutsche Gesellschaft für Wehrtechnik

4. Andersson, K., Bartelson, N. and Gunners, N-E., Afterbody Drag of Projectiles with Base Drag Reducing Systems *Proc. 4th International Symposium on Ballistics, October 17-19, 1978*

5. Gunners, N-E., Hellgren, R., Liljegren, T. and Nilsson, Y., Basflödesaggregat för 155 mm sgr m/77 för haubits F FOA rapport C 20305-D3, May 1979

6. Liljegren, T., Erfarenheter från tillverkning av krut för basflödesladdningar till 105 mm stgr m40/74 FOA rapport C 20350-D1, April 1980

7. Hellgren, R.V., Range Calculation for Base-Bleed projectiles *Proc. 6th International Symposium on Ballistics Orlando, Florida, Oct 1981*

8. Englund, C-E., Sammanställning av teori och resultat från undersökningar av stabilitet och aerodynamiska koefficienter hos friflygande projektiler FOA rapport C 20690-2.1, Dec 1987

9. Gunners, N-E., Andersson, K., Hellgren, R., Base-Bleed Systems for Gun Projectiles in *Gun Propulsion Technology AIAA Progress in Astronautics and Aeronautics* Vol 109, 1988 ed. L. Stiefel, pp 537-562

COMMENTS

G. Laube, Defense Technology Agency - Switzerland--Regarding Figure 18, is the C_D point measured by doppler radar? If yes, at which distance rep time from muzzle? If no, couldn't this be the effect of combining two different measurement systems (e.g., for V_0 and first doppler) for drag calculation?

Reply: The C_D-curves are deduced from continuous velocity measurements by the same doppler radar system. Measurements start less than 100 m from the muzzle and cover the greater part of the exterior trajectory during which igniter combustion occurs.

**

T. Wijker, Muiden Chemie - Netherlands--1. Have you done any investigations concerning the width of the slots? Can you tell something about the results? 2. You mentioned that the design of base bleed with no inhibitor on one of the ends of the grain gave problems. Can you tell us something more about these problems?

Reply: 1. Throughout the years several investigations concerning slot design and slot width have been performed. For most medium caliber base-bleed systems approx. 1 mm slot width seems to be adequate. 2. The one end non-inhibited grain was used for a very special application commented very shortly during the presentation of our paper. It is not of any great importance for ordinary base-bleed system designs. Unfortunately, we cannot give any further information on these tests.

**

W. L. Chow, Florida Atlantic Univ. - USA--Do you design the combustion process for base-bleed according to trajectory? If so, firing at different elevation angles would result in different trajectories and different performances. It is a very interesting paper.

Reply: As indicated in the question, the performance of the base-bleed system depends on the flight trajectory and thus on the elevation angle. Hence, the base-bleed combustion process has to be designed according to the trajectory. For long range projectiles the process is normally optimized for maximum range. However, other design criteria are also possible, such as minimum flight time to a given range. In any case, a special firing table taking into account the effect of the base bleed has to be prepared.

**

B. Zeller, SNPE - France--Dr. Gunners, did you consider performing tensile tests under pressure as, in the gun, the environmental pressure is very high and may significantly affect propellant mechanical properties?

Reply: Tensile tests are normally performed at atmospheric pressures and are used for comparisons between different propellants. They are also correlated to results from gun firings. Tests at gun pressures would be most interesting. They are however very difficult experimentally and have yet not been performed.

**

K. K. Kuo, Penn State Univ. - USA--Would you please comment on the effectiveness of inhibitors of propellant grain segments? Are propellant grains cast with inhibitor on their surfaces? Would you please comment on the type of inhibitor material used?

Reply: The inhibitor we use is based on the same binder as used in the propellant. It is cast on the cured propellant grain. This choice of inhibitor materials offers a high bond strength and good compatibility and has proven to be very effective.

**

D. Chargelegue, Establishement Technique de Bourges - France--You said you made tests with spinning base bleed units in your supersonic wind tunnel. Did you make tests with different spin rates? If yes, did you see an evolution of the mass flow rate with the spin rates?

Reply: Wind tunnel tests have mainly been performed at spin rates corresponding to actual projectile spin. The wind tunnel test results are in consistency with long range firings as well as spin stand tests performed at these spin rates. Spin stand tests varying from 0 to about 300 r/sec show an evolution of the mass flow with the spin rate.

**

Analysis of Dispersion of Base Bleed Projectile

XIFU GUO
Department of Engineering Thermal Physics and Aerodynamics
East China Institute of Technology
Nanjing, PRC

ABSTRACT

The dispersion of shot is an important performance index of a base bleed projectile. Correctly estimating the characteristic quantity of dispersion of the base bleed projectile, effective reducing the dispersion of the projectile, reasonably determining the dispersion index are problems of great concern for projectile designers and ballisticians.

In this paper, the dispersion of the base bleed projectile has been estimated, and the methods for reducing the influence of nutation and the drag dispersion with base bleeding and the highaltitude wind on the dispersion of projectile have been introduced. Then, the effects of various factors on the dispersion of projectile have been analyzed, and it is pointed out that the initial velocity dispersion, nutation dispersion, gusty wind and drag dispersion with base bleeding are the main factors affecting dispersion of projectile. Because the drag of base bleed projectile is so weak that the effect of the drag dispersion on over all–dispersion will be more sensitive. Factors affecting dispersion of the base drag with base bleeding and the influence of the dispersion of ignition time on over all–dispersion have been discussed. The methods of reducing dispersion have been stated, and the methods of determining the index of dispersion and of making reasonable distribution have been demonstrated. The firing test have proved that the theory and methods given in this paper are verified.

1. INTRODUCTION

A base bleed projectile can increase range efficiently, and the increase of range rate is normally about 20~30%. Under the condition of high initial velocity, the less the wave drag of the projectile, the more the proportion of the base drag to total drag, and also the more noticeable the effect of the increasing range of base bleeding. Hence, the increasing range of base bleeding especially suitable for the low drag projectile with the shape of high fineness ratio head.

Owing to the structure characteristic of the base bleed projectile, it is very probable that slight carelessness in design causes excess large dispersion of shot. The reasons of large dispersion can be summed up in following respects: 1. For large calibre and long–range projectile, charge weight is more, the ignition and flame transfer are not uniform, it is easy to cause the large dispersion of initial velocity; 2. Due to rather large fineness ratio of the projectile, long gun barrel, high bore pressure, spin rate of projectile, it is not easy to centre the projectile in bore so that the considerable yaw dispersion of projectile during launch period is caused; 3. Due to extended range, high altitude and large time of flight, a long–range projectile is easy to be affected by random change of meteorological elements; 4. Because of the additional base bleed unit, the stability of propellent gas flow and dispersion of ignition, etc, which are related to the base bleed unit immediately, will cause despersion of base drag; 5. Because base bleed projectile has the shape of low drag, the sensitivity of projectile for the initial velocity and drag dispersion increases as drag decreases. But the relative variation of drag dispersion increases, so that the dispersion of shot further increases.

In this paper, based on theoretical analysis and test data, the estimating method of the dispersion of base bleed projectile has been researched, the factors causing dispersion have been analyzed, the ways of decreasing dispersion have been pointed out, the dispersion index have been discussed. Theoretical foundations and actual methods of reducing the dispersion of base bleed projectile have been provided, in order to realize the overall optimum design of base bleed projectile, and estimate the dispersion elements of the firing table of base bleed projectile.

2. ESTIMATION OF DISPERSION PROBABLE ERROR OF BASE BLEED PROJECTILE

The flight stability of the base bleed projectile is a necessary condition to meet the dispersion index. In estimating the dispersion of shot, the gyroscopic stability, dynamic stability and trailing stability of the base bleed projectile must meet the requirement along overall trajectory. Under these conditions, the dispersion probable error of the falling point of base bleed projectile can be calculated by following formulas:

$$E_x^2 = \sum_{i=1}^{n}\left(\frac{\partial X}{\partial \alpha_i} E_{\alpha_i}\right)^2$$

$$E_z^2 = \sum_{i=1}^{n}\left(\frac{\partial Z}{\partial \alpha_i} E_{\alpha_i}\right)^2 \qquad (1)$$

Where E_x–probable error in range; E_z–probable error in deflection; α_i–factors related with the range and the deflection, such as the initial velocity, v_0, the ballistic coefficient, c, the weight of the projectile, G, the range wind, W_x, the cross wind, W_z, the ignition time of base bleed unit, t_H, etc. $E_{\alpha i}$–the probable errors of dispersion of factors (α_i), such as E_{V_0}, E_G, E_C,E_{wx}, E_{W_z}, etc. $\partial X / \partial \alpha_i$–the sensitivity of the factors α_i for range, $\partial Z / \partial \alpha_i$–the sensitivity of the factors α_i for deflection. Hence the dispersion of range and deflection caused by the dispersion of factor α_i can written as

$$E_{x\alpha i} = \frac{\partial X}{\partial \alpha_i} E_{\alpha_i}$$

$$E_{z\alpha i} = \frac{\partial X}{\partial \alpha_i} E_{\alpha_i} \qquad (2)$$

This is, the dispersion of falling point caused by factor α_i is a product of its sensitivity factor with its probable error. When the dispersion is analyzed, the two respects–the sensitivity factor and dispersion amount must be considered.

3. TREATING METHOD OF FALLING POINT DISPERSION CAUSED BY NUTATION DESPERSION, DRAG DISPERSION WITH BASE BLEEDING AND GUSTY WIND

3.1 Falling Point Dispersion by Nutation

The nutation dispersion causes not only the range dispersion but also the deflection dispersion. They are discussed respectively as follows:

3.1.1 Estimation of range dispersion caused by nutation dispersion. Several methods are described as follows:

a. Method of ballistic equation

Using equation of motion for a modified mass point or ballistic equation of six degrees of freedom, the range X_i corresponding to the individal nutation amplitude δ_{mi} can be calculated and the $E_{x\delta 1}$ can be written as

$$E_{x\delta_1} = 0.6745\sqrt{\frac{\sum_{i=i}^{n}(X_i - \overline{X})^2}{n-1}} \tag{3}$$

$$\overline{X} = \frac{\sum_{i=1}^{n} X_i}{n}$$

b. Method of equivalent initial velocity

The influence of range can be considered as the influence on the initial velocity, v_0, if the change of initial velocity caused by nutation δ_{mi} is Δv_{oi} , the their relation is

$$\Delta V_{oi} = a\left(\frac{\delta_{mi}}{10}\right)^2 \tag{4}$$

then, the probable of initial velocity caused by the nutation dispersion is givn by

$$E_{V_0\delta} = 0.6745\sqrt{\frac{\sum_{i=1}^{n}\left[\Delta V_{oi} - \left(\frac{\sum_{i=1}^{n}\Delta V_{oi}}{n}\right)\right]^2}{n-1}} \tag{5}$$

The range dispersion caused by the nutation dispersion is given by

$$E_{X\delta_1} = \frac{\partial X}{\partial V_0} E_{V0\delta} \tag{6}$$

c. Method of ballistic coefficient

Under the condition of existing nutation, the ballistic coefficient of projectile is given by

$$C_i = C_0\left(1 + K\delta_i^2\right) \tag{7}$$

$$\delta_i^2 = \frac{\int_0^{t_1} \delta^2 dt}{T} \tag{8}$$

where C_0 is the ballistic coefficient when nutation is zero; K is the coefficient related with the projectile shape and Mach number, M; t_1 is the time when nutation is damped to zero; T–total time of flight.

The dispersion of the ballistic coefficient caused by the nutation dispersion is given by

$$E_{C\delta} = 0.6745\sqrt{\frac{\sum_{i=1}^{n}(C_i - \bar{C})^2}{n-1}}$$

$$\bar{C} = \frac{\sum_{i=1}^{n} C_i}{n} \tag{9}$$

The probable error in range is given by

$$E_{X\delta_1} = \frac{\partial X}{\partial C} E_{C\delta} \tag{10}$$

d. Method of nutation deviation

Firstly, calculate

$$\frac{\partial X}{\partial \delta_m} = \sqrt{\frac{\sum_{i=1}^{n}\left(\frac{\partial X}{\partial \delta_{mi}}\right)^2}{n}} \tag{11}$$

Secondly, calculate

$$E_{\delta_m} = 0.6745\sqrt{\frac{\sum_{i=1}^{n}(\delta_{mi} - \bar{\delta}_m)^2}{n-1}}$$

$$\bar{\delta}_m = \frac{\sum_{i=1}^{n}\delta_{mi}}{n} \tag{12}$$

finally, calculate

$$E_{X\delta_1} = \frac{\partial X}{\partial \delta_m} E_{\delta_m} \tag{13}$$

3.1.2 Dispersion of aerodynamic jump caused by nutation dispersion.
The aerodynamic jump caused by the nutation is

$$\psi_i = \frac{\dot{\delta}_{oi}}{V_{oi}} \frac{R_A^2}{h} \tag{14}$$

$$E_\psi = 0.6745\sqrt{\frac{\sum_{i=1}^{n}(\psi_i - \bar{\psi})^2}{n-1}} \tag{15}$$

$$\bar{\psi} = \frac{\sum_{i=1}^{n}\psi_i}{n}$$

where $\dot{\delta}_{oi}$ – the initial nutation angle velocity: R_A – the equator radius of the gyration of projectile; h–the distance of drag centre to mass centre of the projectile.
The range and deflection dispersion caused by E_ψ is respectively given by

$$E_{X\delta 2} = \frac{\partial X}{\partial \theta_0} E_\psi \tag{16}$$

$$E_{Z\delta} = XE_{\psi} \tag{17}$$

finally, the falling point dispersion caused by the nutation is yielded by

$$E_{X\delta} = \sqrt{E_{X\delta 1}^2 + E_{X\delta 2}^2}$$
$$E_{Z\delta} = XE_{\psi} \tag{18}$$

3.2 Range and Deflection Dispersion Caused by Gusty Wind

To calculate the influence of the gusty wind at every point on trajectory on the falling point dispersion, firstly, the method of calculating the falling point dispersion caused by the gusty wind of every point on trajectory need to be discussed. For the point i, the mean wind $\overline{Wi}$ and the mean square error of the gusty wind σ_{wi} have a relation in the following

$$\sigma_{wi} = \psi_{wi} \overline{W}_i \tag{19}$$

where ψ_{wi}–the relative intensity of wind.
If the falling point dispersion caused by the gusty wind at every point along the trajectory is equal to the mean square error of the qusty wind, that is considered the same along the trajectory, hence, following formula can be yielded

$$\left(\frac{\partial X}{\partial W}\sigma_{wb}\right)^2 = \sum_{i=1}^{n}\left(\frac{\partial X}{\partial W_i}\sigma_{wi}\right)^2$$
$$\sigma_{w_b} = \sqrt{\sum_{i=1}^{n}(q_i\sigma_{wi})^2}$$
$$q_i = \frac{\partial X}{\partial W_i} \Big/ \frac{\partial X}{\partial W} \tag{20}$$

where σ_{wb}–mean square error of ballistic gusty wind; $\partial X / \partial W$ –the change of range caused by a unit wind along the trajectory; $\partial X / \partial W_i$ – the change of range caused by unit wind at point i , q_i –the layer–weight of layer i.
If ψ_{wi} is constant along the trajectory, or $\psi_{wi} = \psi$, hence

$$\sigma_{wb} = \psi\sqrt{\sum_{i=1}^{n}(q_i\overline{W}_i)^2} \tag{21}$$

generally, $\psi = 0.1$
for the range wind,

$$\sigma_{wxb} = \sqrt{\sum_{i=1}^{n} (q_i \psi_{wi} \overline{W}_{x_i})^2} \tag{22}$$

for the cross wind,

$$\sigma_{WZb} = \sqrt{\sum_{i=1}^{n} (q_i \psi_{w_i} \overline{W}_{z_i})^2} \tag{23}$$

The probable errors of ballistic range wind and cross wind are respectively given by

$$E_{Wxb} = 0.6745 \ \sigma_{wxb} \tag{24}$$

$$E_{Wzb} = 0.6745 \ \sigma_{wzb} \tag{25}$$

Hence, the probable errors of the falling point caused by the gusty wind are given by

$$E_{XWX} = \frac{\partial X}{\partial W_x} E_{Wxb}$$

$$E_{ZWZ} = \frac{\partial Z}{\partial W_Z} E_{Wzb} \tag{26}$$

3.3 Falling Point Dispersion Caused by Dispersion of Base Bleed Drag

The influence of the drag dispersion of the base bleed projectile with base bleeding on the falling point dispersion is concerned with different times and positions on trajectory. To simplify calculating procedure, the concept of ballistic probable error of the base bleed drag coefficient has been presented. Based on following principle: the falling point dispersion caused by the ballistic deviation of the base bleed drag coefficient is equal to that caused by the dispersion of the drag coefficient of every point on trajectory with base bleeding, the ballistic deviation of drag coefficient can be determined by

$$\left(\frac{\partial X}{\partial C_D} E_{CDb}\right)^2 = \sum_{i=1}^{n} \left(\frac{\partial X}{\partial C_{D_i}} E_{CDi}\right)^2 \tag{27}$$

$$E_{CDb}^2 = \sum_{i=1}^{n} q_{CDi}^2 E_{C_{Di}}^2$$

$$q_{C_{Di}} = \frac{\partial X}{\partial C_{Di}} / \frac{\partial X}{\partial C_D} \tag{28}$$

TABLE 1. Falling point dispersion caused by factors of dispersion

α_i	$E_{x\alpha_i}$ (m)	$E_{z\alpha_i}$ (m)
V_0	184	6.6
G	28.1	0
W_x	58.65	2.14
W_z	0	22.5
ρ	0	0
τ	0	0
h	0.046	0.022
A	0.118	0.43
B	0	0
t_H	2.56	0.23
δ_M	280.6	38.6
C_{DBB}	436.5	1.1
Z	0	39
ψ_0	6.31	3.9
ω_{BB}	6.31	0
E_Σ	544.4	59.6

where E_{CDb} −the ballistic probable error of the drag coefficient with base bleeding; $\partial X / \partial C_D$ −the change of range caused by an unit change of the drag coefficient with base bleeding; $\partial X / \partial C_{Di}$ −the change of range caused by an unit change of the drag coefficient of point i with base bleeding; q_{CDi} −the layer weight of drag coefficient at point i.

During base bleeding, if E_{CDi} is constant $\overline{E}_{CD}$, then

$$E^2_{CDb} = \overline{E}^2_{CD} \sum_{i=1}^{n} q^2_{CDi} \tag{29}$$

The range dispersion caused by the dispersion of drag coefficient with base bleeding is given by

$$E_{XCD} = \frac{\partial X}{\partial C_D} E_{CDb} \tag{30}$$

TABLE 2. Compare of theoretical result with experimental result

	$E_x(m)$	$E_z(m)$	E_x / X
theoretical result	544.4	59.5	1 / 72
experimental result	505.8	65.6	1 / 77

4. ESTIMATION AND ANALYSIS OF INFLUENCE OF DISPERSION FACTORS ON DISPERSION OF BASE BLEED PROJECTILE

4.1 Step of Estimation Dispersion of Base Bleed Projectile

a. Calculate sensitivity factors;

b. Calculate dispersion;

c. Calculate falling point dispersion caused by dispersion factors

d. Composition of probable error of dispersion.

4.2 An Example

For a projectile, the diameter is 155mm, the weight is 47.8Kg, the initial velocity is 850m / s, the angle of departure, is 52 ° , the burning time is 24s, how to estimate the dispersion of the projectile.

Based on above–mentioned formulas and steps of calculation, the probable error of falling point dispersion by the dispersion of factors can be calculated, the results are shown in table 1.

TABLE 3. Falling point dispersion in proportion to overall dispersion caused by the dispresion of any factor

αi	$E^2_{X\alpha i} / E^2_X$ 100%	$E^2_{Z\alpha i} / E^2_Z$ 100%
V_0	11.4	1.2
G	0.266	0
W_x	1.16	0.12
W_z	0	14.17
t_H	0.02	0
C_{DBB}	64.29	0.003
δ_M	26.57	41.16
ψ_0	0.013	0.042
Z	0	42.44

TABLE 4 The influence of base bleed units on the range

change of factor	change of range (m)
ω_B(1g)	5.34
T_B(1℃)	0.12
ρ_B(1%)	32.74
r_f(1%)	22
A_f(1%)	75
S_B(1%)	7.5

4.3 Example Analysis

Above–estimated theory results and test results are shown is table 2. They are in very good agreement. The falling point dispersion in proportion to overall dispersion caused by the dispersion of any factor is shown in table 3.

From table 3, main factors that cause the range dispersion of base bleed projectile can be shown. They are the drag dispersion with base bleed, nutation dispersion, dispersion of initial velocity and the dispersion of range wind, main factors that cause the deflection dispersion are drift dispersion, nutation dispersion and the dispersion of cross wind. The law obtained by lots of calculations and tests from this example is of universal significance. The formulas of the probable error of falling point considering above main factors are given by

$$E_x^2 = E_{XBB}^2 + E_{Xv_0}^2 + E_{XWX}^2 + E_{X\delta}^2$$

$$E_Z^2 = E_{ZZ}^2 + E_{Z\delta}^2 + E_{ZWZ}^2 \qquad (31)$$

Next, the causes of formation of main factors will be further analyzed, and the key points are the drag dispersion with base bleeding, nutation dispersion and drift dispersion.

(1) Drag dispersion with base bleeding. By using wind tunnel and flight test, it has been known that the drag dispersion with base bleeding is larger, but its order of magnitude is close to that of the drag dispersion of common high explosive projectile, and some time is larger. In the wind tunnel test, the relative drag dispersion of base bleed projectile is around 1.5%, the relattive drag dispersion measured by Doppler radar is 1.6 to 1.7%, the corresponding relative deviation of ballistic coefficient is 0.6% to 0.7%, the value of common high exposive projectile is mostly less than 0.5%, and the value of 175mm M483 high explosive projectile is only 0.2%. To analyze the cause of forming above phenomenon, the influence of each factors of base bleeding on the range has been calculated, and concrete data is shown in table 4.

TABLE 5 The influence of spin rate on the burning rate

spin rate (RPM)	r_f(mm / s)	r_f / r_{fo}
0	1.29	1.0
6000	1.47	1.14
7200	1.48	1.15
9000	1.51	1.17
10200	1.56	1.21
10800	1.60	1.24

It is shown from calculated results that the dispersion of burning rate and burning surface are two main factors that affect the drag dispersion of base bleed projectile, and the reasons for the existence of such phemomenon can be divided into interior and external factors, the stability of burning and ejecting of propellant, burning law of burning surface, spin rate and velocity dispersion of projectile, random swashing effect of gas flow of ignition propellant on bleed propellant, and sudden disturbance of meteorologic elements on trajectory, etc, all these would randomly affect the burning of bleed propellant and flowing of gas, and cause base drag dispersion, these have been proved by tests, it is shown from firing test that the dispersion of falling point is also large when the dispersion of burning time is large, the burning test is 28s and that in firing test is 24s, even decreases to 13s. In ground spin test, the burning rate would increase when the spin rate increases, the data is shown in table 5. Based on above tests and analysis, it is shown that the controlling of stability of burning rate and surface is very important to reduce the drag dispersion of base bleed.

(2) The nutation dispersion. Factors that cause the nutation are very complicated. From the moment when a projectile is charged into bore to the end after effect period, or, in overall launch process, factors that cause the nutation would exist all along. The malalignment of band–engraving, imbalance of projectile, the clearance between bourrelet and tube, non–stableness of in bore motion of projectile, the jump of pressure in bore and the vibration of barrel, etc, all these would cause random yawing of projectile in bore. For the low drag high explosive projectile of extended range with bourrelet nub, the position of nub on body, shape of nub, canted angle of nub would all affect greatly the motion of projectile in bore. For example, for some projectile, the intial maximum extent of the nutation is 15 ° , after improving the condition of the motion of projectile in bore, it is reduced to 5 ° . By using firing test and theoretical calcultation, it is proved that the precision would be improved if the nutation angle is controlled to be lesss than 5 ° .

(3) Drift dispersion. For a spin–stabilized projectile, the drift is caused by the repose angle, the dispersion of repose angle is the reason that affects directly the drift dispersion, and the dispersion of spin rate velocity would also cause the drift dispersion, the setting error of nub, especially the error of canted angle of nub would affect the dispersion of spin rate, and cause the

TABLE 6 The ballistic coefficient of different projectiles

projectile	ballistic coefficient
M107 155	0.51
M46 130	0.472
M62 152	0.529
ERFB 155	0.38
ERFB–BB 155	0.28

drift Dispersion, formula of the drift dispersion is given in the form.

$$E_{zz} = \gamma_z X tg\theta_0 \tag{32}$$

where E_{zz} –the probable error of drift dispersion; X–range; θ_0 –the angle of deqarture; γ_z –drift coefficient, its value is 0.0005 to 0.0002, generally adopted 0.001.

5. DISPERSION CHARACTER OF LOW DRAG PROJECTILE

To extend range, the drag must be reduced in a large extent. The drag of low drag extended range projectile is about half as large as that of old common high explosive projectile. If the ballistic coefficient is used as standard, its variation is shown in table 6.

To extend range, initial velocity also must be increased, however, sensitivity factors are increased largely, the data is given in table 7.

Compare with M107155mm projectile, $\partial X / \partial V_0$ and $\partial X / \partial c$ of ERFB 155 and ERFB–BB 155 respectively is increased 2.2 times and 3.1 times.

Secondly, the relative change of the drag coefficient or the ballistic coeffieient of low drag projectile is increased largely. Define relative chang as

$$E_C = \frac{E_{C1}}{C} 100\% \tag{33}$$

where E_{C1} –the probable error of ballistic coefficient. c–ballistic coefficient; E_c–the relative change of the dispersion of ballistic coefficient.

Because the ballistic coefficient c of base bleed projectile is reduced, even if $E_{c1}$is maintained constant, E_c would be increased besides, because of the base bleed unit, the drag dispersion of base bleed is increased, E_{c1} of base bleed projectile is also larger than that of common high explosive projectile, and would be still larger.

$$E_{XC} = \frac{\partial X}{\partial C} E_C$$

$$E_{XV0} = \frac{\partial X}{\partial V_0} E_{v_0} \tag{34}$$

TABLE 7 The sensitivity factors of different projectile

projectile	$\frac{\partial X}{\partial V_0}$	$\frac{\partial X}{\partial c}$
M107 155	28.5	99.5
M46 130	39	107
M62 152	29	116
ERFB 155	62	220
ERFB–BB–155	92	310

From above fomulas, because the low drag projectile is added the base bleed unit, E_C, $\partial X / \partial C$, $\partial X / \partial V_0$ are increased. This problem is caused by the low drag projectile itself, designers must correctly deal with the relation of extended range with dispersion, they would attend to one and neglect the other.

6. DEMONSTRATION AND DISTRIBUTION OF INDEX OF PRECISION OF BASE BLEED PROJECTILE

Under the condition of a strict controlled precision test, the influence of natural condition must be about the same. For the extended range high explosive, the deflection
dispersion is the major concern at the stage of preliminary design main attention must be paid to the range dispersion. Use following formula

$$E_X^2 = \left(\frac{\partial X}{\partial V_0} E_{V_0}\right)^2 + \left(\frac{\partial X}{\partial c} E_C\right)^2$$

6.1 Under the Condition of Precision Index of Common High Explosive Known, Determination of the Index of Precision of Base Bleed Projectile.

For example, following data of common high explosive projectile: E_{Voa},E_{Ca},
$(\partial X / \partial c)_a$,$(\partial X / \partial V_0)_a$, X_a, E_{xa}, and the data of base bleed projectile: $(\partial X / \partial V_0)_b$, $(\partial X / \partial C)_b$, X_b, are known. How to determine E_{xb} / X_b of base bleed projectile.

adopt:

$$\left(\frac{\partial X}{\partial C}\right)_b = K_1 \left(\frac{\partial X}{\partial C}\right)_a$$

$$\left(\frac{\partial X}{\partial V_0}\right)_b = K_2 \left(\frac{\partial X}{\partial V_0}\right)_a$$

$$E_{Cb} = K_3 E_{Ca}$$
$$E_{V_0 b} = K_4 E_{V_0 a}$$
$$X_b = K_5 X_a \tag{35}$$

yield

$$\frac{E_{xb}}{X_b} = \frac{\sqrt{\left[K_1 K_3 \left(\frac{\partial X}{\partial C}\right)_a E_{C_a}\right]^2 + \left[K_2 K_4 \left(\frac{\partial X}{\partial V_0}\right)_a E_{V_0 a}\right]^2}}{K_5 X_a} \tag{36}$$

For convenience to discussion, firstly, assume: $K_1 = K_2 = K, K_3 = K_4 = 1$, then

$$\frac{E_{xb}}{X_b} = \frac{K\sqrt{\left[\left(\frac{\partial X}{\partial C}\right)_a E_{C_a}\right]^2 + \left[\left(\frac{\partial X}{\partial V_0}\right)_a E_{V_0 a}\right]^2}}{K_5 X_a}$$

$$\frac{E_{Xb}}{X_b} = \frac{K}{K_5}\frac{E_{Xa}}{X_a} \tag{37}$$

from (37), it is shown that

$K = K_5, \frac{E_{xb}}{Xb} = \frac{E_{xa}}{Xa}$, or the precision index of base bleed projectile is same as that of high explosive projectile.

$K > K_5, \frac{E_{xb}}{Xb} > \frac{E_{xa}}{Xa}$, or the precision index of base bleed projectile is larger than that of high explosive projectile.

$K < K_5, \frac{E_{xb}}{Xb} < \frac{E_{xa}}{Xa}$, or the precision index of base bleed projectile is less than that of high explosive projectile.

where, K denotes the amplification of sensitivity factors $\partial X / \partial V_0$, $\partial X / \partial c$; K_5 denotes exended range rate; K_3 and K_4 denote amplification of E_{V0} and E_c. Hence, some conclusions can be obtained: Under the condition that E_{V0} and E_c of two kinds of projectile are kept constant, if the amplification of sensitivity factors is equal to the extended range rate. the precision index may be kept constant. Taking ERFB155 and ERFB–BB155 for example, $K = 1.5$, $K_5 = 1.3$, or, $K > K_5$, yield: $E_{xb} / Xb > E_{Xa} / Xa$

Above–mentioned discussion is made under, $E_{voa} = E_{vob}$, $E_{ca} = E_{cb}$. In fact, $E_{v0a} = E_{v0b}$ may be obtainable but E_{cb} is not equal to E_{ca} generally, its reason has been discussed above. When demonstrating, correct estimating E_{cb} is very important. In terms of wind tunnel tests and experience, it is reasonable to assume $E_{Cb} = (1.2\sim 1.4)\ E_{Ca}$ for preliminary design, hence

$$E_{C1b} = (1.2 \sim 1.4)\frac{C_a}{C_b} E_{C1a} \tag{38}$$

take ERFB155 for example, $(\partial X/\partial V_0)_a = 62$, $(\partial X/\partial C)_a = 220$, $C_a = 0.38$, $E_{v0a} = 1.62$, $E_{Ca} = 0.5$, $X_a = 3000$m, $(\partial X/\partial C)_b = 310$, $(\partial X/\partial V_0)_b = 92$, $C_b = 39000$m, result are given by calculation as follows:

$$\frac{E_{Xa}}{X_a} = \frac{1}{201}, \qquad \frac{E_{Xb}}{X_b} = \frac{1}{131}$$

6.2 Distribution of Precision Index and Design

Next, put main factors that affect precision as standards, to distribute the precision index, and to design structure under the condition of ensuring precision index.

If $E_x, E_z, c, V_0, \theta_0$, X are known, steps to distribute the index and to design given as follows:

(a) Distribution of index and determination of dispersion probable error E_{xai}, E_{zai} caused by every factor, Or.

$$E^2_{XBB} = C_1 E^2_X \qquad E^2_{zz} = C_5 E^2_z$$

$$E^2_{X\delta m} = C_2 E^2_X \qquad E^2_{Z\delta} = C_6 E^2_Z$$

$$E^2_{XV2} = C_3 E^2_x \qquad E_{zwz} = C_7 E^2_z$$

$$E^2_{XWX} = C_4 E^2_x$$

where, coefficient C_1—C_7 can be determined by lots of statistic data and experience. So E_{XBB}, $E_{x\delta}$, E_{XVo}, E_{XWx}, $E_{zz}, E_{z\delta}$, E_{ZWz} can be yielded.

(b) Determination of sensitivity factors of every factor. With c, v_0, θ_0, sensitivity factors $\partial X/\partial c$, $\partial X/\partial \delta_m$, $\partial X/\partial V_0$, ······ of every factor can be yielded.

(c) Determination of probable error of dispersion factor

$$\frac{\partial X}{\partial C} E_{CBB} = E_{XBB} \qquad \frac{\partial Z}{\partial \delta_p} E_{\delta_p} = E_{ZZ} = \gamma_Z X tg\theta_0$$
$$\frac{\partial X}{\partial \delta_m} E_{\delta_m} = E_{X\delta_m} \qquad \frac{\partial Z}{\partial \psi} E_{\delta_p} = E_{Z\delta} \tag{39}$$
$$\frac{\partial X}{\partial V_0} E_{V_0} = E_{X_{V_0}}$$

from above, E_{CBB}, $E_{\delta m}$, E_{v0}, γ_z, E_ψ, can be determined.

(d) Determination of probable error of dispersion of every factor assumed to satisfy needs.

and determination of the structure of gun system, or, presentation as advanced index such as E_{V0}, E_{CBB}. According to the probable error of the dispersion of every factor, when the base bleed unit, propellant structure, projectile / gun structure, etc, are considered, as the result of respective error of every factor is reached, E_x / X and $E_z$that statisfy needs would be obtained. This is the process of combining theory and experiment, among other some theoretical problems are not completely solved and satisfactory results would be obtained when modified design and test are taken again and again.

NOMENCLATURE

A_f	burning area
C_{DBB}	drag coefficient of base bleed projectile
C	ballistic coefficient
$\overline{C}$	average ballistic coefficient
E_X	probable error in range
E_Z	probable error in deflection
$E_{X\delta 1}$	probable error in range caused by the nutation dispersion
$E_{V0\delta}$	probable error of initial veloeity caused by the nutation
$E_{\delta m}$	probable error in δ_m
$E_{x\delta z}$	the range dispersion caused by E_ψ
$E_{Z\delta}$	the deflection dispersion caused by E_ψ
E_{CDb}	the ballistie deviation of drsg coefficient
E_{CDi} E_ψ	the deviation of drag coeffictent
$E_{C\delta}$	the dispersion of the ballistic coefficient caused by the nutation dispersion
E_{Wxb}	the probable error of ballistic range wind
E_{Wzb}	the probable error of ballistic cross wind
n	the number of round
q_i	the layer weight of layer i
q_{CDi}	the layer weight of drag coefficient at point i
R_A	the quator radius of the gyration of projectile
r_f	burning rate
t_i	the time when nutation is damped to zero
T	total time of flight
V_{oi}	the inifial velocity caused by nutation δ_{mi}
X_i	range corresponding to the individal nutation amplitude δ_{mi}
$\overline{X}$	average range
α_i	factors related with the range and the deflection
$\dot{\delta}_{0i}$	the initial nutation angle veloeity
δ_{mi}	nutation amplitude
δ	nutation angle
$\overline{\delta}$	the average of nutation angle
σ_{wi}	mean square error of the gusty wind
σ_{wb}	mean square error of ballistic gusty wind

ψ_i the aerodynamic jump caused by the nutation
ψ the average of ψ_i
ψ_{wi} the relative intensity of wind
ψ_0 initial deflection angle
ω_B the charge of bleed propellant
ρ_f density of base bleed propellant
$\partial X / \partial \alpha_i$ the sensitivity of the factors α_i for range
$\partial Z / \partial \alpha_i$ the sensitivity of the factor α_i for deflection
$\partial X / \partial V_0$ the sensitivity of the factor V_0 for range
$\partial X / \partial c$ the sensitivity of the factor c for range
$\partial X / \partial \delta_m$ the sensitivity of the factor δ_m for range
$\partial X / \partial \theta_0$ the sensitivity of the factor θ_0 for range
$\partial X / \partial W$ the sensitivity of the factor factor W for range
$\partial X / \partial W_i$ the sinsitivity of the factor W_i for range
$\partial X / \partial W_x$ the sinsitivity of the factor W_x for range
$\partial Z / \partial W_z$ the sensitivity of the factor W_z for deflection
$\partial X / \partial C_D$ the sensitivity of the factor C_D forrange
$\partial X / \partial C_{Di}$ the sensitivity of the factor C_{Di} for range

REFERENCE

1. Pu Fa, Exterior Ballistics, National Defence Industry Publishing House, 1980.
2. Guo, Xifu Etc, the Exterior Ballistics of base Bleed Projectile, Ballistic Reasearch Lab. of EICT, 1986.

Analysis of Gun-Launched, Solid Fuel Ramjet Projectiles

ALON GANY
Faculty of Aerospace Engineering
Technion, Israel Institute of Technology
Haifa 32000, Israel

ABSTRACT

The feasibility and performance of self-propelled, gun-launched projectiles by means of a solid fuel ramjet (SFRJ) motor is studied. It is demonstrated that the SFRJ can provide remarkable improvement of the projectile performance without changing the weapon itself. Consuming fuel in a flow rate comparable to that of a base bleed round, the SFRJ can generate thrust-equal-drag situations resulting in more than doubling the effective range compared to the modest range increase of the order of 20%, offered by base bleed. The much higher terminal velocity and kinetic energy along with the shorter flight time and flatter trajectory can make the SFRJ projectile significantly more effective, particularly for air defense systems.

INTRODUCTION

Increasing the terminal velocity and the range of a projectile can be achieved by several methods: (a) by increasing the muzzle momentum of the projectile, (b) by reducing the aerodynamic drag, and (c) by providing active propulsion to the projectile during its flight. Increased muzzle momentum can be obtained by either increasing the muzzle velocity and/or the projectile mass. Both imply changes in the gun system to accommodate the higher pressure and the larger propellant amount involved, which actually mean a new design of the weapon. Drag reduction can be accomplished by either the use of smaller cross-section, high density material, sub-caliber projectiles, or by raising the base pressure to a level closer to the ambient pressure using an injection of a fluid (burning or non-burning) near the projectile base zone (so called "base bleed"). Base drag is a significant portion of the total aerodynamic drag, (whose other components are skin friction drag and form or wave drag), and the most amenable to change. Nevertheless, a maximum reduction of about 25% of the total drag can be achieved due to the use of base bleed, hence the projectile range increase is typically limited to approximately 5 to 20% (and, exceptionally, as high as 30%).

Data and correlations on the total projectile drag and the different drag components exhibit quite a spread although reflecting similar typical behavior. The total drag coefficient C_D (based on the free stream dynamic pressure and the maximum cross-section of the projectile) has a relatively low value (approximately 0.1) in the subsonic range, which increases sharply attaining a maximum of 0.3-0.4 in the transonic range (around M=1.2-1.3) and then decreases monotonically in the supersonic flight range. The base drag coefficient C_{Db} demonstrates similar behavior. Hudgins Jr. [1] gives the following empirical correlation for C_{Db} versus Mach number

$$C_{Db} = 0.306\, e^{-0.353\, M} \quad , \qquad M = 1\text{–}8 \tag{1}$$

while Murthy and Osborn [2] correlate some experimental data in the Mach number range of 1.5 to 3, using another approximation:

$$C_{Db} \cong \frac{0.7}{M^2} \quad , \qquad M > 1.5 \tag{2}$$

Summarizing the data and analyses in Refs. [1]-[5] (Fig. 1) one can see that the base drag component is indeed a large portion of the total drag (of the order of one third). The correlation used in this work to represent the (total) aerodynamic drag coefficient is given by Eq. (3):

$$\left.\begin{aligned} C_D &= 0.35\, M^2 - 0.27\, M + 0.14 \quad , & M &< 1.2 \\ C_D &= 0.32 \quad , & 1.2 &\le M \le 1.3 \\ C_D &= 0.3876 - 0.052\, M \quad , & M &> 1.3 \end{aligned}\right\} \tag{3}$$

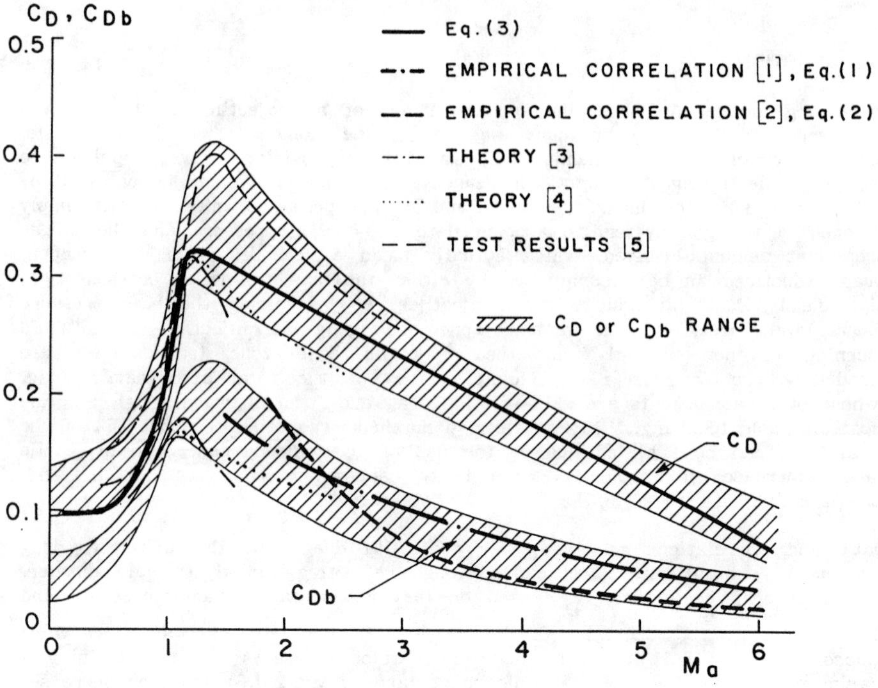

FIGURE 1. Typical variations of the total and base drag coefficients of projectiles with Mach number.

Investigations on base drag reduction by base bleed (usually by means of a solid propellant gas generator, so-called "fumer"), have shown a decrease in C_{Db} to 50% and even less compared to its original value [1], [4], [5]. The effect is more pronounced in the supersonic flight range ($M \gtrsim 1.5$). The optimal situation (maximum base drag reduction) requires, however, an injection of gas amounting to 0.030 of the reference flow rate $\dot{m}_{ref}$ for cold gas (air) and to about $0.013 \cdot \dot{m}_{ref}$ for hot gas [1]. The precise optimal value depends on a number of parameters including the flight Mach number, bleed gas composition, and base and bleed nozzle geometries. Practical base bleed systems often operate at lower $\dot{m}_{ref}$ ranges of approximately 0.002 to 0.008 [6]. $\dot{m}_{ref}$ is defined as

$$\dot{m}_{ref} = \rho_a u_a A_{max} \tag{4}$$

A projectile with base bleed is illustrated in Fig. 2.

Active propulsion for projectiles can be provided by two main methods: rocket motors (rocket assisted projectiles, RAP) or solid fuel ramjet (SFRJ) motors. The rocket motor is a simple and reliable propulsion means, which is independent of flight conditions. Its main disadvantage is its high propellant consumption rate and low specific impulse, resulting in a limited total momentum that can be provided for the projectile and, hence, a limited improvement in the overall performance. In addition, the RAP operation inherently introduces larger trajectory errors and inaccuracies. The RAP concept is shown in Fig. 3.

The solid fuel ramjet seems to be "naturally tailored" to the specific use of propelling projectiles. Muzzle conditions provide the high velocity necessary for the aerodynamic compression, which is essential for the SFRJ operation. This way the gun system serves as an economical, high accuracy accelerator, avoiding the need for a special rocket-type booster stage. The ramjet motor then starts to operate upon leaving the gun barrel. The main advantage of the ramjet over a rocket is in its much higher specific impulse (of the order of 800-1000 seconds versus 200 seconds, approximately) resulting in a much smaller fuel consumption rate. Hence, it can provide a much higher total impulse to the projectile, applied as a specified thrust for the required amount of time.

The SFRJ is the simplest air breathing engine. It does not contain any moving parts, valves, storage tanks, or injection systems. It consists of a diffuser, a combustion chamber, and a nozzle. The solid fuel grain, typically of a hollow cylinder shape, is placed in the combustion chamber, and the incoming air flows through its port. It should be noted, however, that the SFRJ is a more complex motor than a rocket, and its operation depends on several factors that do not affect the rocket motor, such as flight Mach number, density and temperature of the atmosphere, and the angle of attack. In addition, the SFRJ projectile may exhibit more problems in accommodating a payload. Figure 4 presents the concept of an SFRJ projectile of two different inlet types: a simple pitot diffuser and a higher performance, external compression center-body diffuser.

It is desirable that a self-propelled projectile operates at thrust-equal-drag situations. Such conditions imply a pure (vacuum) ballistic trajectory, enabling high precision and self-correcting action against winds [7]. In general, the SFRJ motor compensates for variations in atmospheric density as a result of the similar effect of the density on the drag and the thrust.

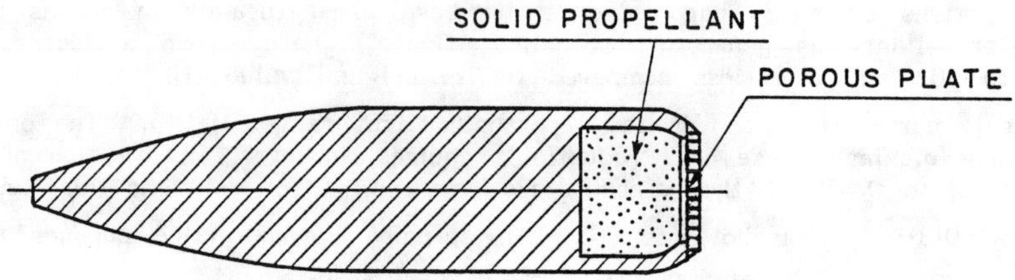

FIGURE 2. Projectile with base bleed.

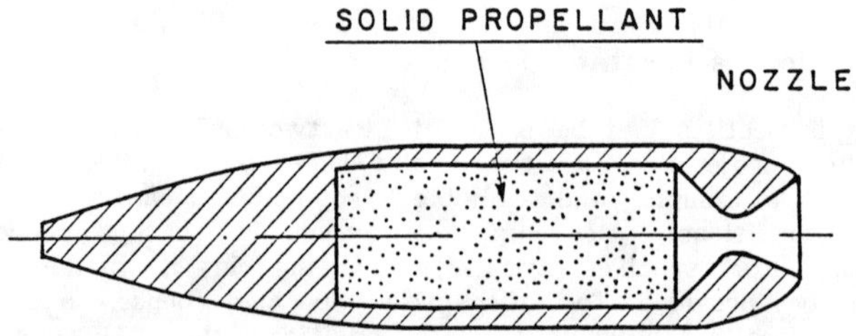

FIGURE 3. Rocket assisted projectile (RAP).

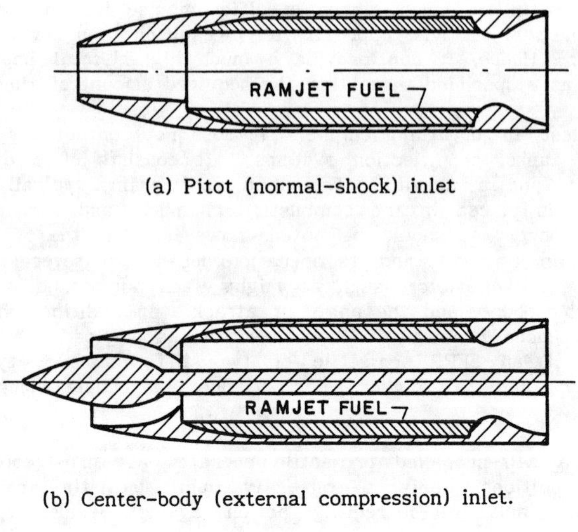

(a) Pitot (normal-shock) inlet

(b) Center-body (external compression) inlet.

FIGURE 4. Solid fuel ramjet (SFRJ) projectiles.

The objective of this work is to present the operational characteristics of gun-launched, self-propelled SFRJ projectiles. It will be shown that for fuel flow rates and effective specific impulse comparable to those of the base bleed system, the superiority of the SFRJ system in terms of projectile performance may be essential, considering the inherent limitations of the base bleed concept. The thrust-equal-drag condition implies a more-or-less constant flight velocity, equal to the muzzle velocity. This situation results in a number of significant advantages: (a) the terminal kinetic energy of the projectile remains equal to that at the muzzle exit, implying the maximum penetration potential. The improvement is more pronounced in the case of small caliber projectiles because of the larger ratio between drag and mass. (b) The trajectory is flatter than in regular rounds, permitting easier aiming and better precision. (c) The time in flight may be much shorter, a major advantage particularly for air defense systems, also reducing wind effects. (d) The range may be increased by a factor of 2 and even more [7], compared to a maximum range increase of as high as 30% expected for based bleed projectiles [6].

Figure 5 demonstrates the striking superiority of a thrust-equal-drag operation, which should supposedly be achieved for an SFRJ projectile, compared to regular and base bleed rounds of the same caliber, mass, and initial velocity, regarding the velocity, kinetic energy, and time along the path. Calculation and data [8], [9] are presented for 5.56 mm, 3.6 gram projectiles, with a muzzle velocity of 990 m/s.

THE SFRJ PROJECTILE - GENERAL CHARACTERISTICS

Detailed configuration and general characteristics of an SFRJ projectile having a pitot-type inlet are shown in Fig. 6. A similar presentation of a 75 mm caliber SFRJ projectile is given by Nusca [10]. An $\dot{m}_a$ mass flow rate of air at supersonic flight speed u_a and Mach number M_a enters the diffuser through a normal shock wave. The resulting subsonic flow is then expanding from the inlet cross section A_{in} into the combustion chamber, flowing through the cylindrical port (cross section A_p) of the fuel grain. The inlet step, represented by the area ratio A_p/A_{in}, is essential for flame stabilization, especially at low air temperatures corresponding to low flight Mach numbers (up to about M_a=3) [11]-[13]. The fuel, typically a hydrocarbon (HC) polymer, burns with the air, heating the gas mixture to its stagnation combustion temperature T_c, while the surface regresses radially at a typical regression rate $\dot{r}$, affected by the air mass flux through the port, the inlet air temperature, the fuel type, and some other operational parameters (e.g., combustor geometry and spinning rate [10], [14]).

Optimal operation implies that the exit pressure p_e is equal to the ambient pressure p_a (adapted nozzle). Maximum thrust at given flight conditions results from a stoichiometric fuel/air ratio (equivalence ratio, ϕ=1).

A general theoretical plot of the specific thrust $F/\dot{m}_a$ and the specific drag

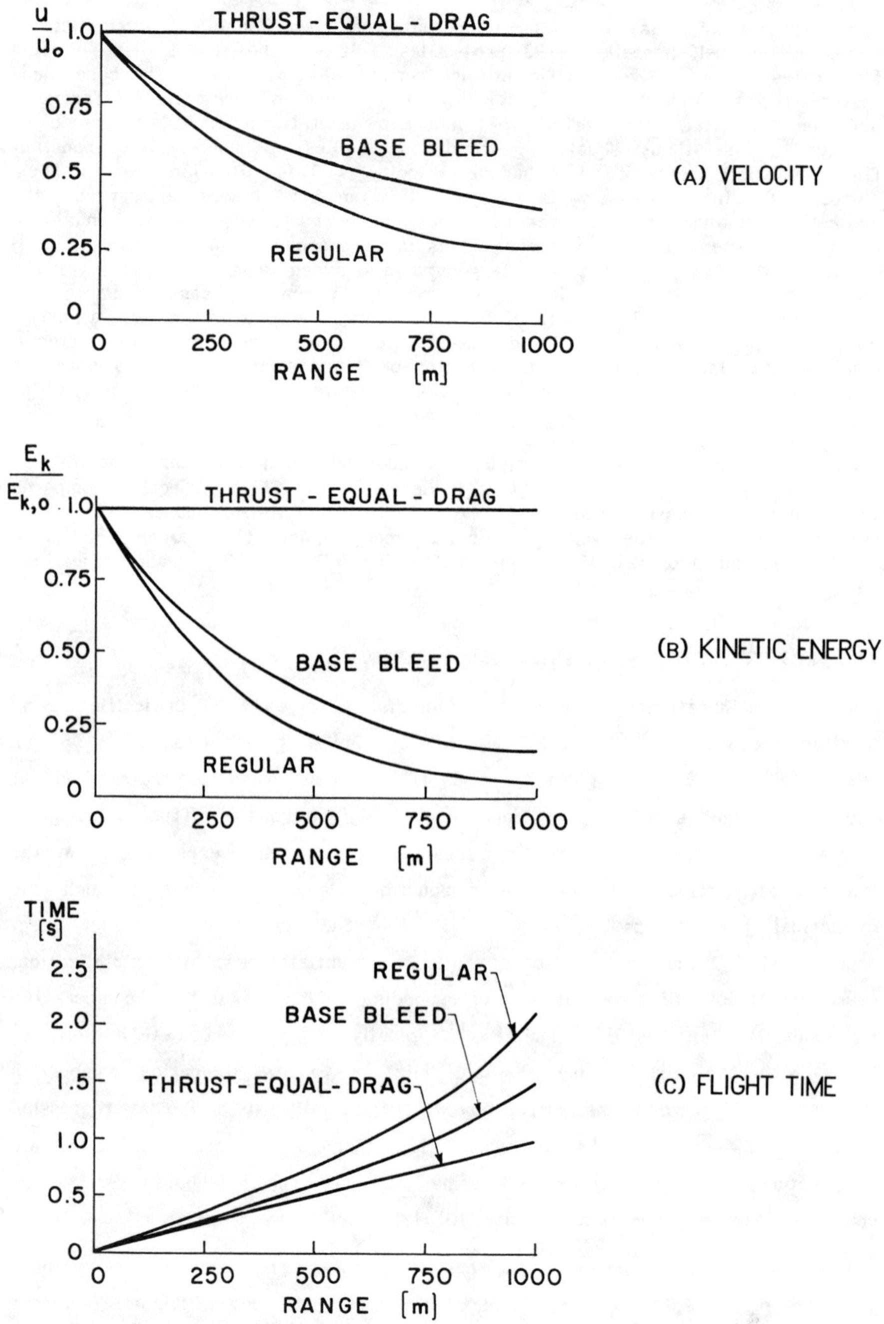

FIGURE 5. Variations of the (a) velocity, (b) kinetic energy, and (c) flight time along the path of regular, base bleed, and thrust-equal-drag, 5.56 mm projectiles.

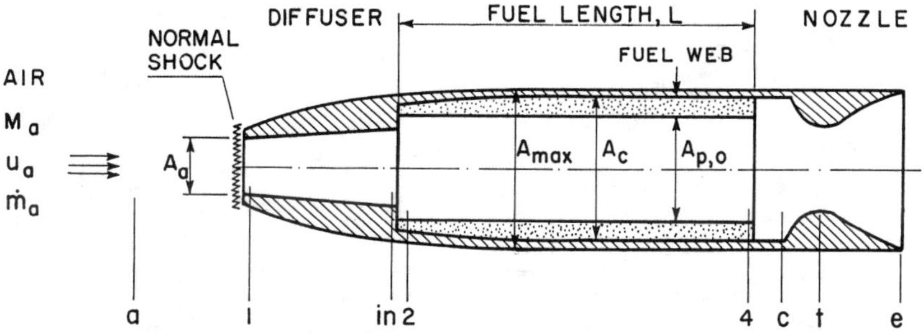

FIGURE 6. Geometry and notation of an SFRJ motor with a pitot diffuser.

$D/\dot{m}_a$ versus Mach number can be drawn, using the ratio between the air capture area (A_a=A_1) and the maximum projectile cross-section area A_{max} as a parameter. The effect of combustion efficiency on the thrust can also be shown (see Fig. 7). In the calculation of Fig. 7, a hydroxyl-terminated-polybutadiene (HTPB) fuel was assumed at sea level flight conditions with a pitot (normal-shock) diffuser. Stagnation pressure losses resulting from the normal shock wave in addition to estimated 5% losses from other processes (e.g., heat addition) were assumed as well. Somewhat better performance can be obtained from an improved (yet conservative) center-body diffuser (Fig. 8), whose pressure recovery [15] is compared to that of a normal-shock diffuser in Fig. 9. Mattingly et al. [16] present the ram recovery of the military specification MIL-E-5008B, which seems to be even higher.

The thrust equation (adapted nozzle conditions) is:

$$F = \dot{m}_a[(1+f)u_e - u_a] \quad , \qquad p_e = p_a \tag{5}$$

where f is the fuel-to-air ratio:

$$f = \frac{\dot{m}_f}{\dot{m}_a} \tag{6}$$

The exit jet velocity was calculated from:

$$u_e = \sqrt{2c_p T_c\left[1-\left(\frac{p_e}{p_c}\right)^{\frac{\gamma-1}{\gamma}}\right]} \tag{7}$$

p_c is the stagnation pressure at the end of the combustion chamber, calculated from the ambient pressure, flight Mach number, and stagnation pressure losses, as described above. T_c is the stagnation combustion temperature calculated by a thermochemical computer code [17] (along with other gas properties such as c_p and γ), accounting for the combustion efficiency defined as

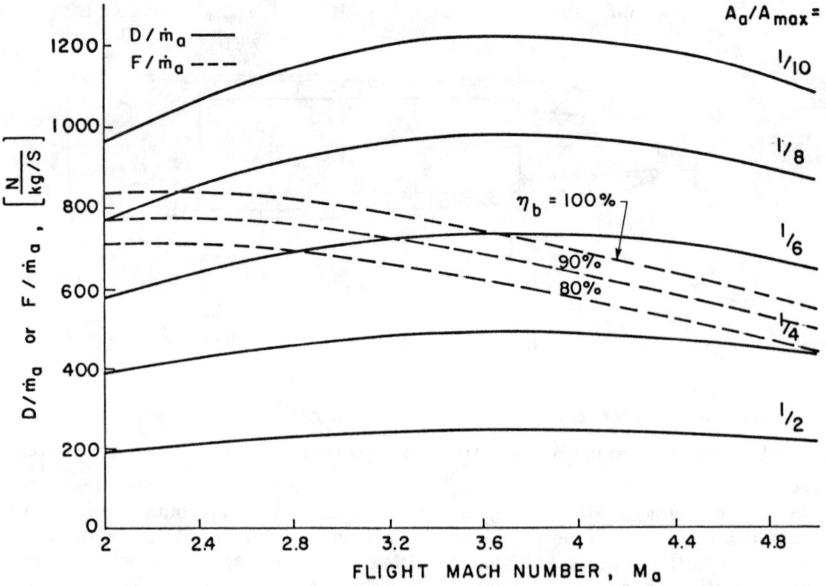

FIGURE 7. $F/\dot{m}_a$ and $D/\dot{m}_a$ versus flight Mach number M_a for different A_a/A_{max} and η_b. Sea level, HTPB fuel, pitot (normal-shock) diffuser.

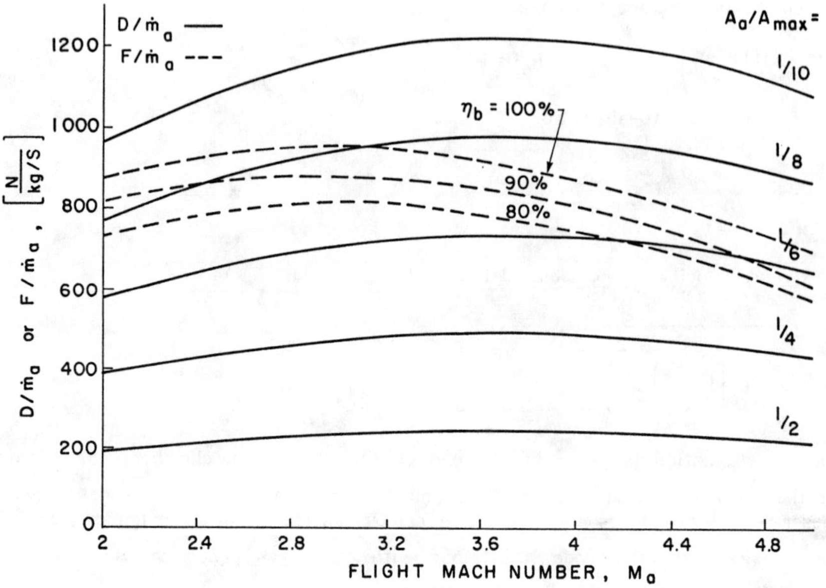

FIGURE 8. $F/\dot{m}_a$ and $D/\dot{m}_a$ versus flight Mach number M_a for different A_a/A_{max} and η_b. Sea level, HTPB fuel, improved (center-body) diffuser.

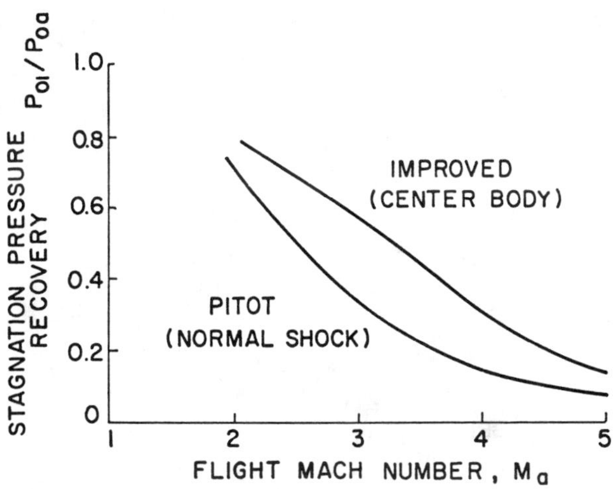

FIGURE 9. Stagnation pressure recovery in a pitot (normal-shock) diffuser and in a typical improved diffuser.

$$\eta_b = \frac{T_c - T_{oa}}{T_{c,th} - T_{oa}} \tag{8}$$

where

$$T_{oa} = T_a\left(1 + \frac{\gamma-1}{2} M_a^2\right) \tag{9}$$

Aerodynamic drag was calculated by

$$D = C_D \cdot \frac{1}{2} \rho_a u_a^2 \cdot A_{max} \tag{10}$$

where the value of C_D was obtained from Eq. (3). From Figs. 7 and 8 one can draw the minimum A_a/A_{max} ratio necessary to obtain thrust-equal-drag conditions at each Mach number (Fig. 10). Note that lower combustion efficiencies imply larger A_a/A_{max} ratios. Figure 10 demonstrates the superiority of an improved (center-body) diffuser compared to a pitot diffuser.

Constraints on A_a/A_{max} ratio may be relieved to some extent by adding some amount of oxidizer to the SFRJ fuel, thereby reducing the required amount of air. Between the two limiting cases, pure SFRJ (only fuel) and pure rocket - RAP - (no air), lie the different (increasing) oxidizer percentages implying decreasing I_{sp} values. Figure 11 shows the variations of the minimum A_a/A_{max} ratio required for a thrust-equal-drag operation versus M_a for different fuel (or propellant)/air ratios associated with different oxidizer fractions in the fuel, maintaining overall stoichiometric fuel (+ oxidizer)/air ratio. HTPB fuel and ammonium-perchlorate (AP) oxidizer as well as a pitot-type diffuser were used in the calculations.

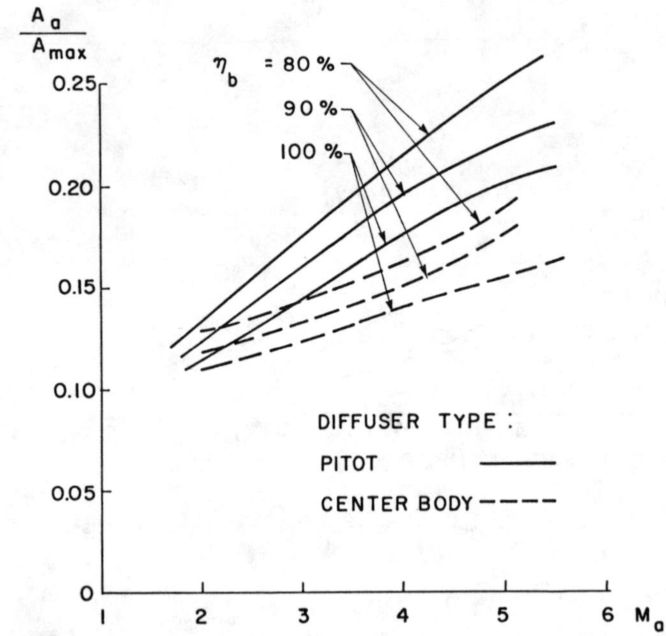

FIGURE 10. Minimum ratio of A_a/A_{max} necessary to achieve thrust-equal-drag conditions versus flight Mach number.

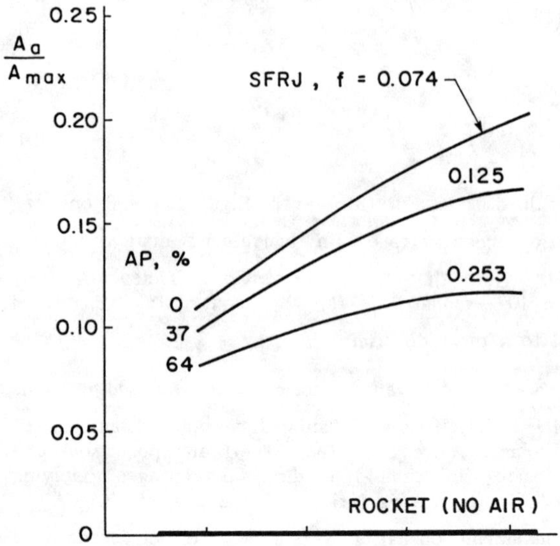

FIGURE 11. A_a/A_{max} ratio required for thrust-equal-drag operation versus Mach number at different oxidizer (AP) fractions.

As has been mentioned before, along with increasing the specific thrust $F/\dot{m}_a$, the use of an oxidizer in addition to the fuel results in a penalty of a reduced specific impulse [10] defined as:

$$I_{sp} = \frac{F}{\dot{m}_f g_o} \tag{11}$$

Figure 12 demonstrates the variations of $F/\dot{m}_a$, I_{sp}, and f with the oxidizer fraction in the fuel for the specific flight conditions of M_a=3 at sea level with a pitot diffuser and η_b=90%.

One should be aware of the fact that a higher specific impulse means more economical use of the fuel (or propellant), i.e., smaller fuel consumption rate $\dot{m}_f$ for the same thrust level. For a given mission, say, a thrust-equal-drag operation, it means longer burn time (and range) per given amount of fuel. The fuel flow rate necessary to maintain a thrust-equal-drag operation of a projectile at Mach number of 3 is given in Fig. 13 (normalized by a reference flow rate) as a function of the oxidizer mass fraction in the fuel or propellant. The two extreme cases, SFRJ with pure fuel (zero oxidizer) and pure rocket (no air) are obvious. A more complete idea on the fuel (or propellant) flow rate necessary to maintain thrust-equal-drag, can be revealed by Fig. 14, showing the variation of the fuel (or propellant) flow rate

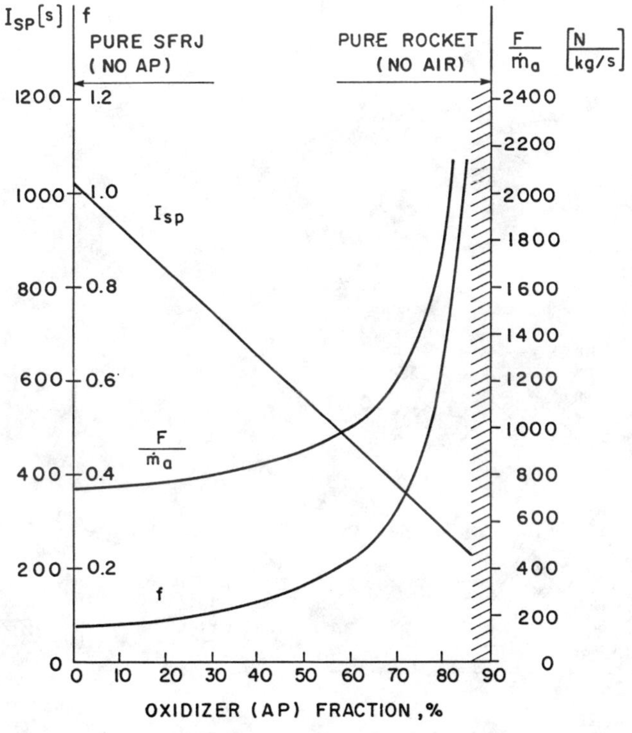

FIGURE 12. Variations of I_{sp}, $F/\dot{m}_a$, and f with the oxidizer (AP) mass fraction in the solid propellant. Pitot diffuser, M_a=3, η_b=90%.

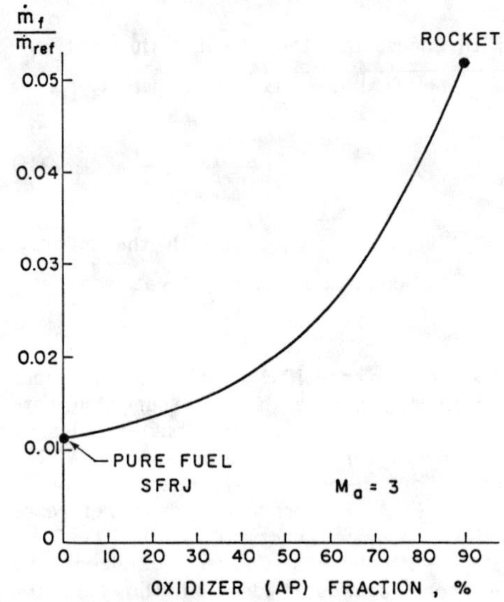

FIGURE 13. Variation of the fuel (or propellant) flow rate necessary for thrust-equal-drag operation as a function of the oxidizer mass fraction. Pitot diffuser, η_b=90%.

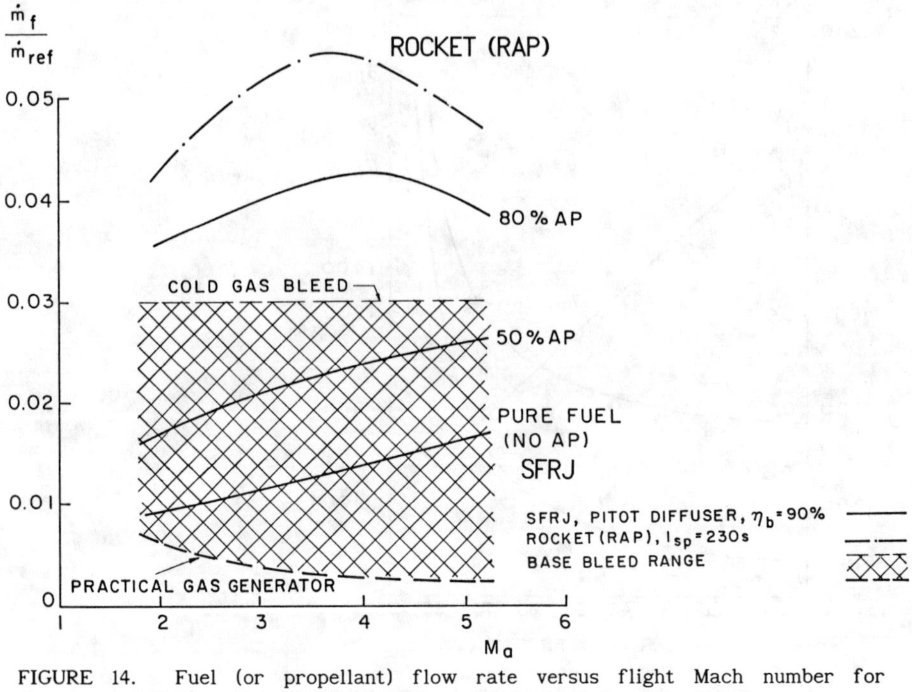

FIGURE 14. Fuel (or propellant) flow rate versus flight Mach number for thrust-equal-drag rocket (RAP) or SFRJ projectiles and for base bleed rounds.

(normalized by a reference flow rate) with the flight Mach number for different fuel-oxidizer combinations between a pure SFRJ (f=0.074, zero oxidizer) and a pure rocket (RAP), having specific impulse of 230 s. For comparison, the typical base bleed operation range (which does not even come close to thrust-equal-drag conditions) is also shown in Fig. 14. The figure demonstrates the remarkable superiority of the SFRJ projectile versus RAP with regard to fuel consumption rate (factor of approximately 4). In addition, it reveals that the fuel flow rate of an SFRJ projectile giving thrust-equal-drag, is comparable to the typical base bleed flow rate range, although actual base bleed projectile systems operate at somewhat lower flow rates [6], particularly at the higher Mach numbers. Note, however, that the base bleed improvement with regard to projectile performance is only a small fraction of the SFRJ's.

INTERNAL BALLISTICS AND CONSTRAINTS

The general requirements from an SFRJ projectile, as presented in the previous chapter, should be examined in the light of the internal ballistics of the SFRJ motor. The internal flow, combustion process, heat addition, and flameholding characteristics, in addition to geometric, construction and other constraints may, in some cases, contradict the mission requirements (such as a specified thrust at given velocity and burning time), hence, questioning the SFRJ as a practical solution for self-propelled projectiles in certain conditions.

The basic operational limit is due to flow choking in the combustion chamber (M_4=1) as a result of the heat addition. This should set the lowest value of the fuel port area to the air capture area ratio, A_p/A_a. The operational limit can be found from the continuity equation:

$$\dot{m}_4 = (1+f)\dot{m}_a \tag{12}$$

taking into account the heat addition in the constant cross section combustor, and the stagnation pressure losses resulting from the inlet shock waves and from the combustion process. The calculations were made for a pitot (normal-shock) diffuser, using a thermochemical code [17] to compute the theoretical stagnation combustion temperature. Combustion efficiency of 90% was assumed [18].

Table 1 shows calculated values of temperatures and pressures for a pitot-diffuser SFRJ motor operating at sea level with ambient temperature of 288 K, and M_4=1. HTPB fuel at stoichiometric ratio was assumed.

TABLE 1. Air and Combustion Stagnation Temperatures and Stagnation Pressure Ratio Versus Flight Mach Number.

M_a	T_{oa}[K]	$T_{c,th}$[K]	$T_{c(\eta_b=90\%)}$[K]	p_c/p_{oa}
2	518	2471	2276	0.59
3	806	2623	2441	0.27
4	1210	2856	2691	0.12
5	1728	3077	2942	0.054

Figure 15 presents the operational limits of the SFRJ motor in terms of the minimum A_p/A_a ratio versus the flight Mach number M_a, based on the data of Table 1. As can be seen from Fig. 15, the operational limits may be even narrower for small caliber projectiles at the lower Mach number range due to the deteriorating flameholding capability, requiring larger inlet step heights under these conditions, as reported by Netzer and Gany [13]. Note that, in general, a smaller ratio A_p/A_a is desirable, since it means that a larger amount of solid fuel can be placed in the combustor volume.

The operational limits in terms of A_p/A_a can determine the required fuel burning rate and the burning time and, hence, the projectile range with active, thrust-equal-drag propulsion, provided that certain geometric parameters are specified.

Substituting

$$\dot{m}_f = f \cdot \dot{m}_a \tag{13}$$

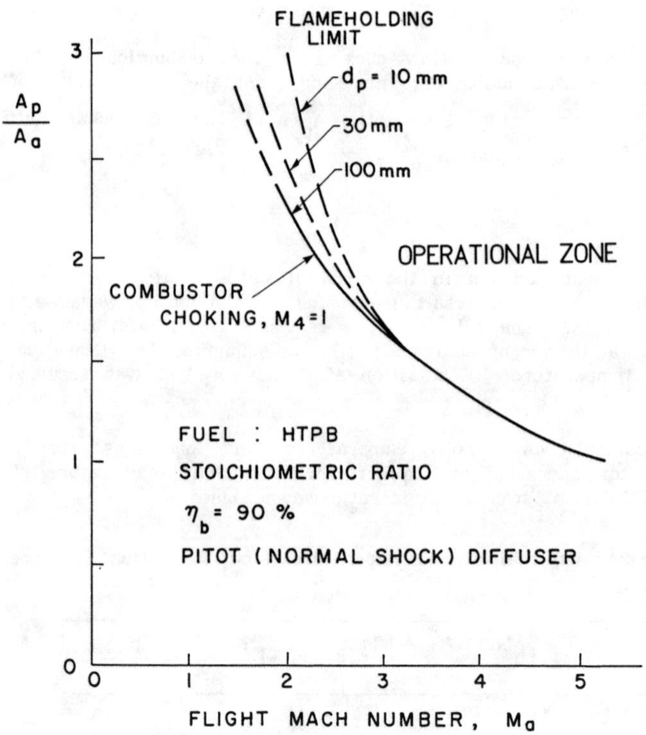

FIGURE 15. Operational limits of an SFRJ motor due to combustor's flow choking and flameholding requirements.

and

$$\dot{m}_f = \rho_f \pi d_p L \dot{r} \tag{14}$$

one obtains:

$$\dot{r} = \frac{f \cdot \rho_a \cdot u_a \cdot \left(\frac{A_a}{A_{max}}\right)}{4\rho_f \cdot \left(\frac{d_p}{d_{max}}\right) \cdot \left(\frac{L}{d_{max}}\right)} \tag{15}$$

It is convenient to relate the fuel length L and the different diameters to the projectile diameter d_{max} as a basis for geometric similarity.

The requirement for constant thrust at constant flight speed implies a fixed fuel consumption rate, i.e.,

$$\dot{r} \cdot d_p = \text{const} = \dot{r}_o \cdot d_{p,o} \tag{16}$$

The fuel burning time can then be found from the web thickness, by integrating Eq. (16) for d_p from $d_{p,o}$ to d_c (external fuel diameter):

$$t_b = \frac{d_c^2 - d_{p,o}^2}{4\dot{r}_o d_{p,o}} \tag{17}$$

It should be noted that Eqs. (15) and (17) contain two free parameters: L/d_{max} and d_c (or by some algebraic manipulations, d_c/d_{max}). Those are design, geometric (construction) parameters. For a given flight Mach number and similar geometric ratios, the required initial value of $\dot{r}$ is found to be independent of the projectile diameter, and inversely dependent on L/d_{max}:

$$\dot{r}_o = \frac{K_1}{(L/d_{max})} \tag{18}$$

where K_1 is a constant. In the specific conditions of M_a=3 and d_c/d_{max}=0.8 (namely, the combustor metal wall thickness is 10% of the projectile diameter d_{max}), when using HTPB fuel of ρ_f=0.95 g/cc and η_b=90%, the value of K_1 is found to be approximately 8 mm/s. $\dot{r}_o$ can be controlled by changing the ratio of L/d_{max}. For instance, L/d_{max}=4 gives $\dot{r}_o$=2 mm/s. The value of $\dot{r}_o$ should increase when increasing the flight Mach number (e.g., to about 5 mm/s at M_a=5 for the same geometry and fuel). The experimental results and empirical correlations of the regression rate of an HTPB-based fuel [13] indicate that the required regression rates mentioned above are achievable in the typical conditions, particularly for small caliber projectiles (e.g., 20 mm). Furthermore, data on small port diameter SFRJ combustors [18], [19] reveal that a fixed fuel consumption rate (according to the desirable situation presented by Eq. (16)) is possible. Hence, constant fuel/air ratio f during the combustion can be achieved, permitting optimal performance throughout the combustor operation. Typical fuel regression rates for larger port diameter combustors were found to be lower [12], [19], [20], hence, fast burning fuels should be used in such cases. In addition, regression rate correlations for large port diameter combustors [11], [20] indicate that the fuel consumption rate as well as the fuel/air ratio may vary a little during the combustion, because of the increase in the port diameter. Thus, some deviations from optimal operation conditions may be expected.

The burning time t_b, which is equivalent to the duration of the propelled-stage of the flight, is one of the most important factors for evaluating the SFRJ projectile system. For a given flight speed, t_b determines the propelled-stage range. In the case of projectiles flying at a specified Mach number and having geometric similarity, t_b is found to be directly proportional to L/d_{max} and to the projectile diameter d_{max}:

$$t_b = K_2\left(\frac{L}{d_{max}}\right) \cdot d_{max} \tag{19}$$

K_2 is a constant which, under the specific conditions giving K_1=8 mm/s, has a value of 0.023, approximately, when t_b is in seconds and d_{max} in millimeters. Table 2 presents some examples of calculated burn-time results based on the combustor operational limits shown in Fig. 15.

Equation (19) suggests that under the same conditions, where the firing angle is relatively small, the propelled-stage range, R, is approximately:

$$R \cong K_3\left(\frac{L}{d_{max}}\right) \cdot d_{max} \tag{20}$$

The specific conditions of M_a=3 and d_c/d_{max}=0.8 yield $K_3 \cong 23.5$, where R is in meters and d_{max} in millimeters.

Although the burn-time is significantly affected by the flight Mach number, the effect of Mach number on the range is only secondary, as the range is a product of t_b and u_a.

Note that more severe constraints are imposed on the SFRJ combustor for $d_p \lesssim 10$mm

TABLE 2. Example of Calculated Burn-Times of SFRJ Propelled Projectiles at Different Mach Numbers and Normalized Fuel Lengths.

caliber d_{max} [mm]	burn-time, t_b [s]			
	M_a=3		M_a=5	
	L/d_{max}=4	L/d_{max}=6	L/d_{max}=4	L/d_{max}=6
20	1.8	2.8	1.0	1.5
40	3.7	5.5	2.0	3.0
105	9.7	14.5	5.2	7.9

because of flameholding deterioration in smaller port diameter motors [13]. In such cases shorter burn-time and range are expected. d_p of 10 mm corresponds to projectile caliber of about 20 mm (calculations based on operational and geometric constraints revealed that $d_{p,min} \approx 0.5 d_{max}$). Flameholding constraints associated with the inlet air temperature effect could also cause difficulties in achieving the performance predicted by the combustor choking limit in the lower Mach number range. For instance, at Mach 2, even combustors with fuel port diameter as large as 100 mm may yield somewhat lower performance.

Figure 16 presents the predicted propelled-flight-range versus the projectile caliber for SFRJ projectiles of different fuel length-to-projectile-diameter ratios flying at Mach 3 (see Eq. (20)). Note that operational constraints can be eased when using larger L/d_{max} ratios. Both propelled-stage time and range grow linearly with the fuel length-to-projectile-diameter ratio. In this respect it is desirable that SFRJ projectiles have as large as possible length-to-diameter ratio.

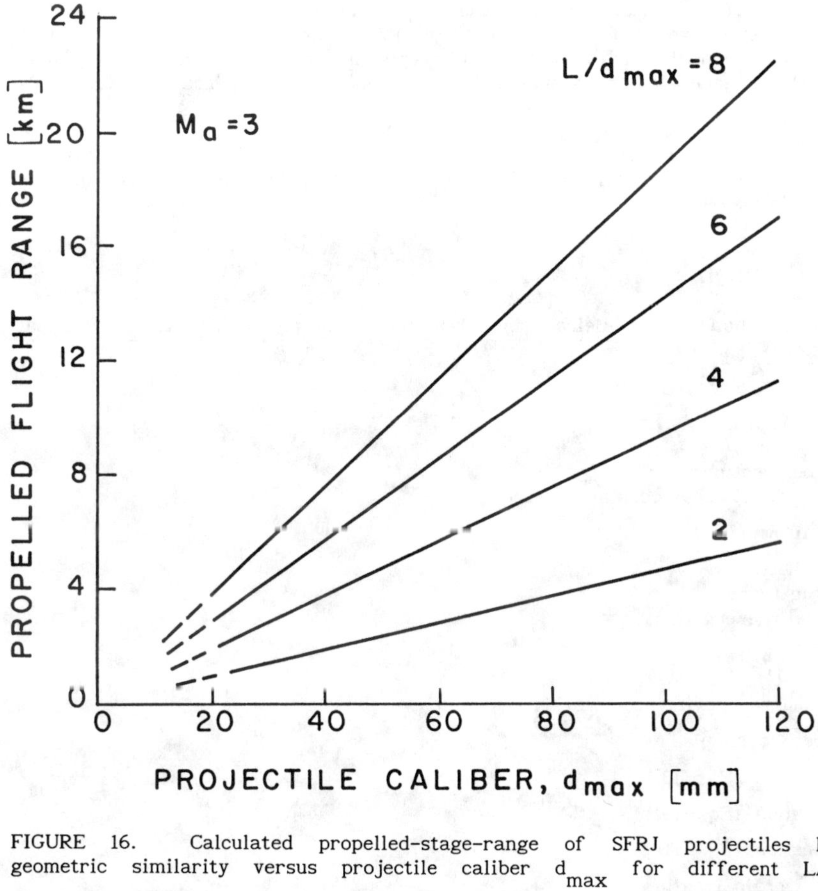

FIGURE 16. Calculated propelled-stage-range of SFRJ projectiles having geometric similarity versus projectile caliber d_{max} for different L/d_{max} ratios. M_a=3, pitot diffuser, η_b=90%.

CONCLUDING REMARKS

Solid fuel ramjet (SFRJ) propulsion can offer a remarkable improvement in projectile performance in terms of terminal velocity, kinetic energy, and range. It can also shorten the flight time and flatten the trajectory for a given range, thus providing easier aiming and increasing hit probability, particularly in air defense operations. The SFRJ propulsion can provide a thrust-equal-drag operation for much longer flight time and range than those encountered in rocket-assisted-projectiles (RAP). The performance superiority of the SFRJ is especially striking when being compared to base bleed projectiles. For comparable fuel (or propellant) mass flow rate and effective specific impulse, the former can double or even triple the effective projectile range, while the latter can only cancel approximately 25% of the overall drag by reducing the base drag, thus offering a modest increase of about 20%-30% in the projectile range.

It should be noted, however, that the SFRJ projectile is not only more complicated, but also has some inherent operational constraints resulting in poor volumetric efficiency of the fuel. This is due to the empty port volume required as well as because of the much lower fuel density in the SFRJ. As a result, the effective fuel volume in the SFRJ is expected to be about three times larger than that of a base bleed propellant having similar mass. Thus, larger than usual length-to-diameter ratios may be required in the SFRJ projectiles.

One should also bear in mind, that more severe operational constraints may be encountered in small caliber SFRJ projectiles due to deterioration of the combustor's flameholding capability.

ACKNOWLEDGMENT

The author thanks Miss Meira Ben-Zaken for assisting in the thermochemical calculations.

NOMENCLATURE

A cross-section area
c_p specific heat at constant pressure
C_D drag coefficient
d diameter
D drag force
E_k kinetic energy
f fuel/air ratio
F thrust
g_o standard gravity acceleration
I_{sp} specific impulse
K_1 constant (Eq. (18))
K_2 constant (Eq. (19))
K_3 constant (Eq. (20))
L fuel grain length
$\dot{m}$ mass flow rate

M	Mach number
p	pressure
$\dot{r}$	fuel regression rate
R	range
t	time
u	velocity
γ	specific heat ratio
η_b	combustion efficiency
ρ	density
ϕ	equivalence ratio

Subscripts

1	station behind inlet normal shock
2	station at the beginning of combustor
4	station at the end of fuel grain
a	ambient (or flight) conditions; air
b	base conditions; burning
c	stagnation conditions at combustor end
e	nozzle exit plane
f	fuel
in	combustor inlet
max	maximum; largest projectile cross section
min	minimum
o	stagnation (total) conditions; initial
p	fuel port
ref	reference (Eq. (4))

REFERENCES

1. Hudgins, Jr., H. E., Range Increase of Projectiles by Heat and/or Mass Addition to Base or External Flow, in *Aerodynamics of Base Combustion*, *Progress in Astronautics and Aeronautics*, Vol. 40, eds. S.N.B. Murthy, J.R. Osborn, A.W. Barrows, and J.R. Ward, pp. 495–521, AIAA, New York, NY, 1976.

2. Murthy, S. N. B. and Osborn, J. R., Base Flow Phenomena With and Without Injection: Experimental Results, Theories and Bibliography, in *Aerodynamics of Base Combustion*, *Progress in Astronautics and Aeronautics*, Vol. 40, eds. S.N.B. Murthy, J.R. Osborn, A.W. Barrows and J.R. Ward, pp. 7–210, AIAA, New York, NY, 1976.

3. Chow, W. L., Base Pressure of a Projectile within the Transonic Flight Regime, *AIAA J.*, Vol. 23, No. 3, pp. 388–395, March 1985.

4. Schilling, H., A Simple Theoretical Approach for BAse Bleed Calculations, *Proc. 9th Int. Symposium on Ballistics*, pp. 2-59 to 2-66, 1986.

5. Elmendorf, T. A. and Trifiletti, R. A., Gas Generators for Base Drag Reduction (Fumers), in *Aerodynamics of Base Combustion, Progress in Astronautics and Aeronautics*, Vol. 40, eds. S.N.B. Murthy, J.R. Osborn, A.W. Barrows and J.R. Ward, pp. 471–486, AIAA, New York, NY, 1976.

6. Gunners, N.E., Andersson, K. and Hellgren, R., Base Bleed Systems for Gun Projectiles, in *Gun Propulsion Technology, Progress in Astronautics and Aeronautics*, Vol. 109, ed. L. Stiefel, pp. 537-562, AIAA, Washington DC, 1988.

7. Fink, M.R., Aerodynamic Properties of an Advanced Indirect Fire System (AIFS) Projectile, *Journal of Spacecraft and Rockets*, Vol. 19, No. 1, pp. 36-40, 1982.

8. Barrows, A. W., Drag Reducing Fumers, in *Aerodynamics of Base Combustion, Progress in Astronautics and Aeronautics*, Vol. 40, eds. S.N.B. Murthy, J.R. Osborn, A.W. Barrows and J.R. Wood, pp. 1-6, AIAA, New York, NY, 1976.

9. Kwatnoski, R., Drag-Reducing Fumer as a Tool in Ammunition Design, in *Aerodynamics of Base Combustion, Progress in Astronautics and Aeronautics*, Vol. 40, eds. S.N.B. Murthy, J.R. Osborn, A.W. Barrows and J.R. Ward, pp. 487-494, AIAA, New York, NY, 1976.

10. Nusca, M., Steady Flow Combustion Model for Solid Fuel Ramjet Projectiles, AIAA Paper 89-2797, 25th Joint Propulsion Conference, July 1989.

11. *Solid Fuel Ramjet*, United Technologies Chemical Systems, San Jose, CA.

12. Schulte, G., Fuel Regression and Flame Stabilization Studies of Solid Fuel Ramjets, *Journal of Propulsion and Power*, Vol. 2, No. 4, pp. 301-304, 1986.

13. Netzer, A. and Gany, A., Burning and Flameholding Characteristics of a Miniature Solid Fuel Ramjet Combustor, Journal of Propulsion and Power (forthcoming). Also, AIAA Paper 88-3044, 24th Joint Propulsion Conference, July 1988.

14. Duesterhaus, D.A. and Högel, A., Measurements in a Solid Ramjet Combustion with Swirl, AIAA Paper 88-3045, 24th Joint Propulsion Conference, 1988.

15. Kinroth, G. D. and Anderson, W. R., *Ramjet Design Handbook, CPIA Publication 319*, p. 2-17, 1980.

16. Mattingly, J.D., Heiser, W.H. and Daley, *Aircraft Engine Design*, p. 107, AIAA Education Series, AIAA, New York, NY, 1987.

17. Cruise, D. R., Theoretical Computations of Equilibrium Compositions, Thermodynamic Properties, and Performance Characteristics of Propellant Systems, NWC-TP-6037, Naval Weapons Center, China Lake, CA, 1979.

18. Zvuloni, R., Levy, Y. and Gany, A., Investigation of a Small Solid Fuel Ramjet Combustor, *Journal of Propulsion and Power*, Vol. 5, No. 3, pp. 269-275, 1989.

19. Zvuloni, R., Gany, A. and Levy, Y., Geometric Effects on the Combustion in Solid Fuel Ramjets, *Journal of Propulsion and Power*, Vol. 5, No. 1, pp. 32-37, 1989.

20. Mady, C.J., Hickey, P.J. and Netzer, D.W., Combustion Behavior of Solid Fuel Ramjets, *Journal of Spacecraft and Rockets*, Vol 15, No. 3, pp. 131-132, 1978.

COMMENTS

D. Lyster, Space Research Corporation, Belgium. What do you see as the practical lower limit for the gun caliber?

Author's Reply. Motor burning time as well as propelled-stage range decrease with decreasing the projectile caliber (see general analysis, Eqs. (19) and (20), and Fig. 16). Furthermore, for calibers below, say, 20 mm, deterioration in flameholding capability becomes a significant problem. The situation may be improved for higher stagnation air temperatures (resulting from higher muzzle velocities). While it is hard to define the "practical lower limit", it should be noted that the use of ramjet projectiles in the light arms caliber range presents severe limitations and design problems that are much less pronounced in large caliber guns.

*

T. Wijker, Muiden Chemie, The Netherlands. How do you launch such a ramjet projectile, considering the relative large outlet endport (nozzle)? You have to realize the relative high g-forces to realize a muzzle velocity of approximately Mach 3.

Author's Reply. The SFRJ projectile should be launched like a regular bullet, thus acquiring the high muzzle velocity necessary to commence the ramjet motor operation. A special cap at the projectile base can be used to block the nozzle exit, while the projectile is moving in the barrel. The cap will separate from the projectile after leaving the gun as a result of the ram pressure.

*

N.-Y. Wang, East China Institute of Technology, People's Republic of China. The flight stability has to be considered in the projectile design. It will be very difficult to use spin-stabilized structure in an SFRJ; the yaw angle (angle of attack) will change the head shock wave, thus the air mass flow rate.

Author's Reply. It is believed that the effect of the small yaw angle expected for spin-stabilized SFRJ projectiles should be similar to that encountered in other supersonic inlets.

*

Author Index

Subject Index